Astrophysics and Space Science Proceedings

Volume 57

More information about this series at http://www.springer.com/series/7395

Mário J. P. F. G. Monteiro • Rafael A. García •
Jørgen Christensen-Dalsgaard • Scott W. McIntosh
Editors

Dynamics of the Sun and Stars

Honoring the Life and Work of Michael J. Thompson

Springer

Editors
Mário J. P. F. G. Monteiro
Universidade do Porto
Porto, Portugal

Rafael A. García
IRFU/DRF/CEA
Saclay, France

Jørgen Christensen-Dalsgaard
Aarhus University
Aarhus, Denmark

Scott W. McIntosh
NCAR
Boulder, CO, USA

ISSN 1570-6591 ISSN 1570-6605 (electronic)
Astrophysics and Space Science Proceedings
ISBN 978-3-030-55335-7 ISBN 978-3-030-55336-4 (eBook)
https://doi.org/10.1007/978-3-030-55336-4

Cover illustration: Michael J. Thompson, photographed in Nov. 2015, in Taiwan, by Ying-Hwa "Bill" Kuo (UCAR)

This Springer imprint is published by the registered company Springer Nature Switzerland AG.
The registered company address is: Gewerbestrasse 11, 6330 Cham, Switzerland

Foreword

Dear friends and colleagues of Michael,

Thank you for allowing me to add my voice to this volume, honouring the life and work of Michael J. Thompson, my beloved husband. And thank you for your friendship and community over the last 35 years. I am not a scientist and so it is with a different, entirely personal, perspective that I write.

I first met Michael when he was a Part III student in Cambridge. He was a shy and intense young man, a typical 'Cambridge mathematician' with that combination of introversion and confidence. At the end of that year, he won something called The Tyson Medal, described by Wikipedia as being *'awarded for the best performance in subjects related to astronomy at the University of Cambridge, England. It is awarded annually for achievement in the examinations for Part III of the Mathematical Tripos when there is a candidate deserving of the prize'*. Deserving of the prize, indeed. Being an undergraduate in English Literature, I had no idea what this meant or who Michael was in his world.

Some of the contributors to this volume would have known Michael at least as long as I did, some perhaps longer. He loved collaborating with you and being a member of the solar and stellar physics community. I had the privilege to travel with Michael to places where I met many of you. In the last 35 years, you, Michael and I have shared many stimulating conversations, rich cultural experiences, natural disasters, food and drink.

As I write, I do not know exactly what the other contributions to this volume are, other than a research article by Robin Thompson, our son. But, attending the conference, I heard talks from those who had known Michael, speaking personally about what he meant to them and how his work had influenced them. I am, of course, deeply saddened that we gathered to honour Michael's memory and not to celebrate his 60th birthday. Despite this, he would have been pleased and perhaps surprised to hear about the many positive impacts that he had on people, both through his work and through other aspects of his life and person. We heard from numerous individuals about how he had been so generous to them—the cafeteria staff, the IT department, the early career and senior scientists and many more. W.B. Yeats wrote that

> The intellect of man is forced to choose
> Perfection of the life, or of the work.

As tributes here show, Michael, like so many of us, aspired to some combination of those.

The language of this subject—the Dynamics of the Sun and Stars—has become somewhat familiar to me over the decades; I understand many of the words (from helioseismology onwards) if not the concepts they are used to describe. The meeting allowed me to hear this language spoken inside and outside the conference room, and although it is not really a language I speak, I enjoyed hearing it this way. It gave me a deeper sense of a part of Michael's life that was always important to him.

I know that you will continue this science and this language, and so Michael's work and his contributions to the subject will also live on. He would be pleased to have been useful and to have shared in this. As the research continues, your collaborations and calculations will bring new understanding of the forces and objects in our universe. This contributes to the sum of human knowledge. Who could not be proud of this? I know Michael would want to have been a part of that and also to know that you and those who come later will be taking the research in new directions. This volume represents the continuation of this field and the start of those future discoveries.

Boulder, CO, USA
March 2020

Kate L. Thompson

Preface

This volume is a collection of original articles resulting from the contributions presented at the conference on

DYNAMICS OF THE SUN AND STARS:
HONOURING THE LIFE AND WORK OF MICHAEL J. THOMPSON
24–26 September 2019, NCAR–USA
Website: https://www2.hao.ucar.edu/MJTWorkshop2019

Rationale for the Conference and Proceedings

This meeting celebrated Michael Thompson's seminal work on solar and stellar physics, as well as his major contributions to the development of the National Center for Atmospheric Research, and marks his untimely death in October 2018. Michael played a key role in the development of helioseismology and its application to the study of the structure and dynamics of the solar interior, and he provided a strong foundation for the extension of seismic studies for other stars. After focussing for several years on more administrative activities, he was returning to leading the seismic studies of solar interior rotation and he was deeply involved in the understanding of the dynamics of the core of stars, when his life was tragically lost.

The conference focussed on dynamical aspects of the Sun and stars, based on the large amount data available on solar and stellar oscillations, and the extensive and detailed modelling now becoming feasible. By combining observations, seismic analysis and modelling, we hope that this will be a fitting memorial for a close colleague and friend, much missed.

Organisation and Committees

The conference was organised by the *National Center for Atmospheric Research (NCAR)* located in Boulder, Colorado, USA.

The scientific programme occupied three working days, complemented by the social programme. On the first day, there was a specific session dedicated to remember and honour Michael's work and role in science administration. It was a very dense programme with a long list of communications, including a total of 36 oral communications (two of these were public talks), a discussion session and 37 posters. The social programme included a concert.

Scientific Organising Committee

Jørgen Christensen-Dalsgaard	Chair—Aarhus University, DK
Sarbani Basu	Yale University, USA
Rafael A. García	Astrophysics Division, CEA, Saclay, FR
Douglas O. Gough	University of Cambridge, UK
Frank Hill	NSO, USA
Rachel Howe	University of Birmingham, UK
Rekha Jain	University of Sheffield, UK
Scott McIntosh	NCAR, USA
Mário J. P. F. G. Monteiro	University of Porto, PT
Jesper Schou	Max Planck Inst. for Solar System Research, DE
Takashi Sekii	NAOJ, JP
Juri Toomre	University of Colorado, USA

Local Organising Committee

Steve Tomczyk	Chair—NCAR, USA
Sheryl Shapiro	NCAR, USA

A special thank you from the SOC and the participants must go to the local organisation, for the excellent work done in ensuring we all had a perfect meeting. Also, we are very grateful to Kate and Robin Thompson for their contributions to the conference, including the social programme.

Sponsors

The organisation of the conference and the publication of the present volume were supported by:

- *National Science Foundation, USA*
- *National Center for Atmospheric Research (NCAR), USA*, through funding of the local organisation, events and participant support.
- *Stellar Astrophysics Centre, Aarhus University, Denmark*, by funding the support of some participants.
- *Instituto de Astrofísica e Ciências do Espaço—Centro de Astrofísica da Universidade do Porto* (IA U.Porto), and the *Departamento de Física e Astronomia—Faculdade de Ciências da Universidade do Porto*, through funding of the publication of this volume of proceedings.
- *Centre National d'Etudes Spatiales* (CNES) through the GOLF and PLATO grants at the Astrophysics Division of the CEA/Saclay, France, by funding the color pictures of this volume.
- *Private donations* in support of student travel.

Porto, Portugal	Mário J. P. F. G. Monteiro
Gif-sur-Yvette, France	Rafael A. García
Aarhus, Denmark	Jørgen Christensen-Dalsgaard
Boulder, CO, USA	Scott W. McIntosh
July 2020	

Contents

Part I
The Life and Work of Michael J. Thompson

Michael Thompson's Legacy in Solar and Stellar Physics

William J. Chaplin

Abstract Professor Michael J. Thompson died on 15 October 2018. Michael made long-lasting contributions to the international research community as both a research pioneer in the field of solar and stellar astrophysics and as an administrator who held major leadership positions in the UK and the USA. In this review we summarize his outstanding contributions to research.

It was a huge honour to be asked to give the opening talk of this conference, commemorating Michael's numerous and very significant achievements. He left a significant and long-lasting legacy, not only in terms of the results and outputs from his research in solar and stellar physics, but also from the way he achieved that impact. He made his mark through work in theory, in analysing and interpreting data, and in developing the methodology and techniques of analysis applied to seismic data on stars.

In his research Michael tackled a wide variety of subjects, reflecting a broad range of interests and an inquiring disposition. This breadth is captured by the word cloud in Fig. 1, created from the titles of research papers on which Michael was a co-author.

Michael liked to work with other people, and had a range of collaborators who without fail enjoyed working with him. Figure 2 is a visual representation of his collaborative network. It shows co-authors on scientific papers, with the font size denoting the number of common co-authorships. Aside from the number of collaborators, what is most strking about this diagramme is that it is not just a few names that dominate; many collaborators share a similar-sized representation on the diagramme, reflecting Michael's interest in working with new people. He was

W. J. Chaplin (✉)
School of Physics & Astronomy, University of Birmingham, Birmingham, UK
e-mail: w.j.chaplin@bham.ac.uk

M. J. P. F. G. Monteiro et al. (eds.), *Dynamics of the Sun and Stars*, Astrophysics and Space Science Proceedings 57,
https://doi.org/10.1007/978-3-030-55336-4_1

Fig. 1 A word cloud created from the titles of research papers on which Michael was a co-author (ADS)

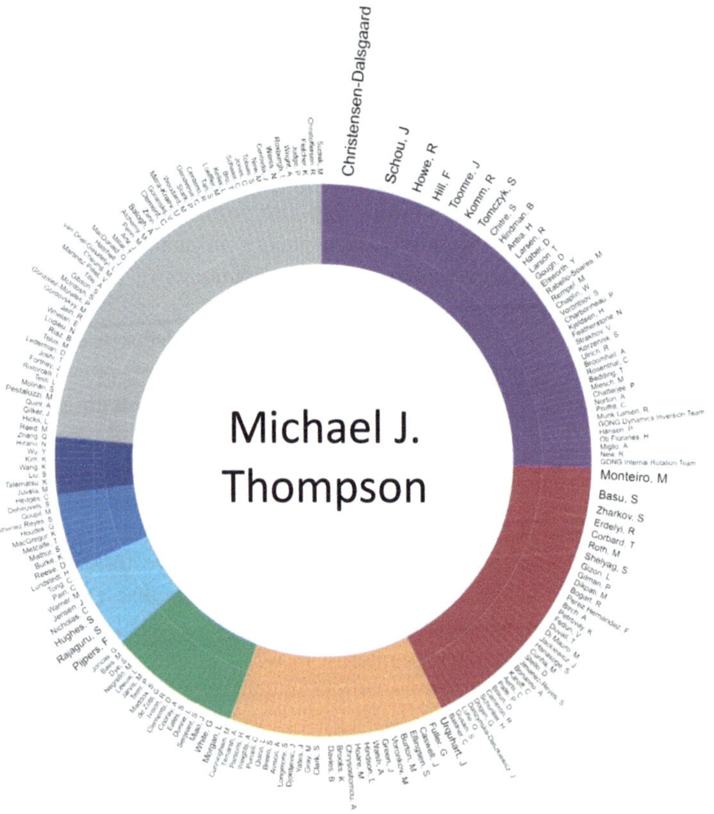

Fig. 2 A visual representation of Michael's network of research collaborators. The diagramme shows co-authors on scientific papers, with the font size denoting the number of common co-authorships (ADS)

always encouraging, particularly with younger colleagues, and treated everyone the same, irrespective of their position and standing.

Michael was an unfailing supporter of theoretical investigations of the solar interior. He was a strong advocate of research excellence, believing implicitly that long-term understanding could only arise through a sound theoretical underpinning.

Michael took his undergraduate degree at Cambridge from 1978 to 1981. Following a brief period in industry, he returned to Cambridge to complete a Certificate of Advanced Studies in Mathematics (Part III Mathematics) in 1984. He is remembered as an outstanding student, who won the Tyson Medal and Prize for his scholastic achievements. He excelled in exams, including one in particular that would shape the future direction of his career: astrophysical fluid dynamics, taught by Professor Douglas Gough. Michael was convinced he had performed badly in the exam to the extent that he aired his frustrations and resolved never to speak to Douglas again, although he had in fact come top by some margin. Michael was so taken by the course that he decided to pursue research in this field. Wanting also to stay in Cambridge, he approached Douglas to supervise and a life-long friendship and working relationship developed from there, and so Michael added his name to a prestigious academic family tree (Fig. 3).

For his PhD research Michael worked with Douglas in the nascent field of helioseismology. This was a time when helioseismology, the study of the Sun's interior through observation of its oscillations, was becoming established as a field in its own right. Michael's research developed around using seismology as a tool to further understanding of the internal structure and physics of stars, the Sun being a focus throughout much of his career.

After completing his PhD, in 1987 Michael began a postdoctoral fellowship at Aarhus University, working with another former student of Douglas, Jørgen Christensen-Dalsgaard. He then joined the High Altitude Observatory (HAO) in Boulder in 1988, where one day he would become its Director. During this period he began building a body of work on investigating the internal rotation of the Sun, the research topic he is probably best known for (see Fig. 4 for a photo from this period). It is fitting that Michael's first lead-author paper [4] in a refereed journal reported results on an inversion for rotation, which he performed using data from Ken Libbrecht's instrument at the Big Bear Solar Observatory.

Michael was a theoretician, but displayed a strong interest in developing the techniques needed to analyse data. Indeed, he was in the forefront of developing so-called inverse analysis techniques—which had been used in the Geophysics community—for application to helioseismic data, allowing the Sun's internal rotation and structure to be inferred in great detail. After Michael moved to Queen Mary University of London in 1990—initially as a Science and Engineering Research Council Fellow, and as a member of the faculty from 1992—he worked with Frank Pijpers on major improvements to the Optimally Localized Averages (OLA) method of inversion.

As its name suggests, the technique utilizes optimal combinations of sensitivity kernels of individual modes to produce so-called averaging kernels that isolate information on the structure within a specific layer or region of the stellar interior. In

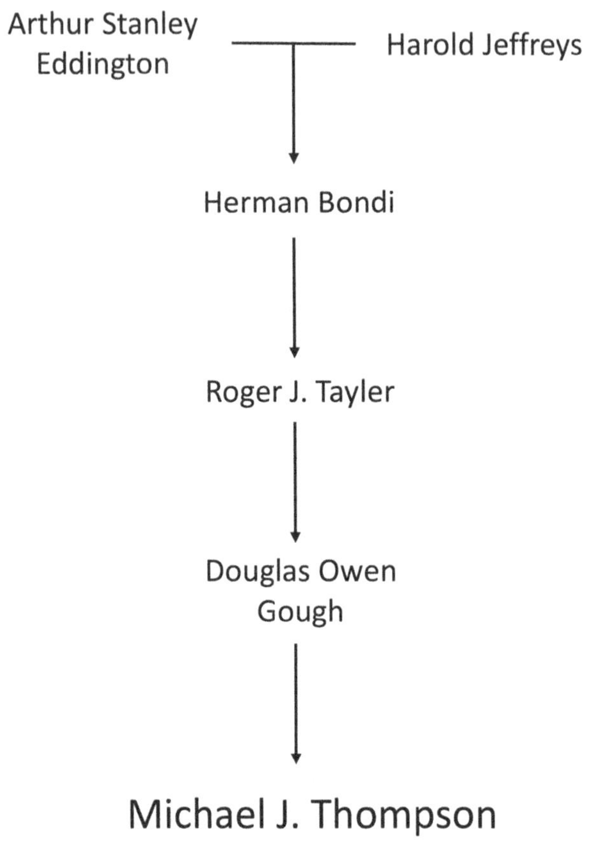

Fig. 3 Michael's academic family tree

the usual approach, one minimizes the integral of the square of the kernel multiplied by some "cost function" that penalizes the kernel having significant amplitude away from the target radius: hence multiplicative OLA, or MOLA. This calculation can be computationally expensive, and Michael and Frank proposed an alternative, more efficient approach [2]. Here, the trick was to instead compare the integral with some desired target function (e.g., a Gaussian), implying a subtraction in the minimization rather than a multiplication, hence Subtractive OLA or SOLA. In a second paper, they then discussed how to choose a suitable width for the target function depending on the location in the interior [3].

The year 1996 marked an important landmark for the growing field of helioseismoloy, with the publication of a topical edition of Science Magazine, carrying first-light papers on results from the Global Oscillations Network Group (GONG). Tellingly, and a testament to the leadership he had established in the field, Michael was the lead author of the paper on inference on the rotation and dynamics of the interior [5]. Its results revealed with greater clarity than before the differential

Fig. 4 Michael (right-hand side, in the blue check shirt) and colleagues listen intently at the IAU 121 "Inside the Sun" conference, Versailles, France 22–26 May 1989

rotation of the convective envelope, the pronounced rotational sheer at the base of the envelope, in addition to shear in the near-surface layers.

Results on rotation did not fit with prevailing theory. Whilst they showed that the pattern of differential rotation observed at the Sun's surface—with equatorial regions rotating more rapidly than the poles—penetrated the Sun's convective envelope, the patterns failed to match those predicted by models. Deeper down the rotation was observed to transition across a thin layer called the tachocline to that expected of a solid body, again a surprise, and one that continues to challenge models today. Michael would continue to publish key results in this area. This included further work on the tachocline [1], and studies of the variations of rotation in the outer envelope over time [7]. These variations were revealed to carry signatures of the Sun's activity cycle and have provided important observational constraints for those working to model the internal evolution of the Sun's global magnetic field, which has far-reaching implications for understanding the influence the Sun has on the Earth (what we now call "space weather").

In 2003, it was fitting that Michael was approached to lead a review in the prestigious Annual Reviews of Astronomy and Astrophysics on the then state-of-play of knowledge on the internal rotation and dyanamics of the Sun [6].

Whilst researches in rotation were a common theme throughout his career, it is important to recognise Michael's contributions across a range of different areas of solar and stellar physics. There was work on the physics of the solar interior. Important examples included: the impact of introducing diffusion and settling into stellar evolutionary models of the Sun, which brought the models into better agreement with observations from helioseismology; and constraining overshooting at the base of the convective envelope from subtle differences in the frequencies that carry signatures of the change in stratification at the base of the convective envelope. Michael performed important calculations on the internal structure and physics of the Sun, and on evaluating different ways of performing local helioseismology (which carries many similarities with terrestrial seismic studies). Later in his research career, Michael also worked on transferring helioseismic techniques to other Sun-like stars, notably using data from the NASA *Kepler* Mission, where the reduced quantity and quality of data present additional challenges.

In 2001 Michael moved to Imperial College, London, to take up a professorial Chair. Michael's professional interests went much wider than physics and astronomy, and in 2004 he moved to the University of Sheffield to take on his first major administrative role. There, he led the successful creation of a new School of Mathematics and Statistics.

Throughout this period Michael maintained strong links with HAO, including regular research trips, and in 2003 he became a National Center for Atmospheric Research (NCAR) affiliate scientist at HAO. In 2010 those links became more permanent when Michael moved with his wife Kate to Boulder to become the Director of HAO and Associate Director of NCAR. In 2013, he was named Deputy Director and Chief Operating Officer of NCAR and interim Director for thirteen months in 2015/16.

Very recently Michael had been in the process of becoming research active once more, and was leading an international project to provide timely updates on the Sun's internal rotation. That work will be completed, under the supervision of friend and colleague Jørgen Christensen-Dalsgaard. The results will serve as a fine testament to Michael's research legacy.

Michael will be much missed by all who knew him.

Acknowledgments The author expresses his thanks to Robin Thompson, Jørgen Christensen-Dalsgaard, Steve Tobias, and Rachel Howe, and to the organizing committee of the conference for the opportunity to share memories of Michael's notable achievements.

References

1. Howe, R., Christensen-Dalsgaard, J., Hill, F., et al. (2000). *Science, 287*, 2456.
2. Pijpers, F. P., & Thompson, M. J. (1992). *Astronomy & Astrophysics, 262*, 33.
3. Pijpers, F. P., & Thompson, M. J. (1993). *Astronomy & Astrophysics, 281*, 231.
4. Thompson, M. J. (1990). *Solar Physics, 125*, 1.
5. Thompson, M. J., Toomre, J., Anderson, E. R., et al. (1996). *Science, 272*, 1300.
6. Thompson, M. J., Christensen-Dalsgaard, J., Miesch, M. S., & Toomre, J. (2003). *Annual Review of Astronomy and Astrophysics, 41*, 599.
7. Vorontsov, S. V., Christensen-Dalsgaard, J., Schou, J., Strakhov, V. N., & Thompson, M. J. (2002). *Science, 296*, 101.

Michael Thompson in Sheffield

Rekha Jain

Abstract Michael Thompson gave huge contributions to the field of Helioseismology and inspired many young researchers in the discipline with his insight and new ideas. He spent six years in Sheffield after being appointed as the Head of Applied Mathematics Department in the University of Sheffield (UK) where colleagues hold fond memories of him, both as a committed academic and a very effective leader. His hard work and sincere approach was infectious to all those who worked around him. This tribute is an opportunity to share a little part of his work and life during his time in Sheffield.

1 As a Head

Michael Thompson was appointed as the Head of Department of Applied Mathematics in 2004 at the University of Sheffield. Over the next four years, he led the change from three separate Departments, Applied Maths, Pure Maths and Probability and Statistics each with very different cultures into the single School of Mathematics and Statistics (SoMaS) becoming its first head in 2008. The event marking the launch of SoMaS was organized on May 21, 2008 where distinguished external speakers from other Universities, Michael Atiyah (Edinburgh), Douglas Gough (Cambridge), Jon Keating (Bristol) and Terry Lyons (Oxford), were invited to present their life-long academic work in their respective field. Figure 1 shows Mike Thompson introducing the meeting on the Launch day of SoMaS on what one imagines as one of the busiest and happiest days for Mike. Figure 2 shows him enjoying the occasion with the dignitaries.

R. Jain (✉)
School of Mathematics and Statistics, University of Sheffield, Sheffield, UK
e-mail: R.Jain@sheffield.ac.uk

M. J. P. F. G. Monteiro et al. (eds.), *Dynamics of the Sun and Stars*, Astrophysics and Space Science Proceedings 57,
https://doi.org/10.1007/978-3-030-55336-4_2

Fig. 1 Michael Thompson: presenting the vision for SoMaS on May 21, 2008. Source: http://www.maths.dept.shef.ac.uk

2 Opportunity for All

As the Head of School of Mathematics and Statistics (2008–2010), Mike believed in participation and progress done through forming various official committees and groups. Colleagues felt Mike was a very patient listener, always providing staff members with the opportunity to speak up.

He supported staff in acquiring various skills encouraging their training through various workshops in and outside Universities. On his recommendation, I also attended one such session run by the University and gained huge benefit from it. He was particularly keen to involve staff members in the decision-making process and train the young upcoming staff for Leadership roles. One of my colleagues Professor E. Winstanley had this to say about Mike:

Fig. 2 Michael Thompson with Fields medallist and Abel prize winner Sir Michael Atiyah, then President and Vice-Chancellor of UoS Prof. Sir Keith Burnett CBE, FRS, FLSW, on May 21, 2008; source: http://www.maths.dept.shef.ac.uk

> He was always extremely kind, supportive and encouraging to me as a younger colleague when he was Head. I am very grateful for his support and encouragement, and also for the opportunities he gave me to be involved in decision-making within the department and school, which was invaluable experience. Mike set an example in leadership which one should strive to emulate!

3 Good Citizenship

One of the greatest aspects of Mike's leadership was his belief for staff to be engaged with SoMaS and not just hide in their offices doing their own teaching and research. Mike was himself a good citizen and expected everyone to be so. My colleague, Prof. D Applebaum from Probability and Statistics mentioned to me that

> he has never forgotten this thinking of Mike's. Mike valued being a good citizen so much that when we were considering promotion of academic staff, he would always ask "is he/she a good citizen?" meaning does he/she do their bit for the school.

4 A Committed Academic

As is well known, Mike was a dedicated researcher and was remarkably driven; his contribution to the field of Helioseismology will always be recognized around the world. Mike gave great credit to his PhD students, postdocs, collaborators and research visitors with whom he had scientific discussions on a daily basis. As reseachers and academics ourselves, we all know that to be able to work so efficiently and fruitfully with folks of different personalities and abilities requires a very special skillset, one that Mike certainly had!

Although as the Head of SoMaS, he had many important and time-consuming administrative tasks, he was a teacher at heart. He took his fair share of teaching duties and taught large undergraduate classes. He somehow found time to meet his tutees several times during the semester to check their progress and to sort out their problems, and took support for his PhD students very seriously. It demonstrated to staff around him that he indeed practiced what he preached. His seminal work on Helio and Astero-seismology is discussed by many in this proceedings so I will not repeat those but I will briefly mention some of the work that he involved me in during the supervision of his PhD student Kara Burke at the University of Sheffield.

His expertise in analytical method was obvious to me when I saw the process of his supervision to Kara Burke who submitted her PhD thesis in 2011 to the University of Sheffield. The thesis work was about the investigation of the Effects of Rotation on the Frequencies of Stellar Oscillations by treating rotation as a perturbation to the non-rotating star. They derived the rotational splitting coefficients, correct to second order in rotational velocity and added to the non-rotating frequencies (see Burke [1]).

$$\omega_{nlm} = \omega_0 + \omega_1 + (2I\omega_0)^{-1} \langle\, \xi_0^* \cdot \left(\mathcal{N}_0 - \mathcal{L}_\Omega - \rho_\Omega P_2 \omega_0^2\right) \xi_0 \,\rangle$$

$$-\frac{\omega_1^2}{2\omega_0} - \frac{\omega_1}{I} \langle\, \rho_0 \xi_0^* \cdot \mathcal{M}_0 \xi_1 \,\rangle + \frac{1}{2I\omega_0} \omega_1 \langle\, \rho_0 \xi_0^* \cdot \mathcal{M}_0 \xi_0 \,\rangle \qquad (1)$$

Here, the subscripts '0' and '1' denote the equilibrium and perturbed quantities respectively. ξ is the displacement (eigenfunction) with * denoting the complex conjugate. ω_0 is the frequency of the non-rotating case, ω_1 is the correction due to the first order perturbation theory and ρ_0 is the equilibrium density. I is an integral involving the spherical harmonic and P_2 is a Legendre function. $\mathcal{M}_0\xi_{0,1}$ contains all the first order rotation terms. Similarly, $\mathcal{L}_\Omega$ denote the remaining perturbed terms in the linearized momentum equation. The new terms arising due to second order rotation effects are contained in $\mathcal{N}_0$ and are quadratic in velocity u_0. The subscript Ω denotes the perturbed quantity due to rotation.

The corresponding perturbations to the eigenfunctions were also calculated. These were then implemented in the EVOLPACK stellar evolution and adiabatic pulsations package and splitting coefficients of various models calculated. The splitting coefficients were found to converge with increasing radial order for solar-like models, regardless of the mass or evolutionary stage of the star. They also concluded that a comparison of eigenfrequencies correct to first and second order in Ω revealed that for first order frequencies the splitting is uniform with the multiplet evenly spaced with the $m = 0$ mode in the middle and that the spacing between the modes increases with rotation rate. The second-order frequencies are evenly spaced at low rotation rates where the first order rotation effects are dominant, however as rotation rate increases the effect of centrifugal distortion causes the shape of the star to change from spherical to that of an oblate spheroid. By comparing the observations of other stars it was concluded that without rotation it is more difficult to constrain the models.

Another project where he showed his interest in bold application to Helioseismology was through numerical simulations. We jointly worked with Pedro A. González-Morales who tested the suitability of FLASH code for simulating subphotospheric wave motions in helioseismology (see, Gonzalez-Morales et al. [2]). In order to check the capability of this code for different sources of waves, Pedro implemented a non-magnetic plane-parallel atmosphere with the energy source as suggested by Rast [3]. The plan was to analyse the numerically generated data using the same methods as used in local helioseismology. However Mike's move to Boulder during the project made it difficult to be completed.

A good researcher also has ability to work with observational data and Mike's knowledge on observations and his ability to formulate a mathematical problem using them was quite unique. He obviously was an excellent mathematician and so solving the resulting equations and interpreting the solutions in the context of these observations made him a successful allrounder Helioseismologist.

He was very much respected and admired by his postdocs as well. When I asked Sergei Zharkov for his memory of Mike, this is what he had to say about Mike

> Mike was a uniquely successful solar physicist and mathematician, an English gentleman of highest personal integrity and a great sense of humour, somebody who believed in Science above individual interest, somebody you can trust. The last time we met was at NAM held at Hull a couple of years ago—he invited me to a dinner with his wife, it was amazing that even with his outstanding career and success he would find time for very personal approach, listening and offering advice and support. I suspect, he was a great family man—he was close with his wife and he talked a lot about his son and was so proud of his son's achievements.

5 A Musical Talent and His Love of Carols

Mike was a great admirer of classical music and he himself was a good pianist. He would play the piano and sing Christmas carols in the annual Christmas celebration of the Applied Maths Department, as we see in Fig. 3 pictured by his postdoc. Daniel Rees.

In addition to this, Mike was also the pianist for the New Year's Day sing in their home village of Grindleford near Sheffield. My colleague Dr Frazer Jarvis, another caroller who took over as pianist on Mike's recommendation after he left Sheffield, has prepared a DVD featuring Mike singing in a number of the carols. Figure 4 shows the back cover of the DVD that was presented to Kate and Robin. Here is an excerpt from the commentary Frazer provided for the DVD:

During his time in Sheffield, Mike was a regular at "Sheffield Carol" events, generally taking place in pubs where the repertoire was singing carols local to the area, and often joined by Kate and Robin.

Fig. 3 Michael Thompson playing keyboard at a Christmas party. Source: Daniel Rees

Fig. 4 Back cover of the DVD containing Mike's Carols

Some of the video recordings of Mike:

1. BBC The Truth About Christmas Carols (2009)
   A segment from the Blue Ball pub in Worrall, with both Mike and Kate.
2. BBC Songs of Praise (2010)
   Another segment from the Blue Ball, this time broadcast as part of another BBC programme.

3 and 4. Festival of Village Carols 2008
   Every two years since 1994, a festival has taken place in Sheffield to celebrate the tradition of the pub carols. Mike first attended in 2008—here are the carols "Tinwood" and "Egypt" from that year. After Mike left for Boulder in 2010, he returned to Sheffield to sing in the festival in 2012, 2014 and 2016.

5. Grenoside: at The Old Red Lion "Top Red" (2014)
   A pub sing just before the 2014 Festival, with Robin joining Mike, singing "Hark, Hark".
6. Mike's last festival in 2016 with Robin
   Here are Mike and Robin in part of "Jesu lover of my soul".
7. 2013 Foolow village tour
   Most carol sings take place in pubs, but there is a small village a little west of Sheffield called Foolow, a village with only about 60 houses, but with the largest collection of unique carols in the UK. Every Christmas evening, there is a village tour, where the singers visit a number of houses, and sing the Foolow carols to the local villagers. Mike, Kate and Robin joined this tradition on a number of occasions during Mike's time in Sheffield. This short clip is from 2013, when they were in the USA, but it gives some idea!

8 and 9. 2014 Festival with Foolow
   Mike had a ticket for the 2018 festival also, but died before it took place. Robin was able to come, and tributes were paid to Mike at the festival. These clips are from the 2014 Festival, where Mike and Robin appeared as part of the Foolow group. You will see Robin–Mike is beside him, but obscured by someone in front. The carols are "Once More" and "Farewell", the latter being sung at the end of every house visit in Foolow on Christmas evening.

10. Mike's last festival in 2016 with Robin
    The carol "Merry Christmas".

In short, Mike created a flourishing and more efficient unit in SoMaS. The ex-head of Applied Maths department, Prof. Lucy Wyatt summed Mike's tenure as:

> It was not without difficulties. Like most University Departments we have our big egos and feifdoms and Mike had to cope with those sometimes successfully, sometimes not so well. No doubt he was glad to leave those battles behind when he went to the USA in 2010 and recently, managing to find a bit more time for research before his untimely death in 2018.

Even after Mike left, his legacy is felt across SoMaS and we reflect upon the great loss our School, Sheffield and the Solar physics community of the world has suffered.

References

1. Burke, K. D. (2011). The effects of rotation on the frequencies of stellar oscillations, Ph.D. Thesis, University of Sheffield.
2. Gonzalez-Morales, P., Jain, R., & Thompson, M. J. (2011). *Journal of Physics Conference Series, 271*(1), 012013.
3. Rast, M. P. (1999). *The Astrophysical Journal, 524*, 462–468.

Michael J. Thompson: A Remarkable Scientist, Leader, and Friend

J. W. Hurrell

The international research community lost a shining star on October 15, 2018. It was on that day that many were shocked and deeply saddened by the news that Michael J. Thompson passed away unexpectedly. Michael was an outstanding solar physicist, known for his exceptional accomplishments in advancing the understanding of the structure and dynamics of the solar interior. Michael was an internationally recognized leader in the field of helioseismology, in large part for his pioneering efforts in developing inversion techniques and applying them to modern data to advance understanding of the physics of the Sun. His many noteworthy achievements as a research scientist were celebrated in September 2019 with the workshop that these proceedings document. Dynamics of the Sun & Stars: Honoring the Life & Work of Michael Thompson was a wonderful gathering of Michael's family, friends and colleagues to discuss the science that was his passion and to commemorate his many professional achievements.

Michael had a wide-ranging influence on essentially all aspects of helioseismology. A hallmark of his research was its extremely high quality, which he always presented with great clarity. His colleagues knew that they could safely rely on his results. Arguably, his best-known and most important contribution was in measuring the rotation profile throughout most of the solar interior. In order to achieve a high level of confidence in his results, Michael devoted considerable attention to developing a framework for understanding and reconciling the properties of different helioseismic inversion techniques. Other contributions in this volume document his scientific achievements in detail. Michael achieved eminence in his field. His intellectual reach was exceptional.

J. W. Hurrell (✉)
Department of Atmospheric Science, Colorado State University, Fort Collins, CO, USA
e-mail: jhurrell@rams.colostate.edu

M. J. P. F. G. Monteiro et al. (eds.), *Dynamics of the Sun and Stars*, Astrophysics and Space Science Proceedings 57,
https://doi.org/10.1007/978-3-030-55336-4_3

With this commentary, I have the honor of saying a few words about Michael's talents as a scientific administrator and the leadership positions he held at my former institution—the National Center for Atmospheric Research (NCAR) in Boulder, Colorado. It was at NCAR that I met Michael. We developed a very close professional relationship and, more importantly, a friendship that I will forever cherish.

Michael first came to NCAR from England as a visiting scientist in 1988, 2 years before I joined the organization. Although he became an NCAR Visiting Scientist in 2003, it was not until he was appointed the Director of NCAR's High Altitude Observatory (HAO) and an Associate Director (AD) of NCAR in 2010 that we met. Around that same time, I also became an AD because of my appointment as Director of another NCAR Laboratory. Together with the Director and Deputy Director of NCAR, the ADs work together on the NCAR Executive Committee (EC). The NCAR EC is responsible for the strategic management and budgetary oversight of NCAR, and it represents the institution to the external community.

Over the next several years, I became fully aware of Michael's scientific management and leadership skills. As Director of HAO, he put great focus on strengthening the observatory's visitor program, enhanced and developed collaborations with universities and the broader solar and geospace research community, and emphasized education and training for early-career scientists. Michael was also deeply committed to working closely with other NCAR laboratories to better understand the Sun's impacts on Earth. Under his leadership, HAO's research thrived, producing new insights into the solar cycle, coronal mass ejections, Earth's outer atmosphere, and other important aspects of the Sun-Earth system. Over this time, Michael and I developed a deep and mutual respect for each other, grounded in shared values and a deep desire to serve NCAR staff. We believed that our primary role as Directors was to create a work environment in which the talented NCAR staff could thrive and conduct cutting-edge research while enabling the broader community to do the same. Michael never sought the spotlight. He was always much more interested in doing whatever he could to help others achieve success.

Thus, when I accepted the responsibility and honor of being appointed NCAR Director in September 2013, I insisted that Michael join me in the Directorate as Deputy Director and Chief Operating Officer (COO). This was a decision I reached even before I applied for the position, having witnessed Michael's stellar leadership of HAO and knowing we shared a common vision of leadership. Over the following five years, Michael and I worked closely together on a daily basis. We spoke on the phone both before and after the workday, over weekends and holidays, and we exchanged thoughts via email no matter the time or in what time zones we found ourselves. This fluid exchange of thoughts and strategies generally was not driven by time imperatives, but rather because of the symbiotic relationship we shared while guiding the institution. Working so closely with Michael was a highlight of my career. He was my strongest ally, but also my most honest critic. He did not hesitate to tell me the truth—even if it was uncomfortable. Because of this, nearly every decision I made or action I took as NCAR Director was improved as a result of Michael's steady, insightful, and reassuring guidance. His advice and leadership provided immeasurable benefits.

As Deputy Director and COO, Michael was deeply devoted to NCAR and its entire staff, and he took great joy from his work with colleagues across the organization on a wide range of topics and issues. His importance to NCAR cannot be overstated. With an unparalleled work ethic and attention to detail, Michael understood the organization in a way that was second to none. His talent as a research scientist was only matched by his talent as an administrator—a statement that can only be made of a special few in any field. Michael had an incredible ability to drill down into complex processes, whether technical, financial, or personnel related, put his finger on what was not working or could be improved, and then deliver a solution that was often transformative in nature. His ability to work on the details while remaining focused on the strategic level was a defining attribute of Michael.

Another defining characteristic was his ability to assume a massive workload, and to do so happily. In fact, he flourished and seemed to be truly happy, professionally, only when his plate was overflowing with responsibilities. As a case in point, despite his tremendous NCAR responsibilities, Michael was appointed Interim UCAR President for 13 months beginning in 2015. UCAR—the University Corporation for Atmospheric Research—is a nonprofit consortium of more than 100 colleges and universities that provide research and training in the atmospheric and related sciences, and it manages NCAR for the National Science Foundation. Amazingly, Michael did all of this while continuing to serve NCAR as its Deputy Director and COO. His brief tenure as Interim UCAR President came during an especially challenging time for the organization, but because of his visionary leadership and his launching of several transformative initiatives, UCAR's ability to support NCAR was strengthened significantly.

In closing, this volume will help document Michael's remarkably distinguished career in research, and I hope my brief commentary also conveys that Michael was equally talented as a scientific leader and administrator. What guided his success was his deep devotion to those he worked with and for—whether at NCAR, UCAR or his previous leadership positions at Imperial College, London and the University of Sheffield. His dedication to NCAR over the last decade of his life certainly left a substantial and enduring impact on the organization.

I am blessed to have worked closely with Michael in the leadership of NCAR, but what I treasure much more was the profound friendship we had. His legacy will live on forever—in his family, in me, and in his many friends and colleagues across the globe. He touched so many lives in profound ways. In his calm and intelligent way, I sense him urging us all onward, wanting us to continue to advance the scientific frontiers that so impassioned him, and to enjoy life. He enriched our lives in immeasurable ways. Like many people, I will forever miss him, but he will never be forgotten.

Michael

Douglas Gough

I was very sorry that I could not attend the conference to help celebrate Michael's life. I certainly would have come had I been able. In these proceedings Bill Chaplin discusses some of Michael's contributions to science, and Rekha Jain and Jim Hurrell consider Michael's impact on the University of Sheffield, and on HAO and NCAR. So I shall write just a few words about the beginning, and the end, of Michael's scientific career.

I begin with his student days. He took a degree in mathematics at the University of Cambridge. But his interests were broader than mere mathematics. And so he left to go into industry—the Big Wide World. But he quickly discovered that this Big Wide World was no less restricting than the university, and much less satisfying. So after two years he returned to the University of Cambridge to take a graduate course in mathematics and theoretical physics: Part III of the Mathematical Tripos, as it's called. It was then that I got to know him. He attended a lecture course that I was giving in astrophysical fluid dynamics. It became evident to me very quickly that he was clearly the brightest student in the class, and also the most curious. He would very often accost me at the end of a lecture to discuss aspects of science that interested him. What also became clear was the caring side of his personality. I recall one day, after having raced up four long flights of stairs to get to the lecture room, encountering him at the door, whereupon he expressed considerable concern that I was somewhat out of breath. He counselled me on looking after the state of my heart and my future wellbeing.

D. Gough (✉)
Institute of Astronomy, and Department of Applied Mathematics and Theoretical Physics, University of Cambridge, Cambridge, UK
e-mail: douglas@ast.cam.ac.uk

M. J. P. F. G. Monteiro et al. (eds.), *Dynamics of the Sun and Stars*, Astrophysics and Space Science Proceedings 57,
https://doi.org/10.1007/978-3-030-55336-4_4

Michael was a man of vision. He could appreciate quickly how far to go, and how to get there. He had strong views, but, unlike many far-sighted strong-viewed people, he was always prepared to adjust them in the light of changing circumstances. I illustrate this by recounting an occasion at the end of an examination that he had sat in Part III Mathematics; it was an examination that I had set. I should point out that Part III of the Mathematical Tripos is regarded by many as one of the most difficult, if not the most difficult, graduate mathematics course in the world; and I had the reputation of being one of the most demanding examiners. I was passing by the examination room at the end of examination that I had set, not entirely by accident, just as Michael was emerging; he was incensed, scarlet with fury, because he thought that he had done very badly, and of course that was my fault. He shouted at me, saying that he would never talk to me again. It turned out that he hadn't performed so badly after all: in fact, he came top of the class by a very large margin. And he performed similarly in other subjects too, for which he was awarded the Tyson Medal by the university for being the best performing candidate that year in astronomically related subjects. It transpired that Michael was more interested in my subject than any other, and he resolved to pursue it into research. But he realized that it was I who was the most suitable potential supervisor. So he rescinded his resolve not to talk to me again.

Michael was not always right. In the beginning of his research days I would see him only very rarely. He would come into my office, and I would discuss his progress and offer suggestions as to how he might best proceed. Then he would hastily depart. Often I had to wait a long time before he returned. What Michael wanted to do was to wait until he had found what he would regard as some substantial concrete result to report to me. And usually that is not very easy, because one always gets stuck on the way. So progress was slow. Then one day he came into my office—this was not long after he had married Kate—and he rather sheepishly told me that Kate insisted that he and I must meet at least once a week. He feared that I would be affronted by this request—or was it a demand?—but that's where he was wrong. I was both delighted and relieved, and we promptly arranged times for future frequent meetings. A further outcome of that encounter was that it left me terrified of Michael's wife, Kate, whom I had not yet met. When we did meet, however, my fear was immediately dissolved. Together with my wife, Rosanne, the four of us became very good friends. And Michael became an excellent scientist who never looked back.

I'll share with you another anecdote from this time: There is a prestigious prize that has been awarded annually for more than a century by the University of Cambridge for the best essay in experimental physics, theoretical physics and mathematics written by a young student. It's called the Rayleigh Prize. "Young" used to be defined in terms of chronological age, and unfortunately, Michael was not eligible because he had intermitted for two years in industry and was now too old. I argued with the university that what matters under these circumstances is academical, not chronological age, and succeeded in having the criterion changed. Michael repaid me by becoming the first recipient of the prize under the new regulation. I hasten to add, incidentally, that I was not one of the adjudicators of the prize that year.

After graduating, Michael went away to hold various postdoctoral positions abroad, and then he returned to the UK into a tenured post at Queen Mary College, University of London, after which he moved on to a professorship at Imperial College. Michael's interests extended beyond the direct science in which he was involved, and so he moved to the University of Sheffield to sort out its mathematics: to establish the School of Mathematics, which amalgamated all the various departments in which mathematics was taught. Having successfully accomplished that, he moved to Boulder to become the Director of the High Altitude Observatory (HAO), at a time which, I believe, was in one of the most difficult periods of the observatory's existence. Jim Hurrell, Director of the National Center for Atmospheric Research, of which HAO is a part, was impressed by Michael's administrative prowess, which led to Michael's promotion to Deputy Director and Chief Operating Officer of NCAR.

During these times, Michael and I embarked on writing a book on helioseismology. Unfortunately, our administrative duties peaked at different times, and we were rarely in a position to collaborate properly. Consequently progress was slow. It is unlikely that the project will ever see the light of day.

In his all-too-short tenure he had almost no time for hands-on research. But he could facilitate research and urge others to pursue it, and at that task he was very good. In fact, he urged me to look again into the possible seismological consequences of a putative black hole in the centre of the Sun, the existence of which had been suggested by my friend Stephen Hawking many decades earlier as a potential solution to the neutrino problem, and which, Michael reminded me, had been the subject of conversation between us when Michael was a research student. Michael was unable to contribute directly to any analysis of that matter, but it was very interesting to me to talk about it with him and learn from his insight into some of the more profound aspects of the subject.

Michael also worked on other projects. In particular, he started an international collaborative project involving many people to reassess the seismological inferences concerning the internal rotation of the Sun using the most up-to-date seismic data, but sadly he didn't live to see it to fruition. Yet the project did not die: Jørgen Christensen-Dalsgaard has kindly stepped in to rescue it, and Rachel Howe reports on its progress in these proceedings.

I must also say that I feel very honoured by Michael for having organized, together with Jørgen and Sylvie Vauclair, a wonderful unforgettable conference to celebrate my sixtieth birthday at Chateau de Mons in the middle of Armagnac country.

Michael had many interests beyond science. He was an accomplished pianist, playing just for relaxation; he loved the outdoors, and after leaving London he and Kate always chose to live in almost isolated houses in the countryside; he loved hiking, and he loved the traditional regional carols of Yorkshire villages; he loved good food, both cooking it and eating it, and, needless to say, he loved good wine.

Latterly, he developed a great affinity for Japan: its people, its culture, and its language, which he was beginning to learn. It was in Japan that he died, prematurely, after a harrowing forced interruption to a month's leave of absence that was to be held in Tokyo to carry out full-time research. Sadly, that research was not completed.

I, and so many of the rest of us, are devastated by the loss of this great colleague who was poised to do many more things in the future. And more importantly, to have lost a good friend. Rosanne and I are still mourning his passing. Our hearts go out to Kate and their son, Robin.

Assessing the Threat of Major Outbreaks of Vector-Borne Diseases Under a Changing Climate

R. N. Thompson, M. J. Thompson, J. W. Hurrell, L. Sun, and U. Obolski

Abstract Michael J. Thompson served as the Director of the High Altitude Observatory at the National Center for Atmospheric Research (NCAR) in Boulder, Colorado, where he was also a Senior Scientist. In September 2013, Michael became the Deputy Director and Chief Operating Officer of NCAR, enjoying a very close working relationship with Director Jim Hurrell. During this time, Michael oversaw an organisation conducting research in a range of fields, including his own topics of solar and stellar physics, as well as others such as atmospheric chemistry and climate science. At the same time, his son Robin was completing a PhD in mathematical epidemiology at the University of Cambridge, UK, after which he was awarded an independent Junior Research Fellowship at the University of Oxford. However, the work conducted at NCAR and Robin's research have more overlap than might at first be expected. Here we present results from a collaboration that was set up following Michael's untimely death in October 2018, between climate scientists (Jim Hurrell and Lantao Sun) and mathematical epidemiologists (Robin Thompson and Uri Obolski). Specifically, we propose a framework for studying the effect of climate variability and change on vector-borne disease risk. We introduce a new quantity—the Instantaneous Outbreak Risk (IOR)—which quantifies the risk posed by an invading pathogen accounting for the climatic conditions when that pathogen

R. N. Thompson (✉)
Christ Church, Mathematical Institute and Department of Zoology, University of Oxford, Oxford, UK
e-mail: robin.thompson@chch.ox.ac.uk

M. J. Thompson
National Center for Atmospheric Research, Boulder, CO, USA

J. W. Hurrell · L. Sun
Department of Atmospheric Science, Colorado State University, Fort Collins, CO, USA

U. Obolski
School of Public Health and Porter School of the Environment and Earth Sciences, Tel Aviv University, Tel Aviv, Israel

M. J. P. F. G. Monteiro et al. (eds.), *Dynamics of the Sun and Stars*, Astrophysics and Space Science Proceedings 57,
https://doi.org/10.1007/978-3-030-55336-4_5

enters the population. We show how the IOR can be used to assess the threat from vector-borne diseases under a changing climate.

1 Introduction

Vector-borne diseases in populations of humans, animals and plants have devastating effects [18, 6, 36]. Future changes in climate are expected to alter the global spatial distribution of vectors, and therefore also of a range of diseases [7, 13, 28]. For example, although there are few reports of *Aedes aegypti* mosquitoes across Europe currently, climate change may allow *Ae. aegypti* to invade countries across the continent, potentially leading to local transmission of pathogens including the Zika, dengue and chikungunya viruses [28, 38, 23].

Epidemiological modelling studies have explored the question of whether or not pathogens are likely to generate large numbers of cases following arrival in a new region or country. In particular, branching processes have been used to estimate the probability of a major outbreak following from $I(0)$ index cases [43, 49, 44, 2, 29, 48, 41]. For simple models describing pathogens that are transmitted directly from host to host, this leads to the approximation

$$\text{Prob(major outbreak)} = 1 - \left(\frac{1}{R_0}\right)^{I(0)}, \tag{1}$$

in which R_0 is the basic reproduction number of the pathogen. This estimate for the probability of a major outbreak can be adapted for more complex epidemiological models, with features such as non-exponential infectious periods [4] and host-vector transmission [46, 24, 1, 5]. Branching process approximations of the major outbreak risk have been considered in the context of diseases such as Ebola virus disease [43, 2, 29, 42] and Zika virus disease [46].

Modelling has been used to forecast vector-borne disease trends using the outputs of climate models (for a review, see [47]). Previous work has included mapping estimated distributions of vectors [25, 19], characterising the potential of different locations for transmission (e.g. estimating reproduction numbers or exploring associations between risk factors and observed cases) [38, 33, 30, 34, 35], and forecasting changes in these assessments [7, 38, 20]. Since climate influences the potential for sustained transmission, the climatic conditions when a pathogen first arrives in a population affects the probability that a major outbreak will follow. It is therefore important to integrate climate forecasts into predictions of future outbreak risks obtained using branching processes [12].

Here, we show how estimates of the probability of a major outbreak can be generated when the values of parameters driving pathogen transmission depend on climate variables. We introduce a new epidemiological concept: the Instantaneous Outbreak Risk (IOR). Specifically, the IOR at any time t is defined as the probability, if the pathogen is brought into the population by a single infected host at time t, that a major outbreak follows—assuming that environmental conditions remain the same after t. In this way, the IOR is a measure of transmissibility at the precise instant that the pathogen first arrives in the population. We use forecasts of the Earth near-surface temperature in Miami from the Community Earth System Model (CESM) [17, 14] to show how climate predictions can be used to forecast vector dynamics and temporal changes in the IOR.

2 Materials and Methods

2.1 Climate Data

We considered a 40-member ensemble of daily near-surface (two metres above the Earth's surface) temperature simulations, conducted with the CESM [17, 14] under a historical and a future radiative forcing scenario (assuming a representative concentration pathway of 8.5) for the period 1920–2100. Each member of the CESM "Large Ensemble" (hereafter CESM-LE) was subjected to the same radiative forcing but began from slightly different atmospheric states in 1920. To demonstrate the principle that climate forecasts can be used to inform estimates of future infectious disease outbreak risks, we used the average of the 40 CESM-LE members in the future period from 2020–2100 for our analysis, so that in the following model 1st January 2020 is represented by the time $t = 0$ days and 31st December 2100 is represented by the time $t = 29{,}564$ days (each year is assumed to have exactly 365 days). By averaging across the range of equally plausible climate projections, we are considering the forced, anthropogenic component of climate change.

2.2 Epidemiological Model

The dynamics of vector-borne pathogen outbreaks are known to depend on climate variables such as temperature [31, 50, 51, 15, 27, 26]. We adapted a host-vector model of Zika virus disease outbreaks [46, 21] so that the vector lifespan was assumed to be a function of the Earth near-surface temperature (τ, measured in °C). We tracked the numbers of the N hosts that are (S)usceptible, (E)xposed, (I)nfectious and (R)emoved, as well as the numbers of vectors that are (S^V)usceptible, (E^V)xposed and (I^V)nfectious. The resulting model is given by

$$\frac{\mathrm{d}S}{\mathrm{d}t} = -\beta I^V \frac{S}{N}, \qquad \frac{\mathrm{d}E}{\mathrm{d}t} = \beta I^V \frac{S}{N} - \alpha_H E, \qquad \frac{\mathrm{d}I}{\mathrm{d}t} = \alpha_H E - \mu I, \qquad \frac{\mathrm{d}R}{\mathrm{d}t} = \mu I,$$

$$\frac{\mathrm{d}S^V}{\mathrm{d}t} = \rho\left(S^V + E^V + I^V\right)\left(1 - \frac{(S^V + E^V + I^V)}{K}\right) - \beta_V S^V \frac{I}{N} - \delta(\tau) S^V,$$

$$\frac{\mathrm{d}E^V}{\mathrm{d}t} = \beta_V S^V \frac{I}{N} - \left(\delta(\tau) + \alpha_V\right) E^V, \qquad \frac{\mathrm{d}I^V}{\mathrm{d}t} = \alpha_V E^V - \delta(\tau) I^V. \tag{2}$$

In this model, the parameter $\delta(\tau)$ describes the mortality rate of each vector when the temperature is τ°C. Consequently, the vector population size ($S^V + E^V + I^V = N^V$) varies in response to temperature. In contrast to the model on which our approach is based [46, 21], here the vector population size is limited by a carrying capacity, K, leading to a decrease in the birth rate due to density dependent effects when the vector population size increases. We set $\rho = \bar{\delta}/(1 - (N_0^V/K))$, where $\bar{\delta}$ is the vector mortality rate corresponding to the mean temperature across the first year of data considered (i.e. the predicted mean temperature in 2020) and N_0^V is the initial vector population size, so that the vector population size would oscillate in an approximately repeating fashion if the climate did not change in future years.

Following and adapting Thompson et al. [46], we define three important epidemiological quantities, assuming that climatic conditions and the vector population size are held fixed at time t (when the temperature is assumed to be τ°C). First, $R_0^{HV} = \frac{1}{\mu}\frac{\beta_V N^V}{N}$ is the expected number of vectors infected (and going on to enter the exposed class) by a single infectious human introduced into the population at time t. Second, the quantity $\rho^{E^V \to I^V} = \frac{\alpha_V}{\delta(\tau) + \alpha_V}$ is the proportion of exposed vectors that become infectious. Third, $R_0^{VH} = \frac{\beta}{\delta(\tau)}$ is the expected number of humans infected by a single infectious vector.

2.3 Model Parameters

As estimated from laboratory experiments by Yang et al. [50], for adult female *Ae. aegypti* mosquitoes the dependence of the vector lifespan on near-surface air temperature is characterised by a quartic polynomial for the mortality rate, $\delta(\tau)$—see black line in Fig. 1. Specifically, $\delta(\tau) = \sum_{i=0}^{4} b_i \tau^i$, where $b_0 = 8.692 \times 10^{-1}$ day^{-1}, $b_1 = -1.590 \times 10^{-1}$ day^{-1} °C^{-1}, $b_2 = 1.116 \times 10^{-2}$ day^{-1} °C^{-2}, $b_3 = -3.408 \times 10^{-4}$ day^{-1} °C^{-3} and $b_4 = 3.809 \times 10^{-6}$ day^{-1} °C^{-4} [50]. The values of the other parameters used in our analysis are given in the caption to Fig. 2.

2.4 Instantaneous Outbreak Risk (IOR)

The IOR can be calculated by considering the analogous stochastic model to the system of equations (2). This quantity is defined as the probability that a single infected human arriving in the population at time t initiates a major outbreak (an outbreak with a large number of hosts ever infected) as opposed to a minor outbreak (an outbreak with few hosts ever infected), assuming that conditions (here, the temperature τ and vector population size N^V) remain constant after time t. The IOR therefore reflects pathogen transmissibility at time t. It is similar to Eq. (1), but accounts for host-vector transmission and varies temporally in response to changing climatic conditions.

To derive the IOR, we denote the probability of no major outbreak starting from i exposed or infectious hosts, j exposed vectors, and k infectious vectors, as q_{ijk}. Starting from a single infectious host, and considering the possibility that the next event is either recovery of the infectious host or infection of a susceptible vector, gives

$$q_{100} = \frac{\mu}{\mu + \frac{\beta_V N^V}{N}} q_{000} + \frac{\frac{\beta_V N^V}{N}}{\mu + \frac{\beta_V N^V}{N}} q_{110}. \tag{3}$$

Similarly, starting instead from a single exposed or infectious vector gives

$$q_{010} = \frac{\delta(\tau)}{\delta(\tau) + \alpha_V} q_{000} + \frac{\alpha_V}{\delta(\tau) + \alpha_V} q_{001},$$

$$q_{001} = \frac{\delta(\tau)}{\delta(\tau) + \beta} q_{000} + \frac{\beta}{\delta(\tau) + \beta} q_{101}. \tag{4}$$

Assuming that infection lineages are independent, we can approximate terms with two exposed or infectious individuals by non-linear terms involving single exposed or infectious individuals, e.g. $q_{110} \approx q_{100} q_{010}$. Noting that $q_{000} = 1$, the three equations above can be solved analytically to give expressions for q_{100}, q_{010} and q_{001}. Taking the minimal solution for q_{100} (for a justification that the minimal non-negative solution should be used when calculating hitting probabilities of Markov chains, see [32]), the probability of a major outbreak starting from a single infected host introduced into the population at time t, i.e. the IOR, is then

$$\text{IOR} = 1 - q_{100} = \begin{cases} 0 & \text{if } \; R_0^{HV} \rho^{E^V \to I^V} R_0^{VH} \leq 1, \\[2ex] \dfrac{R_0^{HV} \rho^{E^V \to I^V} R_0^{VH} - 1}{R_0^{HV} \rho^{E^V \to I^V} R_0^{VH} + R_0^{VH}} & \text{if } R_0^{HV} \rho^{E^V \to I^V} R_0^{VH} > 1. \end{cases} \tag{5}$$

3 Results

In our analysis, we considered predictions of the Earth near-surface temperature in Miami from 2020–2100 according to the CESM-LE [17, 14]. We chose Miami as an example of a location with imported cases of vector borne diseases such as dengue fever, but where the mean predicted annual temperature in 2020 (25.1 °C) is slightly below the optimal temperature for vector survival (Fig. 1).

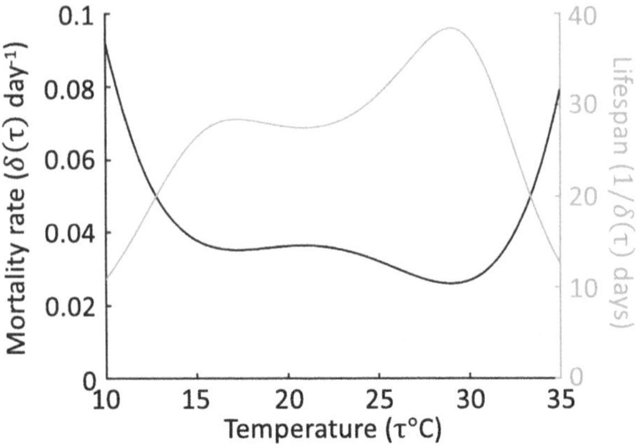

Fig. 1 Assumed dependence of the vector mortality rate (black) and lifespan (grey) on the Earth near-surface temperature, derived from laboratory experiments in adult female *Ae. aegypti* mosquitoes [50]

We first used daily predictions from the CESM-LE (Fig. 2a-black) to predict changes in the average vector lifespan. Numerical solutions of the epidemiological model (2) were then generated in the absence of disease to predict temporal variations in vector abundance, and the IOR was calculated (Fig. 2b-black). In our analysis, we compared predictions under a changing climate to the analogous forecasts if temperatures were assumed unchanged between 2020 and 2100 (i.e. if the daily temperatures in future years were assumed identical to the corresponding daily temperature in 2020)—see light grey lines in Fig. 2.

The general trend over the period 2020–2100 was towards higher temperatures, and therefore larger vector populations and higher IOR values (Fig. 2b-black). It was particularly noticeable that winter conditions were predicted to become more suitable for vector-borne pathogen outbreaks towards the end of this century. A longer vector lifespan leads to an increased IOR for two reasons. First, there are more vectors that can act as vehicles for pathogen transmission. Second, each vector is more likely to generate more successful transmissions between hosts, since each vector has a longer period during which it can transmit the pathogen. However, towards the end of this century, summer temperatures were predicted to increase beyond levels that are optimal for vector survival (Fig. 2a-black, cf. Fig. 1). Temperatures were instead optimal for vector survival twice per year, leading eventually to biannual peaks—a lower peak followed by a higher peak—in the IOR (Fig. 2b-black). There was also a reduction in the maximum IOR between 2070 (maximum IOR in 2070 = 0.277) and 2100 (maximum IOR in 2100 = 0.259).

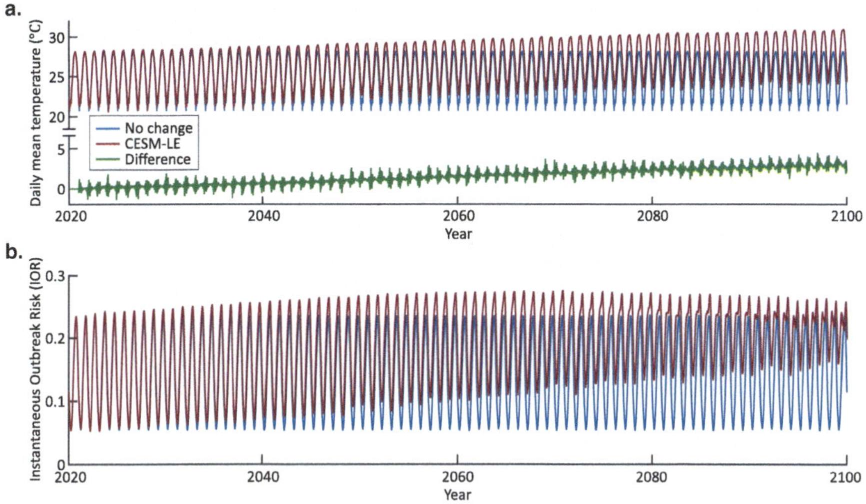

Fig. 2 The impact of temperature forecasts in Miami in the period 2020–2100 on the predicted risk from invading vector-borne pathogens. (**a**) Average temperature predictions from 40 members of the CESM-LE (black) and analogous temperatures if the CESM-LE prediction for 2020 is simply repeated each year (light grey). The difference between the light grey and black forecasts is shown in dark grey. There is a break in the y-axis between 6.5 and 18.5 °C so that the individual lines can be seen more clearly. (**b**) Calculated daily IORs (Eq. (5)) corresponding to the forecasts in panel (**a**). Model parameter values used (where possible, parameter values were obtained from [21]): $N = 10,000$, $N_0^V = 5,000$, $\beta = 0.152$ day^{-1}, $\beta_V = 0.22$ day^{-1}, $1/\alpha_H = 5.9$ days, $1/\mu = 5$ days, $\rho = 0.0419$ day^{-1}, $1/\alpha_V = 10.5$ days, $K = 20,000$

4 Discussion

Assessing and predicting the risk posed by vector-borne pathogens under a changing climate is an ongoing challenge, with the potential to guide climate and outbreak interventions. To honour the life of Michael J. Thompson, here we have presented initial results from our collaboration that aims to forecast vector-borne disease outbreak risks. We have introduced a new quantity, the IOR, which represents the threat of major outbreaks of vector-borne pathogens under a changing climate, and we have shown how temporal changes in this quantity can be predicted using climate forecasts.

The IOR was inspired by another metric, the instantaneous reproduction number, which is used for assessing the transmissibility of pathogens that have already invaded host populations [45, 11, 8]. The instantaneous reproduction number represents the average number of secondary cases of disease arising from a primary case infected at time t, under the assumption that conditions remain identical after t. This is in contrast to the case reproduction number—the equivalent quantity but accounting for transmissibility changes after time t [11]. The IOR could also, in theory, be adapted to account for variations in transmissibility after time t. However, in scenarios in which outbreaks either take off or fade out quickly relative to temporal changes in population sizes or the values of climate variables, this adapted quantity is expected to be very similar to the IOR.

Our goal here was to demonstrate that climate forecasts can be used to inform estimates of the risk that imported cases generate large outbreaks driven by local transmission. To do this, we used a simple model that will require additional features to be included in the future for accurate risk assessments. For example, it will be necessary to consider climate variables other than temperature that affect vector ecology or pathogen transmission, such as precipitation [27, 26, 3] and humidity [9]. Future results from the framework presented here will include the dependence of other features on climatic conditions—such as the effect of temperature on vector egg survival [16], flight distances [37] and virus transmission ability [22, 40]. We will also consider each of the CESM-LE climate simulations individually rather than predicting the IOR under a single, averaged future climate scenario. Such natural variability, driven by atmospheric processes and ocean-atmosphere or land-atmosphere interactions, is expected to contribute significant variation to future epidemic risks.

Nonetheless, our simple model was sufficient to demonstrate how the threat from invading pathogens can be predicted under a changing climate. We note that, while the general trend of short-term climate change suggests increased disease risk, variations in climatic factors may not always exacerbate disease risk. In particular, we found that temperatures in Miami at the end of this century may sometimes increase above those that are optimal for vector survival. In the absence of adaptation of the vector to survive in increased temperatures, this might ultimately lead to reduced disease risk at the warmest times of year, even if the outbreak risk is higher in other seasons [39, 10].

In summary, we have shown that climate forecasts can be used to predict the risk that future imported cases of vector-borne disease will lead to major outbreaks. Providing a metric that can be calculated straightforwardly for assessing the threat from invading vector-borne pathogens under a changing climate—the IOR—is the main contribution of this research. We hope that Michael J. Thompson would be proud of our collaboration.

Acknowledgments This work is dedicated to MJT, who unknowingly set up this research team. RNT is happy to have finally written an article with MJT, who died at the age of 59 on 15th October 2018. RNT and JWH miss him every day. Thanks to Christ Church, Oxford, for funding via a Junior Research Fellowship (RNT).

References

1. Allen, L. J. S., & Lahodny, Jr. G. E. (2016). Extinction thresholds in deterministic and stochastic epidemic models. *Journal of Biological Dynamics, 6*, 590–611.
2. Althaus, C. L., Low, N., Musa, E. O., Shuaib, F., & Gsteiger, S. (2015). Ebola virus disease outbreak in Nigeria: transmission dynamics and rapid control. *Epidemics, 11*, 80–84.

3. Alto, B. W., & Juliano, S. A. (2001). Precipitation and temperature effects on populations of *Aedes albopictus* (Diptera: Culicidae): Implications for range expansion. *Journal of Medical Entomology, 38*, 646–656.
4. Anderson, D., & Watson, R. (1980). On the spread of a disease with gamma distributed latent and infectious periods. *Biometrika, 67*, 191–198.
5. Bartlett, M. S. (1964). The relevance of stochastic models for large-scale epidemiological phenomena. *Applied Statistics, 13*, 2–8.
6. Brooks-Pollock, E., de Jong, M. C. M., Keeling, M. J., Klinkenberg, D., & Wood, J. L. N. (2015). Eight challenges in modelling infectious livestock diseases. *Epidemics, 10*, 1–5.
7. Campbell, L. P., Luther, C., Moo-Llanes, D., Ramsey, J. M., Danis-Lozano, R., & Peterson, A. T. (2015). Climate change influences on global distributions of dengue and chikungunya virus vectors. *Philosophical Transactions of the Royal Society, B: Biological Sciences, 370*, 20140135.
8. Cori, A., Ferguson, N. M., Fraser, C., & Cauchemez, S. (2013). A new framework and software to estimate time-varying reproduction numbers during epidemics. *American Journal of Epidemiology, 178*, 1505–1512.
9. da Cruz Ferreira, D. A., Degener, C. M., De Almeida Marques-Toledo, C., Bendati, M. M., Fetzer, L. O., Teixeira, C. P., et al. (2017). Meteorological variables and mosquito monitoring are good predictors for infestation trends of *Aedes aegypti*, the vector of dengue, chikungunya and Zika. *Parasites & Vectors, 10*, 78.
10. Escobar, L. E., Romero-Alvarez, D., Leon, R., Lepe-Lopez, M. A., Craft, M. E., Borbor-Cordova, M. J., et al. (2016). Declining prevalence of disease vectors under climate change. *Scientific Reports, 6*, 39150.
11. Fraser, C. (2007). Estimating individual and household reproduction numbers in an emerging epidemic. *PLoS One 2*, e758.
12. Guzzetta, G., Poletti, P., Montarsi, F., Baldacchino, F., Capelli, G., Rizzoli, A., et al. (2016). Assessing the potential risk of Zika virus epidemics in temperate areas with established *Aedes albopictus* populations. *Eurosurveillance, 21*, 30199.
13. Hales, S., De Wet, N., Maindonald, J., & Woodward, A. (2002). Potential effect of population and climate changes on global distribution of dengue fever: an empirical model. *Lancet, 360*, 830–834.
14. Hurrell, J. W., Holland, M. M., Gent, P. R., Ghan, S., Kay, J. E., Kushner, P. J., et al. (2013). The Community Earth System Model: A framework for collaborative research. *Bulletin of the American Meteorological Society, 94*, 1339–1360.
15. Johnson, L. R., Ben-Horin, T., Lafferty, K. D., McNally, A., Mordecai, E., Paaijmans, K. P., et al. (2015). Understanding uncertainty in temperature effects on vector-borne disease: A Bayesian approach. *Ecology, 96*, 203–213.
16. Juliano, S. A., O'Meara, G. F., Morrill, J. R., & Cutwa, M. M. (2002). Desiccation and thermal tolerance of eggs and the coexistence of competing mosquitoes. *Oecologia, 130*, 458–469.
17. Kay, J. E., Deser, C., Phillips, A., Mai, A., Hannay, C., Strand, G., et al. (2015). The Community Earth System Model (CESM) large ensemble project: A community resource for studying climate change in the presence of internal climate variability. *Bulletin of the American Meteorological Society, 96*, 1333–1349.
18. Kilpatrick, A. M., & Randolph, S. E. (2012). Drivers, dynamics, and control of emerging vector-borne zoonotic diseases. *Lancet, 380*, 1946–1955.
19. Kraemer, M. U. G., Sinka, M. E., Duda, K. A., Mylne, A. Q. N., Shearer, F. M., Barker, C. M., et al. (2015). The global distribution of the arbovirus vectors *Aedes aegypti* and *Ae. Albopictus Elife, 4*, e08347.
20. Kraemer, M. U. G., Reiner, R. C., Brady, O. J., Messina, J. P., Gilbert, M., Pigott, D. M., et al. (2019). Past and future spread of the arbovirus vectors *Aedes aegypti* and *Aedes albopictus*. *Nature Microbiology, 4*, 854.
21. Kucharski, A. J., Funk, S., Eggo, R. M., Mallet, H.-P., Edmunds, W. J., & Nilles, E. J. (2016). Transmission dynamics of Zika virus in island populations: A modelling analysis of the 2013–14 French Polynesia outbreak. *PLoS Neglected Tropical Diseases, 10*, e0004726.

22. Lambrechts, L., Paaijmans, K. P., Fansiri, T., Carrington, L. B., Kramer, L. D., Thomas, M. B., et al. (2011). Impact of daily temperature fluctuations on dengue virus transmission by *Aedes aegypti*. *Proceedings of the National Academy of Sciences of the United States of America, 108*, 7460–7465.
23. Liu-Helmersson, J., Rocklöv, J., Sewe, M., & Brännström, Å. (2019). Climate change may enable *Aedes aegypti* infestation in major European cities by 2100. *Environmental Research, 172*, 693–699.
24. Lloyd, A. L., Zhang, J., & Root, A. M. (2007). Stochasticity and heterogeneity in host-vector models. *Journal of the Royal Society, Interface. 4*, 851–863.
25. Longbottom, J., Browne, A. J., Pigott, D. M., Sinka, M. E., Golding, N., Hay, S. I., et al. (2017). Mapping the spatial distribution of the Japanese encephalitis vector, *Culex tritaeniorhynchus* Giles, 1901 (Diptera: Culicidae) within areas of Japanese encephalitis risk. *Parasites & Vectors, 10*, 148.
26. Lowe, R., Coelho, C. A., Barcellos, C., Carvalho, M. S., Catão, R. D. C., Coelho, G. E., et al. (2016). Evaluating probabilistic dengue risk forecasts from a prototype early warning system for Brazil. *Elife, 5*, e11285.
27. Lowe, R., Gasparrini, A., Van Meerbeeck, C. J., Lippi, C. A., Mahon, R., Trotman, A. R., et al. (2018). Nonlinear and delayed impacts of climate on dengue risk in Barbados: A modelling study. *PLoS Medicine, 15*, e1002613.
28. Medlock, J. M., & Leach, S. A. (2015). Effect of climate change on vector-borne disease risk in the UK. *The Lancet Infectious Diseases, 15*, 721–730.
29. Merler, S., Ajelli, M., Fumanelli, L., Parlamento, S., Pastore y Piontti, A., et al. (2016). Containing Ebola at the source with ring vaccination. *PLoS Neglected Tropical Diseases, 10*, 1–11.
30. Metcalf, C. J. E., Walter, K. S., Wesolowski, A., Buckee, C. O., Shevliakova, E., Tatem, A. J., et al. (2017). Identifying climate drivers of infectious disease dynamics: Recent advances and challenges ahead. *Proceedings of the Royal Society B: Biological Sciences, 284*, 20170901.
31. Mordecai, E. A., Cohen, J. M., Evans, M. V., Gudapati, P., Johnson, L. R., Lippi, C. A., et al. (2017). Detecting the impact of temperature on transmission of Zika, dengue, and chikungunya using mechanistic models. *PLoS Neglected Tropical Diseases, 27*, e0005568.
32. Norris, J. R. (1998). *Markov chains*. Cambridge: Cambridge University Press.
33. Nsoesie, E. O., Kraemer, M. U. G., Golding, N., Pigott, D. M., Brady, O. J., Moyes, C. L., et al. (2016). Global distribution and environmental suitability for chikungunya virus, 1952 to 2015. *Eurosurveillance, 21*, 30234.
34. Obolski, U., Perez, P. N., Villabona-Arenas, C. J., Thézé, J., Faria, N. R., & Lourenço, J. (2019). MVSE: An R-package that estimates a climate-driven mosquito-borne viral suitability index. *Methods in Ecology and Evolution, 10*, 1357–1370.
35. Perez-Guzman, P. N., Carlos Junior Alcantara, L., Obolski, U., de Lima, M. M., Ashley, E. A., Smithuis, F., et al. (2018). Measuring mosquito-borne viral suitability in Myanmar and implications for local Zika virus transmission. *PLoS Currents, 10*, 1–10.
36. Plantegenest, M., Le May, C., & Fabre, F. (2007). Landscape epidemiology of plant diseases. *Journal of the Royal Society, Interface, 4*, 963–972.
37. Rowley, W. A., & Graham, C. L. (1968). The effect of temperature and relative humidity on the flight performance of female *Aedes aegypti*. *Journal of Insect Physiology, 14*, 1251–1257.
38. Ryan, S. J., Carlson, C. J., Mordecai, E. A., & Johnson, L. R. (2018). Global expansion and redistribution of *Aedes*-borne virus transmission risk with climate change. *PLoS Neglected Tropical Diseases, 3*, e0007213.
39. Semenza, J. C., & Suk, J. E. (2018). Vector-borne diseases and climate change: A European perspective. *FEMS Microbiology Letters, 365*, fnx244.
40. Tesla, B., Demakovsky, L. R., Mordecai, E. A., Ryan, S. J., Bonds, M. H., Ngonghala, C. N., et al. (2018). Temperature drives Zika virus transmission: Evidence from empirical and mathematical models. *Proceedings of the Royal Society B: Biological Sciences, 285*, 20180795.

41. Thompson, R. N. (2020). Novel coronavirus outbreak in Wuhan, China, 2020: Intense surveillance is vital for preventing sustained transmission in new locations. *Journal of Clinical Medicine, 9*, 498.
42. Thompson, R. N., Gilligan, C. A., & Cunniffe, N. J. (2016). Detecting presymptomatic infection is necessary to forecast major epidemics in the earliest stages of infectious disease outbreaks. *PLoS Computational Biology, 12*, e1004836.
43. Thompson, R. N., Jalava, K., & Obolski, U. (2019). Sustained transmission of Ebola in new locations: More likely than previously thought. *The Lancet Infectious Diseases, 19*, 1058–1059.
44. Thompson, R. N., Thompson, C., Pelerman, O., Gupta, S., & Obolski, U. (2019). Increased frequency of travel in the presence of cross-immunity may act to decrease the chance of a global pandemic. *Philosophical Transactions of the Royal Society, B: Biological Sciences, 374*, 20180274.
45. Thompson, R. N., Stockwin, J. E., van Gaalen, R. D., Polonsky, J. A., Kamvar, Z. N., Demarsh, P. A., et al. (2019). Improved inference of time-varying reproduction numbers during infectious disease outbreaks. *Epidemics, 19*, 100356.
46. Thompson, R. N., Gilligan, C. A., & Cunniffe, N. J. (2020). Will an outbreak exceed available resources for control? Estimating the risk from invading pathogens using practical definitions of a severe epidemic. bioRxiv.
47. Tjaden, N. B., Caminade, C., Beierkuhnlein, C., & Thomas, S. M. (2018). Mosquito-borne diseases: Advances in modelling climate-change impacts. *Trends in Parasitology, 34*, 227–245.
48. Whittle, P. (1955). The outcome of a stochastic epidemic - A note on Bailey's paper. *Biometrika, 42*, 116–22.
49. Woolhouse, M. E. J., Brierley, L., McCaffery, C., & Lycett, S. (2016). Assessing the epidemic potential of RNA and DNA viruses. *Emerging Infectious Diseases, 22*, 2037–2044.
50. Yang, H. M., Macoris, M. L. G., Galvani, K. C., Andrighetti, M. T. M., & Wanderley, D. M. V. (2009). Assessing the effects of temperature on the population of *Aedes aegypti*, the vector of dengue. *Epidemiology and Infection, 137*, 1188–1202.
51. Yang, H. M., Macoris, M. L. G., Galvani, K. C., Andrighetti, M. T. M., & Wanderley, D.M.V. (2009). Assessing the effects of temperature on dengue transmission. *Epidemiology and Infection, 137*, 1179–1187.

Touching the Interior Structure and Dynamics of Our Nearest Star

Juri Toomre

Abstract Michael Thompson has had a pivotal and continuing role in developing and refining inversion techniques to be applied to the great blossoming of helioseismic data forthcoming from the GONG, MDI (on SOHO) and HMI (on SDO) observational projects. This has enabled major discoveries about the internal differential rotation of the Sun, revealing both a tachocline of shear at the base of its convection zone and a near-surface shear layer near its surface, and of its temporal variations. It has also guided efforts to map subsurface flows of many scales in the convection zone. In parallel with his abiding interests in helioseismology, Michael was very enthusiastic about recent efforts in solar convection and dynamo theory to address what he saw as the outstanding questions about the dynamics proceeding deep within our nearest star, and thus we touch briefly upon some of these here.

1 Michael's Professional Trajectory and of Helioseismology

The rapid ascent of helioseismology from a tantalizing possibility in the early 1980s to a major research tool, through investments that permit imaging structure and dynamics within the interior of our nearest star, all occurred within a few decades. These were exciting times given the rapid pace of developments and the striking results that emerged at a steady pace. Michael Thompson started his astrophysics career in the beginnings of this subject, and as a mathematical physicist made major and inspired contributions in inverse theory that permitted results from the oscillation frequencies of vast numbers of modes to be interpreted in terms of the underlying physics, such as the detailed differential rotation of the solar convection zone.

J. Toomre (✉)
JILA & Department of Astrophysical and Planetary Sciences, University of Colorado Boulder, Boulder, CO, USA
e-mail: jtoomre@lcd.colorado.edu

M. J. P. F. G. Monteiro et al. (eds.), *Dynamics of the Sun and Stars*, Astrophysics and Space Science Proceedings 57,
https://doi.org/10.1007/978-3-030-55336-4_6

Michael did his doctoral work at the University of Cambridge working with Douglas Gough, partly assessing the role that large-scale magnetic fields could have on the global acoustic oscillations of a star like the Sun. These were forward problems, sorting out how the oscillation frequencies of particular low-degree modes would be shifted or split by the presence of magnetic fields, and also by rotation [10]. He was also very interested in how information from many of the acoustic modes could be combined to get measures of the variation of the sound speed with depth, thus inverting the data, leading to a major discussion of the inversion problem [11], and its applications to deduce the depth of the solar convection zone and of its composition [4, 5]. Some of these inversion developments were carried out in the six-month program in 1990 on helioseismology at the Institute for Theoretical Physics (ITP, now KITP) at UC Santa Barbara. These followed his 1987 postdoc year in Aarhus Denmark with Jørgen Christensen-Dalsgaard, whereas ITP occurred in the second year of his postdoc at the High Altitude Observatory (HAO) of NCAR, starting his yearly (but for one) month-long visits to Boulder as the helioseismology efforts continued to blossom.

The ITP program also led to Michael accepting in fall 1990 what became a lecturer and reader position at Queen Mary, University of London, where his productive scientific work in applying inversions in helioseismology continued in earnest. In 2001 he accepted to be a professor of physics at Imperial College in London, and then in 2004 moved to a professorship at the University of Sheffield and rebuilt the school of mathematics, in addition to continuing his widely collaborative research in helioseismology. In 2010 Michael became the director of HAO, and thus his life within Boulder and NCAR became fully established.

The helioseismology program at ITP in 1990 helped to cement together the major community team efforts to build the analysis procedures necessary to deal with the vast data streams that would be forthcoming from two full-disk helioseismic imaging projects selected for development in the late 1980s. The NSF agreed to the GONG (Global Oscillations Network Group) project to place six ground-based Doppler imaging instruments around the globe, and NASA and ESA selected the MDI (Michelson Doppler Imager) instrument for the SOHO (Solar Heliospheric Observatory) spacecraft. Both had the goal of yielding nearly uninterrupted observation of the full solar disk. GONG became operational in 1995 and is continuing now, and MDI/SOHO started in 1996 and continued observations until April 2011. It was replaced by the higher resolution HMI (Helioseismic Magnetic Imager) instrument on the SDO (Solar Dynamics Observatory) spacecraft, launched February 2010 and observing now. Michael was an active participant in all three projects from their inception and helped to influence much of the analysis and interpretation of the resulting very large data sets.

2 Scientific Surprises Emerging About Dynamics Within the Sun

The helioseismology adventures for Michael involved many threads, but most had as a central theme the application of inversions to determine solar interior structure and of its internal rotation and large-scale flows. Michael has over 250 publications, many focused on the development, application and interpretation of inverse techniques in global and local helioseismology, and then extending to asteroseismology. His early work was on the effects of rotation and magnetic fields on the oscillations, and those continued to be a major interest. The inversions involve using the observed properties of some large fraction of the millions of acoustic oscillations excited within the Sun, such as their frequencies and their splittings, to deduce the differential rotation and magnetic fields that cause these distortions. Yet motivating much of this work dealt with trying to understand how a rotating star like the Sun is able to build its complex and evolving magnetic fields by dynamo action within its deep and turbulent convection zone below its surface. Thus let us now briefly sample some of those dynamical aspects that provided a substantial stimulus to Michael's research interests in probing the solar interior with helioseismology. The topics involved are very broad and involve many active researchers, some of whom have joined this gathering, and here we will only touch very lightly on some of the issues that provided a partial background to Michael's sustained research.

2.1 Nature of Solar Magnetism and Sunspot Mergence

Sunspots have long been a hallmark of solar magnetism, yet much is still unclear about their origins. There is no doubt that a magnetic dynamo is operating within the deep convection zone, and that individual sunspot pairs arise from a toroidal flux loop rising through the surface. But how, and where, is a convective dynamo able to build these toroidal structures? That is very much an open question. More broadly, the Sun as a rotating and magnetic star exhibits a wide variety of activity, ranging from the fairly regular 11-year cycles of major sunspot eruptions with orderly patterns on the global scales to the intense and more chaotic magnetic fields on the small scales. The strong magnetic fields that emerge through the solar surface are pivotal in controlling the evolving structure of the Sun's chromosphere and extended corona. The "butterfly diagram" of the evolving latitudinal dependence of sunspot emergence as the solar cycles advance is the trademark of solar variability. When accompanied with plots in time of the total area occupied by sunspots which waxes and wanes over the course of about 11 years, these two figures exemplify the prominent character of the magnetic Sun (Fig. 1a, b). These are the two "calling cards" of solar physics. Clearly a magnetic dynamo is operating within the convection zone, yet sorting out how this zone's intensely turbulent flows yield orderly magnetic behavior reveals many puzzles. There is considerable uncertainty

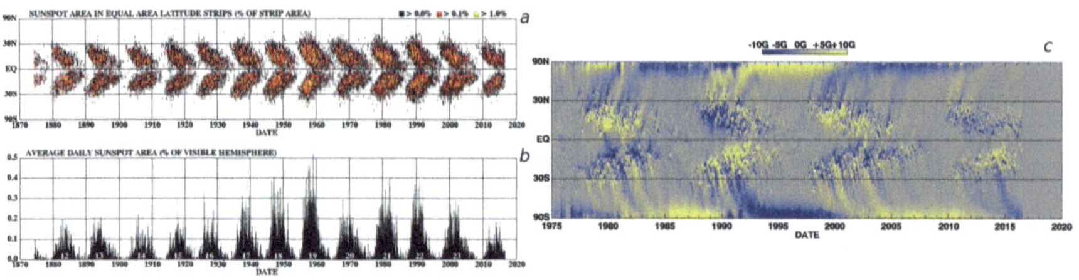

Fig. 1 Solar variability and sunspot butterfly diagrams. (**a**) Emerging sunspots are concentrated in two latitude bands on either side of equator, forming first at mid latitudes, widening, and then moving toward the equator as each cycle progresses. (**b**) The cycle is at a minimum as the sunspot emergence nears the equator, at which time new spots are also beginning to form at mid-latitudes. (**c**) Modern MDI and HMI mapping of evolving radial magnetic field associated with sunspots and active regions, averaged in longitude over successive solar rotations, showing polarity of field

and debate as to where the toroidal flux loops originate that rise through the surface to form individual sunspot pairs, and indeed how they are formed in the first place. Indeed, the sunspot behavior is the primary challenge to all solar dynamo theory. It is of basic significance since the rising flux of sunspots plays a crucial role in sculpting the evolving magnetic topology of the entire solar atmosphere. The advent of SOHO and now SDO has enabled radial magnetic field mappings averaged in longitude over successive solar rotations, as in Fig. 1c, with the butterfly signature of sunspots clearly visible at lower latitudes, along with fainter streaks of new polarity extending to the poles. More importantly, space has given us access to the UV and X-ray portions of the spectrum that reveal the complex loops and arcs of magnetism in the chromosphere and corona that accompany flux emergence in the photosphere.

2.2 Guidance from Helioseismology and Surprise of Tachocline

Helioseismology permits us to probe and assess flows deep within the Sun in ever increasing detail, using many of the inversion methods that Michael helped to develop and refine. It provides important guidance and constraints to dynamo theory in terms of the differential-rotation profile, meridional-circulation patterns, and the large-scale subsurface flows that interact with active regions, as shown in turn in Fig. 2. Indeed, helioseismic probing has revealed that two boundary layers are present, with both the near-surface shear layer (NSSL) and the tachocline at the base of the convection zone (CZ) being sites of enhanced rotational shear that can participate in dynamo action. These findings are now being examined with greater fidelity with HMI on SDO, as it is now providing continuous high-resolution measurements of Doppler velocity and vector magnetic fields across the full solar disk.

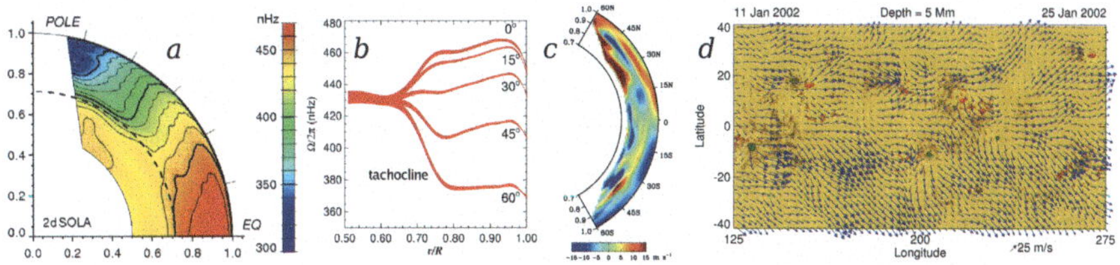

Fig. 2 (**a**) Angular velocity profile Ω from global helioseismology showing differential rotation within the solar interior, with fast rotation (red) at the equator and slow (blue) at poles. (**b**) Radial cuts of Ω at different latitudes. The prominent boundary layers of the tachocline at $r/R = 0.71$ and near-surface shear layer in the outer 5% by radius are evident [26, 27]. (**c**) The meridional flows with depth and latitude as determined with time-distance methods in local helioseismology [28]. (**d**) Synoptic map of horizontal flows of solar subsurface weather (SSW) at a depth of 7 mm deduced from ring-diagram methods, with surface magnetic field overlain [7]

The tachocline had not been anticipated prior to its discovery by helioseismology. This boundary layer of strong rotational shear near the base of the CZ ($r \sim 0.71R$) is a complex transition layer between the prominent differential rotation above and the uniform rotation of the deep radiative zone below. Although the tachocline appears to be about $0.05R$ in thickness, that likely reflects the resolution of the helioseismic inversion at that depth, and the tachocline may well be much narrower. It has been variously proposed that the rotational shear of the tachocline is confined by effects of anisotropic turbulence, by gravity waves, or by magnetic stresses. How and why such a thin rotational boundary layer has arisen in the Sun is still an open research topic, as reviewed by Miesch [17] and Miesch and Toomre [18], and in this meeting by Pascale Garaud. Of considerable interest is the role that the presence of a tachocline has on the operation of magnetic cycles, and on the storage, destabilization, and buoyant rise of coherent toroidal fields.

2.3 *Recent Advances in 3-D Global Dynamo Modeling*

Studying elements involved in the solar dynamo is made challenging by the vast range of length and time scales involved in the highly nonlinear dynamics of the deep convection zone. There has been major progress in modeling the coupling of turbulent convection, shear and magnetism in rotating spherical shells of conducting fluids, representative of the solar CZ. The 3-D global hydrodynamic simulations (Fig. 3), such as with the Anelastic Spherical Harmonic (ASH) code, have achieved solar-like differential rotation profiles that are largely in accord with helioseismic findings (e.g., Miesch et al. [19, 20]). However, although many research groups with other simulation codes have obtained similar differential rotation results, as also discussed here by Sacha Brun, there is an underlying puzzle yet to be resolved, which has come to be called the “convective conundrum”. Namely, as the solar simulations are made to represent ever more turbulent flows by reducing

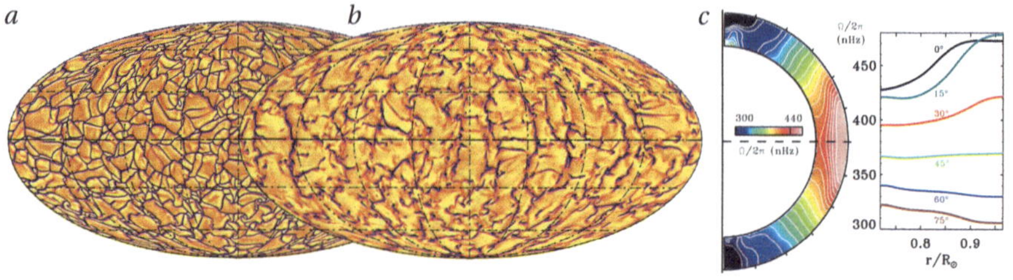

Fig. 3 Global solar convection ASH simulation [20], showing the connectivity of the major downflows with depth as the giant cells become more evident. Snapshot of radial velocity in Mollweide projection (downflows are dark) at (**a**) $0.98R$ near the top of the domain and (**b**) at mid-convection zone ($0.92R$) detailing the largest scale flows. (**c**) Separate ASH modeling with a specified thermal structure in the tachocline [19] yields a solar-like mean angular velocity Ω profile with radius and latitude (along with radial cuts of Ω at indicated latitudes) which make contact with helioseismic deductions

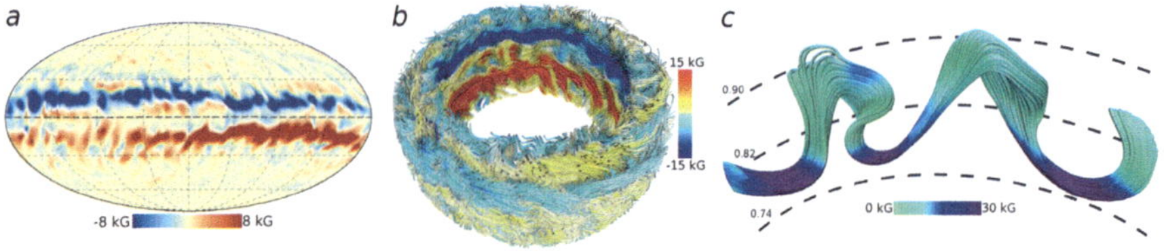

Fig. 4 Global dynamo simulations with ASH in case S3. (**a**) Prominent wreaths of strong toroidal magnetic fields are realized, with opposite polarity above and below the equator in a global Mollweide mapping at mid-depth. (**b**) Volume rendering of magnetic field lines shows the complex structure within these persistent wreaths. (**c**) Buoyant loop segment rising from a magnetic wreath, colored by field strength. (Nelson et al. [21, 22, 23])

the diffusivities inherent in the modeling, there is a tendency for the differential rotation to become anti-solar in the sense of slow equator and fast poles. This can be identified with the Rossby number of the faster and more turbulent convective flows approaching unity, with the effects of rotation through Coriolis forces weakening compared to the nonlinear inertial effects (e.g., Featherstone and Hindman [8]). Thus most related solar dynamo simulations have avoided this behavior by either having the model Sun rotating faster or having a lowered energy flux going through the CZ. These offer ways to use the 3-D simulations to explore how the subtle coupling of rotation and convection leads to differential rotation in a spherical shell geometry, and of its many abilities to operate magnetic dynamos. What causes the conundrum is uncertain, but likely treating the upper boundary of the CZ as a diffusive layer rather than being able to capture the very small scale granulation explicitly may be the culprit, along with having to treat diffusivities by various subgrid scale turbulent representations.

Turning to global MHD simulations with ASH, the simulations have revealed that remarkable wreaths of strong magnetic field (Fig. 4a, b) can be built in the bulk of the CZ by dynamo action (e.g., Brown et al. [2, 3] Nelson et al. [21]). These are

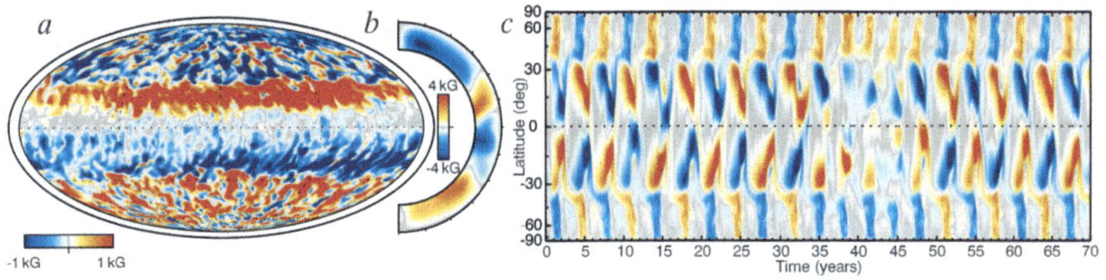

Fig. 5 Nature of toroidal magnetic field B_ϕ in ASH dynamo simulation case K3S. (**a**) Snapshot of B_ϕ in a global view (Mollweide projection) at $0.95R$, showing the connectivity of the magnetic wreaths (field polarity in color). (**b**) Longitude-averaged $\langle B_\phi \rangle$, also averaged in time, shown in a meridional plane, revealing that wreaths at lower latitudes occupy much of the convection zone depth. (**c**) Time-latitude diagram of $\langle B_\phi \rangle$ at $0.95R$, exhibiting repeated cycles of equatorward migration of the wreaths (from the tangent cylinder at latitude 40°), and the poleward propagation of the field at higher latitudes. The cycling enters a grand minimum, after which reversing cycling resumes. (Augustson et al. [1])

quite significant findings, since they are contrary to prior beliefs that such ordered magnetic structures could not be achieved nor survive in the midst of the intense turbulence there due to flux expulsion by the convective gyres. These wreaths of toroidal magnetic fields often have opposite polarity in the two hemispheres, and have a wide range of structure and behavior as discussed briefly in this meeting by Loren Matilsky, and in detail in Matilsky and Toomre [16]. The sense of the magnetic fields can also reverse in a cyclic manner over decade-long time scales. One of the recent ASH simulations (case K3S, Augustson et al. [1]) passed through many reversing cycles (Fig. 5) before experiencing brief quiescent intervals resembling a Maunder Minimum. Interestingly, whereas the primary cycling at low latitudes was interrupted, modulation continued at high latitudes, but without reversals (Fig. 5c). After the hiatus, the model resumed its regular cycling. We are seeing equatorward propagation of toroidal flux in this cycling system, a feature shared with sunspots as the cycle advances.

Another step forward with ASH has been to attain the first global convective dynamo simulation (case S3) to exhibit the spontaneous, self-consistent generation of rising magnetic loops (Nelson et al. [21, 23]), as shown in Fig. 4c. Over 130 such loops have been identified. By "self-consistent" we mean that convection under the influence of rotation generates a differential rotation that in turn generates a coherent toroidal field that in turn destabilizes and rises upward through the CZ. The rise is due to a combination of magnetic buoyancy and advection by the convective motions, which work together to move flux upward. Figures 6a, b illustrate the topology of the field both globally and locally in a buoyant loop. Figure 6c shows the spatial distribution of those rising loops relative to the wreath structure. The distribution of loop tilts near the top of their ascent makes contact with Joy's Law. It is remarkable that neither case K3S nor case S3 has a tachocline. This suggests that a tachocline is not crucial to achieve cyclic magnetic activity or for flux emergence in

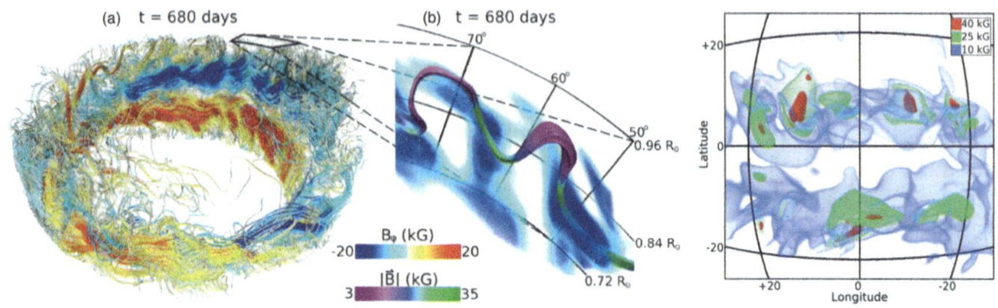

Fig. 6 Details of magnetic wreath structure and ascending loops in case S3. (**a**) Prominent wreaths of strong toroidal magnetic fields are realized, with opposite polarity above and below the equator. (**b**) Buoyant loops of magnetism are self-consistently achieved and transit the convection zone. (**c**) The buoyant loops form in the strongest regions of magnetic field, shown here in volume rendering (Nelson et al. [22])

the form of bipolar magnetic regions. However, its presence may modify the nature of the cycles by providing a second reservoir for strong fields.

In addition to our brief sampling of dynamical results here, there has been striking overall progress in 3-D modeling of the cyclic solar dynamo, using several different global codes including ASH. These studies have attained varied cyclic activity over a range of time scales, with both regular and irregular reversals and some even decadal in length (e.g., Ghizaru et al. [9]; Käpylä et al. [14]; Masada et al. [15]; Fan and Fang [6]; Passos et al. [24]; Guerrero et al. [12]; Hotta et al. [13]; Strugarek et al. [25]). A few of the studies have included a stable region at the base of the CZ to begin to explore the role that a tachocline may have on cycle periods or their regularity. Thus the world of 3-D global dynamo simulations, both solar and stellar, has attracted much competent attention recently, most focused on how to achieve cycling behavior. However, the case S3 in Figs. 4 and 6 is the first global dynamo to attain an abundance of rising loops from the strong magnetic wreaths, thus making detailed contact with the deep origins of sunspots. Due to their anelasticity, our global MHD simulations with ASH cannot track the ascending loops all the way to the surface. Yet by studying their deep origins they can set the stage for where, when and how the flux eruptions will occur. Tracking the final ascent requires fully compressible codes, such as developed by several groups carrying out 3-D radiative MHD modeling of near-surface solar convection and dynamo action in localized domains. A number of such studies are now showing the way forward in coping with the details of flux emergence through the highly turbulent solar surface. This is also the region where granulation serves as the driver of the acoustic oscillations so central to Michael's research interests, emphasizing the complex interplay of topics in solar and stellar physics.

Acknowledgments This project was supported by NASA grants 80NSSC18K1127 and NNX16AC92G.

References

1. Augustson, K. C., Brun, A. S., Miesch, M. S. & Toomre, J. (2015). *The Astrophysical Journal, 809*, 149.
2. Brown, B. P., Browning, M. K., Brun, A. S., Miesch, M. S., & Toomre, J. (2010). *The Astrophysical Journal, 711*, 424.
3. Brown, B. P., Browning, M. K., Brun, A. S., Miesch, M. S. & Toomre, J. (2011). *The Astrophysical Journal, 731*, 69.
4. Christensen-Dalsgaard, J., Gough, D. O., & Thompson, M. J. (1991). *The Astrophysical Journal, 378*, 413.
5. Christensen-Dalsgaard, J., Gough, D. O. & Thompson, M. J. (1992). *Astronomy and Astrophysics, 264*, 518.
6. Fan, Y., & Fang, F. (2014). *The Astrophysical Journal, 789*, 35.
7. Featherstone, N. A. (2011). Ph.D. Thesis, Univ. Colorado Boulder, Fig. 8.7, p. 158.
8. Featherstone, N. A., & Hindman, B. W. (2016). *The Astrophysical Journal Letters, 830*, L15.
9. Ghizaru, M., Charbonneau, P., & Smolarkiewicz, P. (2010). *The Astrophysical Journal Letters, 715*, L133.
10. Gough, D. O., & Thompson, M. J. (1990). *Monthly Notices of the Royal Astronomical Society, 242*, 25.
11. Gough, D. O., & Thompson, M. J. (1991). In A. N. Cox et al. (Ed.), *Solar interior and atmosphere* (p. 519). Tucson: University of Arizona Press.
12. Guerrero, G., Smolarkiewicz, P. K., de Gouveia Dal Pino, E. M., Kosovichev, A. G., & Mansour, N. N. (2016). *The Astrophysical Journal, 819*, 104.
13. Hotta, H., Rempel, M., & Yokoyama, T. (2016). *Science, 351*, 1427.
14. Käpylä, P. J., Mantere, M. J., & Brandenburg, A. (2012). *The Astrophysical Journal Letters, 755*, L22.
15. Masada, Y., Yamada, K., & Kageyama, A. (2013). *The Astrophysical Journal, 778*, 11.
16. Matilsky, L. I., & Toomre, J. (2020). *The Astrophysical Journal, 892*, 106.
17. Miesch, M. S. (2005). *Living Reviews in Solar Physics, 2*, 1.
18. Miesch, M. S., & Toomre, J. (2009). *Annual Review of Fluid Mechanics, 41*, 317.
19. Miesch, M. S., Brun, A. S., & Toomre, J. (2006). *The Astrophysical Journal, 641*, 618.
20. Miesch, M. S., Brun, A. S., DeRosa, M. L., & Toomre, J. (2008). *The Astrophysical Journal, 673*, 556.
21. Nelson, N. J., Brown, B. P., Brun, A. S., Miesch, M. S., & Toomre, J. (2011). *The Astrophysical Journal Letters, 739*, L38.
22. Nelson, N. J., Brown, B. P., Brun, A. S., Miesch, M. S., & Toomre, J. (2013). *The Astrophysical Journal, 762*, 73.
23. Nelson, N. J., Brown, B. P., Brun, A. S., Miesch, M. S., & Toomre, J. (2014). *Solar Physics, 289*, 441.
24. Passos, D., & Charbonneau, P. (2014), *Astronomy and Astrophysics, 568*, A113.
25. Strugarek, A., Beaudoin, P., Charbonneau, P., Brun, A. S., & do Nascimento, Jr., J.-D. (2017). *Science, 357*, 185.
26. Thompson, M. J., Toomre, J., Anderson, E. R., & Berthomieu, G. (1996). *Science, 272*, 1300.
27. Thompson, M. J., Christensen-Dalsgaard, J., Miesch, M. S., & Toomre, J. (2003). *Annual Review of Astronomy and Astrophysics, 41*, 599.
28. Zhao, J., Bogart, R. S., Kosovichev, A. G., Duvall, T. L., Jr., & Hartlep, T. (2013). *The Astrophysical Journal Letters, 774*, L29.

Part II
Solar Interior and Dynamics

Uncovering the Hidden Layers of the Sun

Sarbani Basu

Abstract Professor Michael Thompson had made major contributions in developing techniques that have allowed us to 'invert' helioseismic data. These analyses have allowed us to determine the structure of the inner layers of the Sun. This in turn has led to tests of solar models, and perhaps more importantly, tests of the properties of stellar material. I review some of the interesting results that helioseismic structure inversions have revealed and how today we are still using structure inversions to reveal subtle features of solar structure.

It was an honour to be asked to speak at the workshop honouring the life and work of Michael Thompson. Michael contributed to the development of many inverse techniques in helioseismology (see e.g., [42, 43, 27, 41]), and in particular the method of Subtractive Optimally Localized averages or SOLA [37] is one of the most commonly used techniques. Inverse techniques have allowed us to uncover details about the Sun that are hidden below the photosphere. We know the structure of the Sun very well, and we can use this knowledge to test solar models and the physics that goes into the models. We also know the dynamics of the Sun, and how dynamics changes with solar cycle. I refer you to Thompson [44], Christensen-Dalsgaard [20], Basu and Antia [10], Basu [9], Howe [30] etc., for earlier reviews of results obtained with helioseismic inversions, as well as to Rachel Howe's review of solar dynamics in this volume. I shall limit myself to solar structure.

S. Basu (✉)
Department of Astronomy, Yale University, New Haven, CT, USA
e-mail: sarbani.basu@yale.edu

M. J. P. F. G. Monteiro et al. (eds.), *Dynamics of the Sun and Stars*, Astrophysics and Space Science Proceedings 57,
https://doi.org/10.1007/978-3-030-55336-4_7

1 The Need for Inversions

In astrophysics, inferences are made through "forward modelling" of the data—you make a model of the object (in our case the Sun), calculate the observables (the oscillation frequencies) and compare the data with the observables from the model. A good fit to the data will imply a good model, and the structure and other properties of the model are believed to be the structure of the astrophysical object. This does not work for solar oscillation frequencies.

An issue that plagues all solar models is the fact that we cannot model convection properly. Convection is an inherently three-dimensional, dynamical phenomenon, and we approximate that to a static, one-dimensional equation either through the mixing length formalism or through other approximations. Most of the effect of the lack of proper modelling lies very close to the surface of the model, notably in the so-called super-adiabatic layer where convection is inefficient. The consequence of this is a frequency-dependent differences between frequencies of a model and that of the Sun. This is usually called the 'surface term' [21]. Once corrected for the effect of mode inertia, the frequency differences of modes of different degree collapse to one curve. The surface term also depends on the atmospheric model, and thus models that have essentially the same structure can show different amount of surface term, we show this in Fig. 1. This clearly shows that comparing frequencies will not help us in determining the structure of the Sun.

We therefore need an analysis technique that will allow us to determine solar structure while simultaneously removing the effects of the surface term. An inversion for solar structure generally proceeds through a linearization of the equations of stellar oscillations around a known reference model. The differences between the structure of the Sun and the reference model are then related to the differences in the frequencies of the Sun and the model by kernels. This linearization does not account for the surface term, and the latter has to be added in an *ad hoc* manner. When the oscillation equation is linearized—under the assumption of hydrostatic equilibrium—the fractional change in the frequency can be related to the fractional changes in two of the functions that define the structure of the models. Thus,

$$\frac{\delta\omega_i}{\omega_i} = \int K^i_{1,2}(r)\frac{\delta f_1(r)}{f_1(r)}dr + \int K^i_{2,1}(r)\frac{\delta f_2(r)}{f_2(r)}dr + \frac{F_{\rm surf}(\omega_i)}{E_i}. \qquad (1)$$

Here $\delta\omega_i$ is the difference in the frequency ω_i of the ith mode between the solar data and a reference model. The functions f_1 and f_2 are an appropriate pair of functions like sound speed and density, or, density and adiabatic index Γ_1, etc. The kernels $K^i_{1,2}$ and $K^i_{2,1}$ are known functions of the reference model which relate the changes in frequency to the changes in f_1 and f_2 respectively; and E_i is the inertia of the mode. The surface term is denoted by $F_{\rm surf}$, which is assumed to be a function of frequency alone. Including the surface term by hand works quite well for modes up to about $l = 250$, but does not for higher-degree modes [14]. "Inversions" for solar structure are carried out using different techniques with the

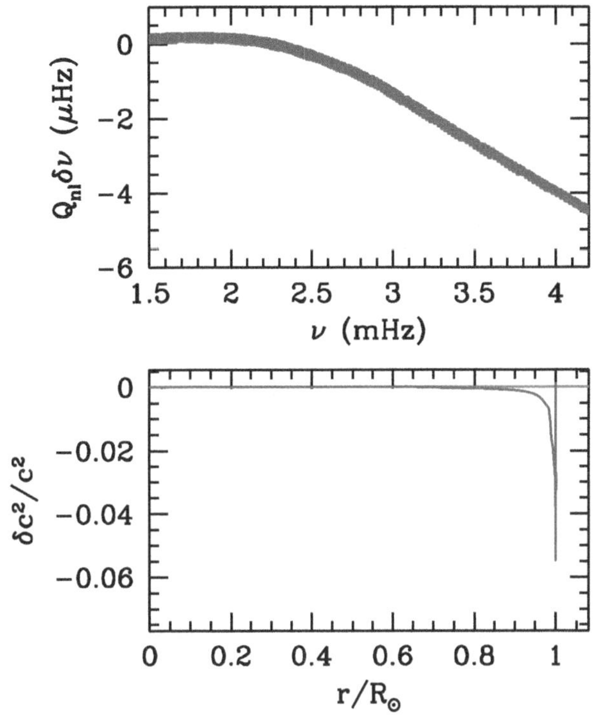

Fig. 1 Top: the frequency differences, scaled with mode inertia, between two solar models constructed with different model atmospheres, but otherwise identical inputs. One model was constructed with the Eddington T–τ relation, the other with the KrisnaSwamy relation. Bottom: The relative difference in the squared sound-speed of the models. Note that differences are concentrated very close to the surface

aim of determining $\delta f_1(r)/f_1(r)$ and $\delta f_2(r)/f_2(r)$ between the Sun and a solar model. Different techniques can be used perform the inversions; how the techniques are implemented are reviewed in [9]. One should note that the SOLA method proposed by Michael Thompson is one of the most popular ways of inverting Eq. (1).

2 Structure Inversion Results

With standard data-sets produced by the GONG, MDI and HMI projects, it is possible to determine the solar sound-speed and density profiles (or more precisely the difference between the sound-speed and density profiles of the Sun and a solar model) between about $0.05R_\odot$ and $0.96R_\odot$. We show inversion results in Fig. 2 for two well-known solar models. The two models are undoubtedly different and the differences are mainly due to the physics and other inputs to the models. Model S

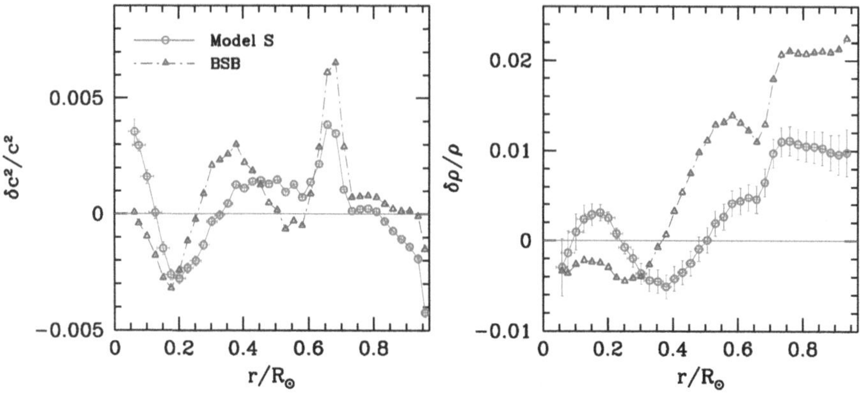

Fig. 2 The relative difference of the squared sound-speed (left panel) and density (right panel) between the Sun and two solar models, Model S is from Christensen-Dalsgaard et al. [23] and BSB refers to model BS05(OP) from Bahcall et al. [7]. Error-bars are shown for only one result for the sake of clarity

was constructed with a higher solar metallicity ($Z/X = 0.024$ from Grevesse and Noels [28]) than model BSB ($Z/X = 0.023$ from Grevesse and Sauval [29]), and that accounts for the sound-speed difference just below the base of the solar convection-zone (known to be at $0.713R_\odot$). The issue of metallicity is discussed further in Sect. 4. Additionally, model BSB used newer reaction rates and an updated equation of state (EOS), that of Rogers and Nayfonov [38], compared with Model S [39], and that accounts for most of the difference between the two models in the core. The error bars show that there are statistically significant differences between the structure of the Sun and the two models. The very localized difference in sound speed below the base of the convection zone ($0.713R_\odot$) is believed to be caused by the lack of rotationally-induced and turbulent mixing in the models.

Density inversion results show a larger difference, particularly in the outer layers. One needs to keep in mind that because the models are constructed to have the same mass as the Sun, a small difference in density in the core (where densities are larger) automatically implies a large difference in the envelope (where densities are much smaller). Unfortunately, density inversion results also depend on the mode set inverted, and change when the number of low-degree modes (modes that penetrate to the core) is increased (see [12]). Thus to determine the density of the different layers of the Sun with better precision, we need more low degree modes, preferably modes with low frequency since their kernels have higher amplitudes in the core. Of course, solar g modes should solve this problem.

Note that for both models the sound-speed difference between the model and the Sun are at the level of fractions of a percent, while density differences are at most a few percent, showing that our solar models are actually very good, in fact most fields of astrophysics would consider the models to be perfect. In both cases, the differences between the Sun and the models in the core are very small. This, hinted that the solution of the so-called "solar neutrino problem" did not lie with modifying solar models. To recapitulate, the solar neutrino problem was a decades long disagreement between the predicted fluxes of neutrinos from solar models, and the observed flux of neutrinos. In all cases, the observed flux was about a third of that predicted. It has been argued that given the relative messiness of a standard solar model compared to the standard model of particle physics, deficiencies in solar models must be the reason for the discrepancy. This led to decades-long efforts to construct solar models with non-standard processes to try to reconcile the neutrino predictions and observations. However, as shown in Fig. 2, our solar models are very good indeed. In fact, it can be shown (see [6]), that constructing a solar model to satisfy the neutrino constraint at the 3σ level will increase the sound-speed discrepancy in the core by a factor of about 15. As we know now, the root of the solar neutrino problem was the assumption that neutrinos are massless. If neutrinos have mass, they can change from being electron-type neutrinos produced in the solar core to μ-type neutrinos. The Sudbury Neutrino Observatory, unlike the other neutrino observatories, can observe both electron-type and μ-type neutrinos, and its observations showed that the predicted flux of neutrinos is very similar to the observed flux [1]. The erstwhile problem and the current situation are illustrated in Fig. 3. This is one of the few cases in astrophysics that a long-standing problem had a clear resolution, so-much-so that the astrophysics community does not even discuss this any more.

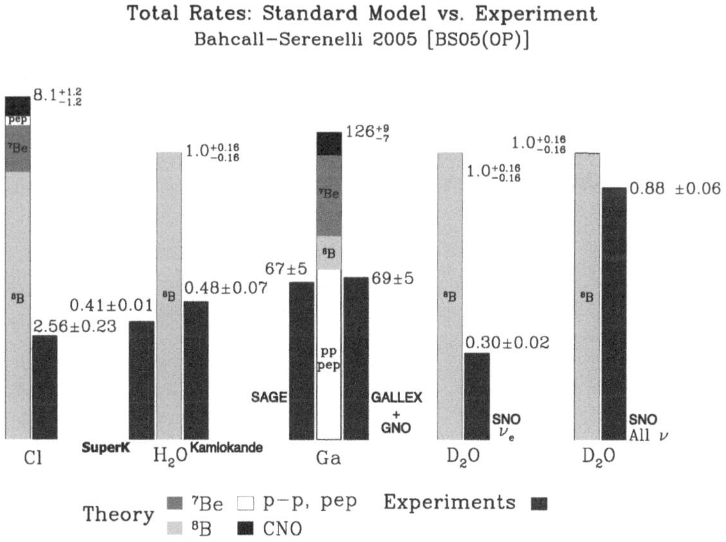

Fig. 3 A comparison of observed (light gray) and predicted (dark grey) neutrino fluxes of different types. Cl, H_2O, Ga, and D_2O refer to the kind of detectors; the name of the experiments are in black bold font. The model predictions are from Bahcall et al. [7]

3 What Else Can We Learn?

The precision with which we can determine the structure difference between the Sun and solar models allows us to test inputs to the models in a relatively simple matter—we change one physics input at a time and see how the resulting model compares with the Sun.

One of the first such tests was to examine whether or not including the diffusion and gravitational settling of helium and heavy elements in solar models is necessary. Christensen-Dalsgaard et al. [22] were the first to demonstrate using inversions that including diffusion improves the match between solar models and the Sun. Models without diffusion have very shallow convection zones and very high convection-zone helium abundances. A comparison of the sound-speed difference between the Sun and two modern models, one with and one without the settling of helium and heavy elements, is shown in Fig. 4. Clearly the model with diffusion is better. The settling of helium and heavy elements is now considered a standard physics process. Inversion results for the hydrogen and/or heavy element abundances also revel the presence of diffusion [33, 40, 3]; these inversion, usually referred to as secondary inversions, required the assumption that the opacity of solar material is known well.

Solar structure results are not just useful for studying physical processes that take place inside the Sun, but also to study miocrophysics inputs such as the equation

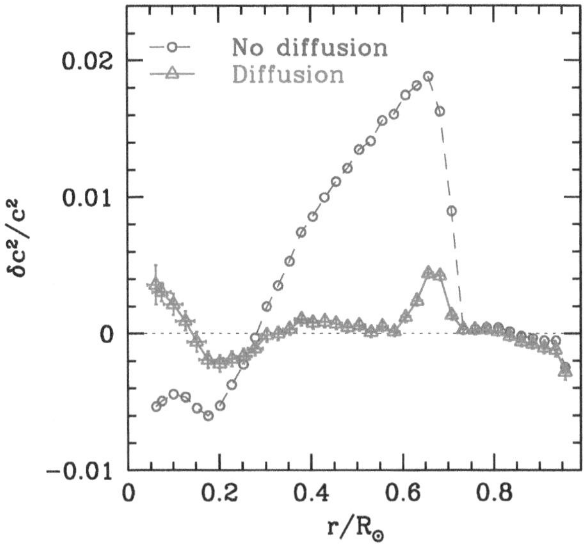

Fig. 4 The relative sound-speed difference between two models and the Sun obtained by inverting helioseismic data. The two models differ in that the model marked by the triangles includes the diffusion and gravitational settling of helium and heavy elements, while the model marked by circles does not. 1σ error bars are shown only on one of the results; uncertainties are the same for both results

of state. The speed of sound is related to the first adiabatic index Γ_1 through the relation $c^2 = \Gamma_1 P/\rho$ and thus the c^2–ρ pair of variables can easily be converted to the Γ_1–ρ pair, which allows us to invert for the Γ_1 profile of the Sun. The Γ_1 profile is essentially determined by the equation of state, though abundances matter too (Γ_1 decreases in ionization zones). In the core one would expect Γ_1 to be constant, about 5/3, the value for a fully ionized non-relativistic perfect gas. Thus one would not expect a Γ_1 difference between models and the Sun near the solar core. Much to our surprise, Elliott and Kosovichev [26] found a significant difference, with Γ_1 in the solar core *lower* than that in the models. We know now that this is because the equations of state available at the time did not account for the fact that the velocity distribution of electrons in the solar core has a relativistic tail, making Γ_1 lower than the expected value of 5/3 (recall that for a perfectly relativistic ideal gas, Γ_1 is 4/3). This oversight has been corrected, and the EOSs available today do not show this discrepancy.

Testing the equation of state in the outer layers on the Sun using Γ_1 is more challenging since abundances, pressure and density play a role in defining the Γ_1 profile. In particular, ionization causes Γ_1 to decrease, but the amount of decrease and the resulting change in the Γ_1 profile depends on the abundance and local conditions, such as pressure and density, in addition to the EOS. Basu and Christensen-Dalsgaard [13] showed that it is possible to separate out the part of the Γ_1 that depends on the EOS alone, they called it the "intrinsic Γ_1" or $\Gamma_{1,\mathrm{int}}$. The difference in $\Gamma_{1,\mathrm{int}}$ between the Sun and models constructed with three different

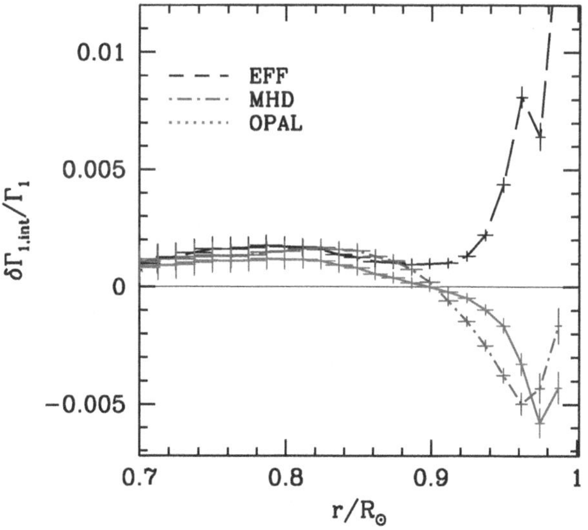

Fig. 5 The relative difference of intrinsic Γ_1 between the Sun and solar models constructed with different EOS. EFF refers to the Eggleton, Faulkner and Flannery [25] EOS, MHD is the Mihalas, Hummer and Däppen EOS [24, 35, 31]. The model labelled OPAL uses the EOS of Rogers et al. [39]

EOSs are shown in Fig. 5. It is clear that none of the EOSs are very good close to the surface, though the MHD and OPAL do much better than the EFF EOS.

Somewhat more indirect means have been used to test opacities and nuclear energy generation rates. See Basu [9] for a review.

4 The Current Issue: The Solar Metallicity Problem

The most contentious issue in terms of studying solar structure today is the question of solar metallicity. It had been believed that the total metallicity as well as the relative abundance of different elements in the Sun (i.e., the "solar mixture") are known well; after all they are used as standards for studying other stars. However, there is a disagreement that has made solar models uncertain.

The abundance problem, as it is often called, started in the early 2000s when in a series of papers, Allende Prieto et al. [2] and others revised the spectroscopic estimates of the solar photospheric composition downwards from the then accepted value of $Z/X = 0.023$ [29]. The solar abundances of carbon, nitrogen and oxygen were lowered by 35% to 45% of those listed in Grevesse and Sauval [29]. These measurements have been summarised by Asplund et al. [4]. The net result of these changes is that Z/X for the Sun is reduced to 0.0165 (or $Z \sim 0.0122$). The difference in their analysis compared with previous efforts was the use of three-dimensional model atmospheres; they also incorporated some non-LTE effects in their analysis. With further improvement in analysis, abundances were updated by Asplund et al. [5] to $Z/X = 0.018$. There were other, independent, attempts to determine solar heavy element abundances using 3D model atmospheres and using non-LTE effects, and this led to $Z/X = 0.0209$ [17, 18].

The heavy-element abundance changes radiative opacities, and a simple rule of thumb is that higher the abundance, the higher is the opacity. The mixture also has an effect, but to a lower degree. The change in opacity changes the position of the base of the convection zone of solar models—low-metallicity models have shallower convection zones—increasing the difference in structure between the Sun and the model; this illustrated in Fig. 6. The changes are not limited to the base of the convection zone—models with lower abundances do not reproduce some of the seismic signatures of the solar cores, namely the small-frequency ratios [19], and the surface helium abundance of the low-Z models are lower than the observed abundance [11]. In fact all helioseismic signatures point to a higher abundance (see, e.g., Basu and Antia [10]), though all the tests assumed that the microphysics inputs to the models, in particular opacity and diffusion rates, are known exactly.

The assumption that radiative opacities are known exactly is not correct, in fact the usual errors are assumed to be around 10%, and even that may be an underestimate. A 30% increase in the opacities localized around the base of the convection zone could actually solve the problem with the model. Thus the position of the convection zone in models is not the best indicator of whether input abundances are correct. However, opacities do not matter at all within the convection

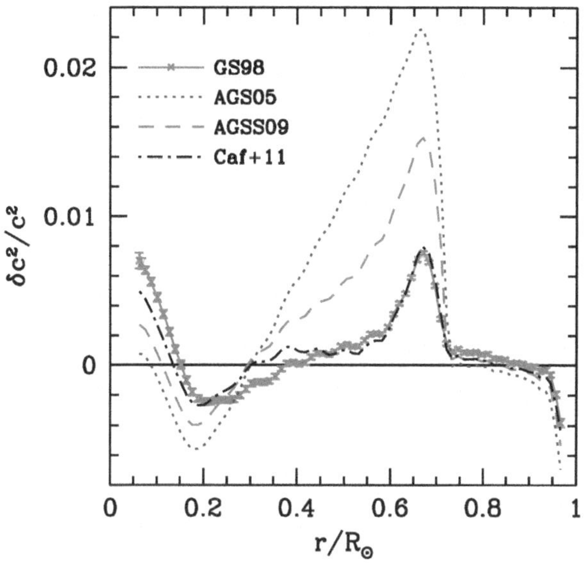

Fig. 6 The relative sound-speed difference the Sun and solar models constructed with different abundances, but otherwise identical inputs. GS98 refers to the abundance of Grevesse and Sauval [29], AGS05 to Asplund et al. [4], AGSS09 to Asplund et al. [5], and Caf+11 to Caffau et al. [17, 18]. Although Z/X of Caf+11 is lower than that of GS98, the difference in the mixture compensates for that

zone, the equation of state and abundances matter instead. Thus this is where one might look at to examine whether the lower abundances work or not. Lin et al. [34] used the signature of the ionization zones on Γ_1 to find that the metallicity has to be high; the only issue is that the signature is EOS dependent, and thus EOS errors may be an issue. More recently Buldgen et al. [15, 16] inverted for the difference in $S_{5/3} = P/\rho^{5/3}$, which is a proxy for entropy, the results are shown in Fig. 7, and again clearly, the higher metallicity models are doing better (GN93 refers to the abundances derived by Grevesse and Noels [28] with $Z/X = 0.0245$), showing that S has an abundance dependence,

Unfortunately, unlike the case of the EOS, one cannot do an inversion to determine opacities that are independent of the metallicity. Opacities can be thought of as having two components, the intrinsic opacity and a metallicity dependence, and inversions cannot separate the two effects. A welcome development in this matter are the experiments to measure opacities at conditions found at the base of the convection zone. Bailey et al. [8] measured Fe opacities at conditions present at the base of the convection zone using the Sandia Z facility to find that the measured opacities were higher than the calculated ones, even far away from lines. At temperatures and densities present near the base of the convection zone, Fe is seventeen-fold ionized; it has a full S shell and one vacancy in the L shell. It has been hypothesized that the increased opacity is due to effects related to the vacancy

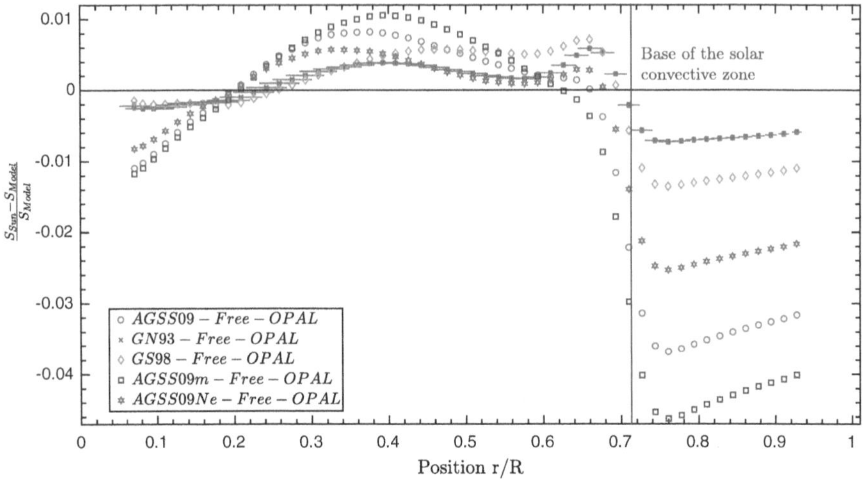

Fig. 7 Inversion results for the entropy proxy S for solar models constructed with different solar metallicities but otherwise identical inputs. GN93 refers to the abundances derived by Grevesse and Noels [28]. "Free" in the legend refers to the FreeEOS of Irwin [32]. Image courtesy of G. Buldgen. Results are from Buldgen et al. [15, 16]

in the L-shell that are not correctly taken into account. To confirm Fe-opacity measurements and test the hypothesis that the L-shell vacancy is the cause of the discrepancy between calculations and observations, Nagayama et al. [36] repeated the experiment and additionally measured the opacity due to Cr (three vacancies in the L shell) and Ne (no vacancy in the L shell). If the L shell vacancy is the cause, the discrepancy between calculations and measurements should be highest for Cr and lowest for Ne. The experiment confirmed the Bailey et al. Fe results, however, it was found that the disagreement between experiment and theory was higher for Fe and not, as was expected, for Cr. Thus clearly a lot more work, both experimental and theoretical, needs to be done before we understand opacities of stellar materials.

5 Summary

Helioseismic inversions, some using the method pioneered by Michael Thompson, have given us a very clear picture of what the structure of the Sun is like. We know solar structure quite well and we have also been able to constrain some of the physical processes that go on inside the Sun. We still have problems with microphysics inputs though, and thanks to the solar abundance problem, opacities are at the forefront of these issues. More direct measurements, along with parallel theoretical work, are needed to understand what is missing in the current opacity calculations.

Acknowledgments I would like to thank the organizing committee for giving me a chance to speak at this workshop honoring Michael Thompson who was my first post-doc supervisor. I would like to thank Gaël Buldgen for drawing Fig. 7 for this article. Figure 3 is based on a similar figure shown by the late John Bahcall.

References

1. Ahmad, Q. R., Allen, R. C., Andersen, T. C., Anglin, J. D., Barton, J. C., Beier, E. W., Bercovitch, M., et al (2002) Direct evidence for neutrino flavor transformation from neutral-current interactions in the sudbury neutrino observatory. *Physical Review Letters, 89*(1), 011301. nucl-ex/0204008.
2. Allende Prieto, C., Lambert, D. L., & Asplund, M. (2001). The forbidden abundance of oxygen in the sun. *The Astrophysical Journal Letters, 556*, L63–L66. astro-ph/0106360.
3. Antia, H. M., & Chitre, S. M. (1998). Determination of temperature and chemical composition profiles in the solar interior from seismic models. *Astronomy & Astrophysics, 339*, 239–251. astro-ph/9710159.
4. Asplund, M., Grevesse, N., & Sauval, A. J. (2005). The solar chemical composition. In T. G. Barnes III, Bash, F. N. (Eds.), *Cosmic abundances as records of stellar evolution and nucleosynthesis*. Astronomical Society of the Pacific Conference Series (vol. 336, p. 25).
5. Asplund, M., Grevesse, N., Sauval, A. J., & Scott, P. (2009). The chemical composition of the sun. *Annual Review of Astronomy and Astrophysics, 47*, 481–522, 0909.0948
6. Bahcall, J. N., Basu, S., & Pinsonneault, M. H. (1998). How uncertain are solar neutrino predictions? *Physics Letters B, 433*(1–2), 1–8. astro-ph/9805135.
7. Bahcall, J. N., Serenelli, A. M., & Basu, S. (2005). New solar opacities, abundances, helioseismology, and neutrino fluxes. *The Astrophysical Journal Letters, 621*, L85–L88. astroph/0412440.
8. Bailey, J. E., Nagayama, T., Loisel, G. P., Rochau, A., Blancard, C., Colgan, J., et al. (2015). A higher-than-predicted measurement of iron opacity at solar interior temperatures. *Nature, 517*(7532), 56–59.
9. Basu, S. (2016). Global seismology of the Sun. *Living Reviews in Solar Physics, 13*, 2. 1606.07071.
10. Basu, S., & Antia, H. M. (2008). Helioseismology and solar abundances. *Phys Rep 457*, 217–283. 0711.4590
11. Basu, S., & Antia, H. M. (2013). Revisiting the issue of solar abundances. In *Journal of Physics Conference Series* vol. 440, p. 012017
12. Basu, S., Chaplin, W. J., Elsworth, Y., New, R., & Serenelli, A. M. (2009). Fresh insights on the structure of the solar core. *The Astrophysical Journal, 699*, 1403–1417. 0905.0651.
13. Basu, S., & Christensen-Dalsgaard, J. (1997). Equation of state and helioseismic inversions. *Astronomy & Astrophysics, 322*, L5–L8. astro-ph/9702162.
14. Basu, S., Christensen-Dalsgaard, J., Perez Hernandez, F., & Thompson, M. J. (1996). Filtering out near-surface uncertainties from helioseismic inversions. *Monthly Notices of the Royal Astronomical Society, 280*, 651.
15. Buldgen, G., Salmon, S. J. A. J., Noels, A., Scuflaire, R., Reese, D. R., Dupret, M. A., et al. (2017). Seismic inversion of the solar entropy. A case for improving the standard solar model. *Astronomy & Astrophysics, 607*, A58. 1707.05138.
16. Buldgen, G., Salmon, S., & Noels, A. (2019). Progress in global helioseismology: A new light on the solar modelling problem and its implications for solar-like stars. *Frontiers in Astronomy and Space Sciences, 6*, 42. 1906.08213.
17. Caffau, E., Ludwig, H. G., Bonifacio, P., Faraggiana, R., Steffen, M., Freytag, B., et al. (2010). The solar photospheric abundance of carbon. Analysis of atomic carbon lines with the CO5BOLD solar model. *Astronomy & Astrophysics, 514*, A92. 1002.2628.

18. Caffau, E., Ludwig, H. G., Steffen, M., Freytag, B., & Bonifacio, P. (2011). Solar chemical abundances determined with a CO5BOLD 3D model atmosphere. *Solar Physics, 268*, 255–269. 1003.1190.
19. Chaplin, W. J., Serenelli, A. M., Basu, S., Elsworth, Y., New, R., & Verner, G. A. (2007). Solar heavy-element abundance: Constraints from frequency separation ratios of low-degree p-modes. *The Astrophysical Journal, 670*(1), 872–884. 0705.3154.
20. Christensen-Dalsgaard, J. (2002). Helioseismology. *Reviews of Modern Physics, 74*, 1073–1129. astro-ph/0207403.
21. Christensen-Dalsgaard, J., & Berthomieu, G., (1991). In A. N. Cox, W. C. Livingston, M. S.Mathews, (eds.), *Solar Interior and Atmosphere* (pp. 401–478). Univ. of Arizona Press: Tuscon.
22. Christensen-Dalsgaard, J., Proffitt, C. R., & Thompson, M. J. (1993). Effects of diffusion on solar models and their oscillation frequencies. *The Astrophysical Journal Letters, 403*, L75–L78.
23. Christensen-Dalsgaard, J., Dappen, W., Ajukov, S. V., Anderson, E. R., Antia, H. M., Basu, S. et al. (1996). The current state of solar modeling. *Science, 272*, 1286–1292.
24. Däppen, W., Mihalas, D., Hummer, D. G., & Mihalas, B. W. (1988). The equation of state for stellar envelopes. III - Thermodynamic quantities. *The Astrophysical Journal, 332*, 261–270.
25. Eggleton, P. P., Faulkner, J., & Flannery, B. P. (1973). An approximate equation of state for stellar material. *Astronomy & Astrophysics, 23*, 325.
26. Elliott, J. R., & Kosovichev, A. G. (1998). The adiabatic exponent in the solar core. *The Astrophysical Journal Letters, 500*, L199.
27. Gough, D. O., & Thompson, M. J., (1991). In A. N. Cox, W. C. Livingston, M. S.Mathews, (eds.), *Solar Interior and Atmosphere* (pp. 519–561). Univ. of Arizona Press: Tuscon.
28. Grevesse, N., & Noels, A. (1993). Cosmic abundances of the elements. In N. Prantzos, E. Vangioni-Flam, M. Casse (Eds.), *Origin and evolution of the elements* (pp. 15–25). Cambridge: Cambridge University Press.
29. Grevesse, N., & Sauval, A. J. (1998) Standard Solar Composition. *Space Science Reviews, 85*, 161–174.
30. Howe, R. (2009). Solar interior rotation and its variation. *Living Reviews in Solar Physics, 6*, 1. 0902.2406.
31. Hummer, D. G., & Mihalas, D. (1988). The equation of state for stellar envelopes. I - an occupation probability formalism for the truncation of internal partition functions. *The Astrophysical Journal, 331*, 794–814.
32. Irwin, A. W. (2012). FreeEOS: Equation of State for stellar interiors calculations, in *Astrophysics Source Code Library*, record ascl:1211.002
33. Kosovichev, A. G. (1995). Determination of interior structure by inversion. In R. K. Ulrich, E. J. Rhodes, Jr., W. Dappen, (Eds.) *GONG 1994. Helio- and Astro-Seismology from the Earth and Space*. Astronomical Society of the Pacific Conference Series (vol. 76, p. 89)
34. Lin, C. H., Antia, H. M., & Basu, S. (2007). Seismic study of the chemical composition of the solar convection zone. *The Astrophysical Journal, 668*(1), 603–610. 0706.3046.
35. Mihalas, D., Däppen, W., & Hummer, D. G. (1988). The equation of state for stellar envelopes. II - Algorithm and selected results. *The Astrophysical Journal, 331*, 815–825.
36. Nagayama, T., Bailey, J. E., Loisel, G. P., Dunham, G. S., Rochau, G. A., Blancard, C., et al. (2019). Systematic study of L - shell opacity at stellar interior temperatures. *Physical Review Letters, 122*(23), 235001.
37. Pijpers, F. P., & Thompson, M. J. (1992). Faster formulations of the optimally localized averages method for helioseismic inversions. *Astronomy & Astrophysics, 262*, L33–L36.
38. Rogers, F. J., & Nayfonov, A. (2002). Updated and expanded OPAL equation-of-state tables: Implications for helioseismology. *The Astrophysical Journal, 576*, 1064–1074.
39. Rogers, F. J., Swenson, F. J., & Iglesias, C. A. (1996). OPAL equation-of-state tables for astrophysical applications. *The Astrophysical Journal, 456*, 902.

40. Shibahashi, H., & Takata, M. (1996). A seismic solar model deduced from the sound-speed distribution and an estimate of the neutrino fluxes. *Publications of the, Astronomical Society of Japan, 48*, 377–387.
41. Sudnik, M., & Thompson, M. J. (2009). Iterative Inversion for Solar and Stellar Internal Rotation. *Astronomical Society of the Pacific Conference Series* vol. 416, p. 411.
42. Thompson, M. J. (1990). A new inversion of solar rotational splitting data. *Solar Physics, 125*(1), 1–12.
43. Thompson, M. J. (1991). Lecture Notes in Physics, 388, p.61
44. Thompson, M. J. (1998). What we have learned from helioseismology. *Astrophysics and Space Science, 261*, 23–34.

Solar Rotation

Rachel Howe

Abstract The study of the solar interior rotation has a long history, in which helioseismology has played a vital role and produced some unexpected results that challenged modellers—such as the nearly-radial profile of the lines of constant rotation in the convection zone, the thin shear layer or tachocline at its base, and the nearly-rigid rotation in the radiative interior. This review will highlight Michael Thompson's important contributions to techniques for helioseismic rotation inversions, and the results that those contributions have enabled. These include the study of the near-surface shear layer and the evolution of zonal flows in the convection zone over the solar cycle—the so-called torsional oscillation, which can be seen in surface observations but which helioseismology reveals to penetrate deep into the convection zone. Finally, I briefly mention the collaborative effort that was initiated by Michael in 2017 with the aim of taking advantage of more than twenty years of high-quality helioseismic observations to refine our picture of the interior rotation profile.

1 Introduction

The study of the solar interior rotation has been an important theme in helioseismology since its early days. Michael Thompson made several important contributions to this field, some of which will be discussed in this brief review.

R. Howe (✉)
School of Physics and Astronomy, University of Birmingham, Birmingham, UK

Stellar Astrophysics Centre (SAC), Department of Physics and Astronomy, Aarhus University, Aarhus C, Denmark
e-mail: r.howe@bham.ac.uk

M. J. P. F. G. Monteiro et al. (eds.), *Dynamics of the Sun and Stars*, Astrophysics and Space Science Proceedings 57,
https://doi.org/10.1007/978-3-030-55336-4_8

2 The Inversion Problem

In global helioseismology, the basic observational data are time series of Dopplergrams or intensity images of the solar disc. Spatial and temporal Fourier analysis is used to generate power spectra each for spherical harmonic, from which the mode frequencies can be obtained by a process colloquially known as "peak bagging." Depending on their radial order n, degree l, and azimuthal order m, different modes are trapped in different radial and latitudinal regions of the solar interior. The inversion problem, simply put, is to infer from the mode frequencies—in the case of inversions for rotation, specifically the m-dependence of the frequencies within a multiplet of a given n and l—the rotation as a function of latitude and radius. In mathematical terms, we have a set of data d_i, which will be either rotational splitting measurements $(\nu_{nlm} - \nu_{nl-m})$, or polynomial coefficients parametrizing the splittings. Each datum represents some weighted linear average of the rotation rate,

$$d_i = \int_0^{R_\odot} \int_0^{\pi} K_i(r,\theta)\Omega(r,\theta)drd\theta + \epsilon_i, \tag{1}$$

where $R_\odot$ is the solar radius, ϵ is an error term, and K is a model-dependent spatial weighting function, the *kernel* [18, 12]. For the two-dimensional rotation inversion, the radial part of K is related to the eigenfunction of the mode and the latitudinal part to the associated Legendre polynomial. We want to find

$$\bar{\Omega}(r_0,\theta_0) = \sum_{i=1}^{M} c_i(r_0,\theta_0)d_i, \tag{2}$$

where (r_0, θ_0) is the location at which the inferred rotation rate $\bar{\Omega}$ is to be found and the c_i are the coefficients to be used to weight the data; in other words, we need to find the best values for these coefficients.

Substituting Eq. (1) into the RHS of Eq. (2) gives

$$\bar{\Omega}(r_0,\theta_0) = \int_0^{R_\odot} \int_0^{\pi} \mathcal{K}(r_0,\theta_0;r,\theta)\Omega(r,\theta)drd\theta + \epsilon, \tag{3}$$

where

$$\mathcal{K}(r_0,\theta_0;r,\theta) \equiv \sum_{i=1}^{M} c_i(r_0,\theta_0)K_i(r,\theta) \tag{4}$$

is the *averaging kernel* for the location (r_0, θ_0), which is independent of the values of the data. Because the uncertainties in the data are used to weight the inversion calculation that generates the coefficients c_i, both the uncertainties and the selection of modes in the data set affect the averaging kernels. Averaging kernels provide a useful tool for assessing the reliability of an inversion inference from a particular mode set [43, 44]. Because of noise and the finite number of data, there is no unique solution, and we need to find a suitable balance between the uncertainties of the inferred rotation rate at each location and the extent to which they can be localized. This is generally implemented as one or more "tradeoff parameters". There are various possible approaches to this. One example is regularized least squares (RLS: e.g. [44]), where the aim is to optimize the match between the data and the corresponding values forward-calculated from the inferred profile, with a penalty term to enforce smoothness. Another approach is known as optimally localized averaging (OLA; e.g. [38]), where the goal is to optimize the match between the averaging kernel and a 'target' kernel such as a Gaussian centered on the target location. An important contribution of Michael Thompson and collaborators [9] was to suggest the use of the tradeoff curve, a plot of averaging-kernel width against error magnification, to assess the best choice of such parameters; this can be applied to any inversion technique where averaging kernels can be calculated.

When interpreting inversion results, it is important to consider the averaging-kernel properties and bear in mind that it is not possible to get well-localized inferences in some locations, particularly at high latitudes and small fractional radius. It can also be useful to consider how the error estimates on the inferred rotation rate at different locations are correlated [22]. We should also remember that the inversion results can be compromised if the data themselves contain systematic errors.

For a more detailed discussion of inversion methods, see [21] and references therein.

3 A Tour of the Rotation Profile

Early attempts to measure the solar interior rotation, such as the one by Thompson [49], used data from instruments such as the one that operated at the Big Bear Solar Observatory (BBSO) between 1986 and 1990. Since the 1990s, we have accumulated many years of high-quality resolved-Sun observations from facilities such as the Global Oscillation Network Group (GONG: [19]), the *Michelson Doppler Imager* (MDI: [42]) onboard the *Solar and Heliospheric Observatory*, and MDI's successor, the *Helioseismic and Magnetic Imager* (HMI: [46]) onboard the *Solar Dynamics Observatory*. In this section we will briefly describe the important features revealed by helioseismic inversions of these data.

The solar interior rotation profile can be divided into a number of zones (Fig. 1): the nuclear-burning core; the radiative interior; the tachocline, or shear layer at the base of the convection zone; the convection zone; and the near-surface shear layer that occupies the outermost few per cent by radius. We will consider each of these in turn.

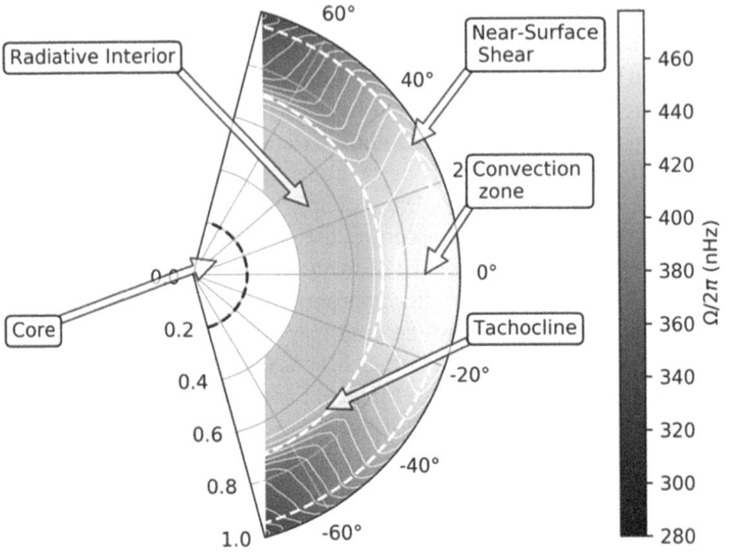

Fig. 1 Solar interior rotation profile from 2dRLS inversions of HMI data, indicating the important areas

3.1 The Core

Only the lowest-degree p-modes penetrate to the innermost region of the solar interior—below 0.2 $R_\odot$—where nuclear fusion takes place, and these modes sample all the higher layers and spend only a small fraction of their time in the core. Furthermore, in Sun-as-a-star observations such as those from the BiSON network, only the low-frequency modes have lifetimes long enough for the rotationally split components of an $l = 1$ mode to be resolved, and these lower-frequency modes have shallower lower turning points. This means that inversions have very little sensitivity to the core rotation, and, conversely, the inferred core rotation rate is very sensitive to systematic errors in the rotational splitting of the low-degree modes (see, for example, [7]). Attempts have been made to constrain rotation close to the core using a combination of low-degree splittings and resolved-Sun observations, for example by Chaplin et al. [8], who used data from BiSON and the LOWL instrument and found a core rotation slightly slower than the surface rate but consistent with a flat rotation profile down to 0.15 $R_\odot$.

While p-modes are of limited use in inferring the solar core rotation, g-modes—oscillations where the restoring force is gravity rather than pressure—are much more sensitive to the core. However, these modes are believed to have very small amplitudes at the solar surface—so small that there are still no widely accepted detections despite numerous attempts to measure them. Intriguing results were reported by Garcia et al. [15], who found a pattern of low-frequency patterns in GOLF (*Global Oscillations at Low Frequencies*) data with the period spacing expected for $l = 2$ g-modes and a splitting corresponding to a core rotation rate three to five times greater than the surface rate, but this result remains unconfirmed. More recently, Fossat et al. [14] also claimed evidence for a rapidly rotating core from a different analysis of the GOLF data that looked for the modulation of p-modes by g-modes in the deep interior. This analysis also remains unconfirmed and controversial (e.g. [47, 3]).

3.2 The Radiative Interior

The radiative interior, between the core and the base of the convection zone, is difficult to study in detail because it is penetrated only by the lower-degree modes, so the resolution of inversions is quite poor. It appears to rotate more or less rigidly; Eff-Darwich et al. [13] used data from GONG, LOWL, and MDI and found an interior rotation rate of 435 nHz, with no significant angular differential rotation below the convection zone.

3.3 The Tachocline

The tachocline, the layer of strong radial shear at the base of the convection zone, is a major feature of the solar rotation profile and plays a role in dynamo models. It cannot be completely resolved in inversions, but its depth and thickness can be inferred by less direct methods, such as forward modelling. Most of the results of such studies (e.g., [11, 6]) find that the tachocline is centered slightly below the 0.71 $R_\odot$ base of the convection zone and has a thickness of just a few percent of the solar radius; see [21] for a table summarizing more results. This presents a challenge to modellers to explain how such a steep shear can be maintained; for a review from a modeller's perspective, see [35]. There is some evidence that the tachocline is shallower at high latitudes [6].

3.4 The Convection Zone

The convection zone—from 0.71 $R_\odot$ up to the surface—exhibits differential rotation, with the equator rotating faster than the poles. That the early rotation inversions ruled out rotation "constant on cylinders"—that is, a rotation rate depending only on the distance from the rotation axis—was something of a surprise [17]. However, improved data showed that neither is the rotation rate quite constant along radial lines. In fact [16, 25] the contours of constant rotation in the bulk of the convection zone ($0.8 \leq r/R_\odot \leq 0.95$) make an angle to the rotation axis of about 25° over a

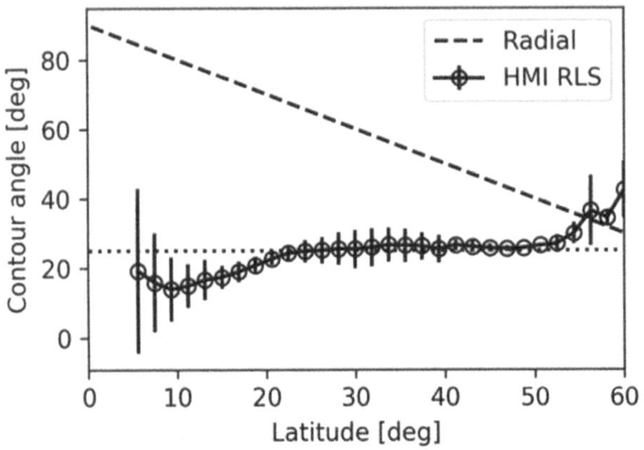

Fig. 2 The angle to the rotation axis of contours of constant rotation, as a function of the latitude at which each contour crosses $r = 0.95\ R_\odot$ for the mean of 2D RLS inversions of HMI data from 2010 to 2019. The dashed line indicates the angle for radial contours and the dotted horizontal line is drawn at 25°. The error bars are shown at 5-sigma for visibility

wide range of latitudes (Fig. 2), which is believed to be a result of the interaction of the Coriolis effect and the meridional flow.

In early inversions of MDI data [45], a jet-like feature was seen at about $0.95R_\odot$ and 75° latitude. However, subsequent work found that this was an artefact of systematic errors in the splitting coefficients derived from MDI data, specifically MDI data that was spatially binned prior to downloading from the spacecraft [33]. This is a good example of the issues that can be caused by problems with the input data, and it strengthens the case for comparing data from more than one source where possible.

3.5 The Near-Surface Shear Layer

In the outer 5 per cent of the solar radius, the rotation rate increases steeply with depth. This shear layer was hinted at in analysis of early observations, for example [52], and it is clearly detected in the first inversion results from GONG [50] and MDI [32, 45]. It was studied in detail by Corbard and Thompson [10] using f-mode data from MDI observations. They found a slope around −400 nHz per solar radius at the equator, or $d \ln \Omega / d \ln R \approx -1$, decreasing with latitude and possibly reversing sign at higher latitudes. Interestingly, the almost linear variation of the rotation rate with depth is not compatible with a simple picture of angular momentum being conserved in parcels of fluid moving relative to the rotation axis. The analysis was revisited more recently by Barekat et al. [5] using data from HMI and an updated analysis of MDI data. They confirmed the Corbard and Thompson results at low latitudes, but found that the gradient did not vary with latitude up to at least 60°.

This suggests that the apparent sign reversal in the shear may have been another artefact of systematic errors in the original MDI splittings [33].

4 Temporal Variations

The solar rotation profile varies slightly during the solar cycle, most obviously in the outer layers of the convection zone but perhaps as deep as its base.

4.1 The Torsional Oscillation

Surface Doppler observations from Mount Wilson [20] revealed a pattern of bands of faster and slower rotation, propagating from mid-latitudes to the equator during the solar cycle, that is known as the torsional oscillation. With the advent of systematic helioseismic measurements, it was found that this pattern is not only a surface phenomenon but penetrates deep into the convection zone [23, 1, 51]. There is also a poleward-propagating branch that (at least in Solar Cycle 23) shows a much stronger angular-velocity variation than the equatorward branch [2].

The bands of faster flow are closely related to the magnetic butterfly diagram, but at mid-latitudes the flow accelerates a year or two before active regions appear. For the cycles covered by Mt. Wilson and helioseismic observations, the activity of a new solar cycle becomes widespread at about the time that the equatorward band of faster flow reaches 25° latitude [26, 27]. As we can see in Fig. 3, the most recent data suggest that this epoch is almost upon us for Cycle 25, with the new equatorward branch well developed and close to 25°.

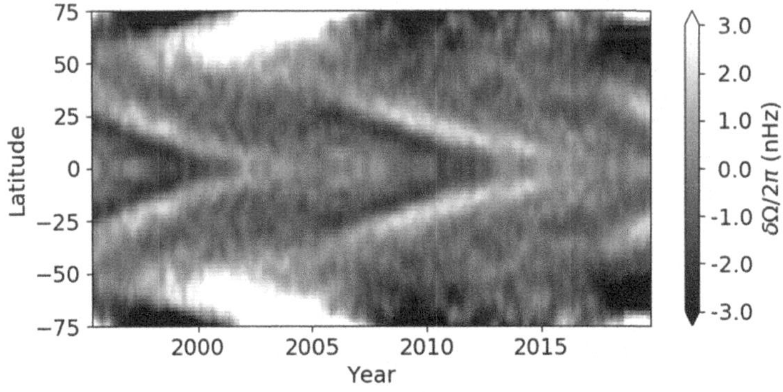

Fig. 3 Time–latitude map of rotation-rate residuals after subtraction of a temporal mean, for 2dRLS inversions of GONG, MDI and HMI data at a target depth of 0.99 $R_{\odot}$

The high-latitude branch of the torsional oscillation pattern was weak during Cycle 24, and the mean rotation profile for Cycle 24 shows significantly slower rotation at high latitudes (and slightly faster at low latitudes) than that for Cycle 23 [28, 29]. This is believed to be related to the weaker polar fields during Cycle 24 [40].

While global inversions can only detect the north–south symmetric part of the flows, local helioseismology can distinguish between the two hemispheres; see, for example, the recent work by Lekshmi et al. [34].

4.2 Tachocline Variations

Howe et al. [24] reported variations in the rotation rate at the bottom of the convection zone in GONG and MDI observations from 1995 – 1999, which appeared to have a 1.3-year period. This was of interest because of the role of the tachocline in the dynamo, and similar periods had been reported in other solar and heliophysical phenomena [48, 41, 37]. However, the result was not reproduced by other authors analyzing the same data [1]. After 2000 the periodic nature of the variations disappeared. Figure 4 shows the rotation-rate residuals at the equator just above and below the tachocline, from a continuation of the Howe et al. analysis. While there are fluctuations that appear in the data from both instruments, there is no strong periodic signal in the recent data.

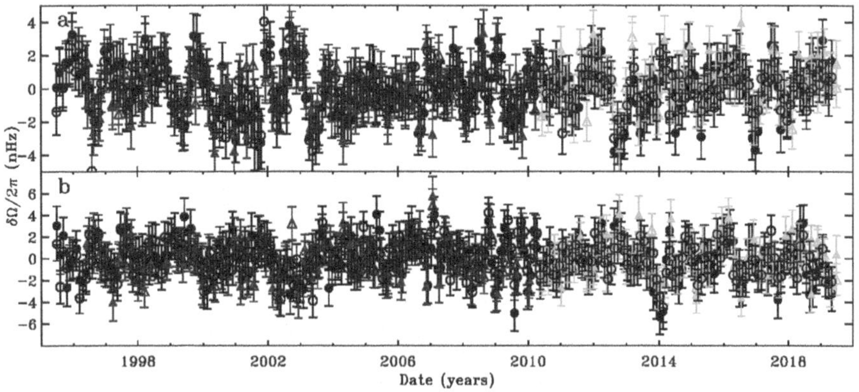

Fig. 4 Rotation-rate residuals from GONG (black circles), MDI (darker gray triangles) and HMI (lighter gray triangles) at 0.72 $R_\odot$ (top) and 0.63 $R_\odot$ (bottom), at the equator, for 2dRLS (filled) and 2dOLA (open) inversions

5 The Rotation Project

Major collaborative efforts studying the rotation profile were carried out in the early days of GONG, led by Michael Thompson [50], and MDI [45], led by Jesper Schou, but these used only a few months of data. We now have nearly 25 years of continous observations from GONG and from the combination of MDI and HMI, which could provide greatly improved inferences of the solar rotation profile. With this in mind, in 2017 Michael Thompson initiated a new collaborative effort to produce a definitive rotation profile, involving over 30 scientists from eight countries. The work had not progressed very far before Michael's untimely death, but it continues under the leadership of Jørgen Christensen-Dalsgaard, and we hope to publish results within the next year or so.

6 Conclusions

Helioseismology, and in particular global-mode inversion, has given us a robust picture of some features of the solar interior rotation, from the radial shear layer in the outer 5% of the solar radius, through the convection zone with its inclined—but not radial—contours of rotation and the tachocline shear layer at its base, to the mostly rigid rotation in the radiative interior. The core rotation is still less well constrained. In interpreting such inversion results, we should always bear in mind the resolution properties inherent in the method and conveniently represented by the averaging kernel. We should also be wary of systematic errors in the input data.

Analysis of a quarter-century of rotation data from GONG, MDI, and HMI reveals that the pattern of migrating zonal flows known as the torsional oscillation penetrates deep into the convection zone and that its appearance in the near-surface layers can help to project the near-term evolution of the solar cycle.

The work of Michael Thompson included several important contributions to this field, and his influence will be felt for many years to come.

Acknowledgments From 1995 to 1997 I was a postdoctoral researcher at Queen Mary, University of London under the supervision of Michael Thompson. In one sense our collaboration ended with his death, but in another sense it is still going on, and I will always be grateful for his mentorship.

I thank the National Solar Observatory for computing support, and the University of Birmingham for computing and travel support.

This work utilizes data obtained by the Global Oscillation Network Group (GONG) program, managed by the National Solar Observatory, which is operated by AURA, Inc. under a cooperative agreement with the National Science Foundation. SOHO is a project of international cooperation between ESA and NASA. HMI data courtesy of NASA/SDO and the HMI science team.

This review has made use of NASA's Astrophysics Data System. The plots were prepared using the Python packages Matplotlib [30], SciPy [31], NumPy [36], and Astropy, a community-developed core Python package for Astronomy [4, 39].

References

1. Antia, H. M., & Basu, S. (2000). *The Astrophysical Journal, 541*(1), 442. https://doi.org/10.1086/309421
2. Antia, H. M., & Basu, S. (2001). *The Astrophysical Journal Letters, 559*(1), L67. https://doi.org/10.1086/323701
3. Appourchaux, T., & Corbard, T. (2019). *Astronomy & Astrophysics, 624*, A106. https://doi.org/10.1051/0004-6361/201935196
4. Astropy Collaboration, Robitaille, T. P., Tollerud, E. J., Greenfield, P., Droettboom, M., Bray, E., et al. (2013). *Astronomy & Astrophysics , 558*, A33. https://doi.org/10.1051/0004-6361/201322068
5. Barekat, A., Schou, J., & Gizon, L. (2014). *Astronomy & Astrophysics, 570*, L12 (2014). https://doi.org/10.1051/0004-6361/201424839
6. Basu, S., & Antia, H. M. (2003). *The Astrophysical Journal, 585*(1), 553. https://doi.org/10.1086/346020
7. Broomhall, A. M., Salabert, D., Chaplin, W. J., García, R. A., Elsworth, Y., Howe, R. et al. (2012). *Monthly Notices of the Royal Astronomical Society, 422*(4), 3564. https://doi.org/10.1111/j.1365-2966.2012.20868.x
8. Chaplin, W. J., Christensen-Dalsgaard, J., Elsworth, Y., Howe, R., Isaak, G. R., Larsen, R. M., et al. (1994). *Monthly Notices of the Royal Astronomical Society, 308*(2), 405. https://doi.org/10.1046/j.1365-8711.1999.02691.x
9. Christensen-Dalsgaard, J., Schou, J., & Thompson, M. (1990). *Monthly Notices of the Royal Astronomical Society, 242*, 353.
10. Corbard, T., & Thompson, M. J. (2002). *Solar Physics, 205*(2), 211. https://doi.org/10.1023/A:1014224523374
11. Corbard, T., Blanc-Féraud, L., Berthomieu, & G., Provost, J. (1999). *Astronomy & Astrophysics, 344*, 696.
12. Cuypers, J. (1980). *Astronomy & Astrophysics, 89*, 207.
13. Eff-Darwich, A., Korzennik, S. G., & Jiménez-Reyes, S. J. (2002). *The Astrophysical Journal, 573*(2), 857. https://doi.org/10.1086/340747
14. Fossat, E., Boumier, P., Corbard, T., Provost, J., Salabert, D., Schmider, F. X., et al. (2017). *Astronomy & Astrophysics 604*, A40. https://doi.org/10.1051/0004-6361/201730460
15. García, R. A., Turck-Chièze, S., Jiménez-Reyes, S. J., Ballot, J., Pallé, P. L., Eff-Darwich, A., et al. (2007). *Science, 316*(5831), 1591. https://doi.org/10.1126/science.1140598
16. Gilman, P. A., & Howe, R. (2003). In H. Sawaya-Lacoste (Ed.), *GONG+ 2002. Local and Global Helioseismology: The Present and Future* (vol. 517, 283–285). ESA Special Publication.
17. Gough, D. O., Kosovichev, A. G., Sekii, T., Libbrecht, K. G., & Woodard, M. F. (1993). In T. M. Brown (Ed.), *GONG 1992. Seismic Investigation of the Sun and Stars. Proceedings of a Conference Held in Boulder, Colorado, August 11–14, 1993* (vol. 42, p. 213). Astronomical Society of the Pacific. ISBN # 0-937707-61-9. LC # QB539.O83 G66 1992.
18. Hansen, C., Cox, J., & van Horn, H. (1977). *The Astrophysical Journal, 217*, 151.
19. Harvey, J., Hill, F., Hubbard, R., Kennedy, J., Leibacher, J., Pintar, J., et al. (1996). *Science, 272*, 1284.
20. Howard, R., & Labonte, B. J. (1980). *The Astrophysical Journal Letters, 239*, L33. https://doi.org/10.1086/183286
21. Howe, R. (2009). *Living Reviews in Solar Physics, 6*(1), 1 (2009). https://doi.org/10.12942/lrsp-2009-1
22. Howe, R., & Thompson, M. J. (1996). *Monthly Notices of the Royal Astronomical Society, 281*, 1385. https://doi.org/10.1093/mnras/281.4.1385
23. Howe, R., Christensen-Dalsgaard, J., Hill, F., Komm, R. W., Larsen, R. M., Schou, J. et al. (2000). *The Astrophysical Journal Letters, 533*(2), L163. https://doi.org/10.1086/312623

24. Howe, R., Christensen-Dalsgaard, J., Hill, F., Komm, R. W., Larsen, R. M., Schou, J., et al. (2000). *Science, 287*(5462), 2456. https://doi.org/10.1126/science.287.5462.2456
25. Howe, R., Christensen-Dalsgaard, J., Hill, F., Komm, R., Schou, J., & Thompson, M. J. (2005). *The Astrophysical Journal, 634*(2), 1405 (2005). https://doi.org/10.1086/497107
26. Howe, R., Christensen-Dalsgaard, J., Hill, F., Komm, R., Schou, J., & Thompson, M. J. (2009). *The Astrophysical Journal Letters, 701*(2), L87. https://doi.org/10.1088/0004-637X/701/2/L87
27. Howe, R., Hill, F., Komm, R., Christensen-Dalsgaard, J., Larson, T. P., Schou, J. et al. (2011). *Journal of Physics: Conference Series, 271*, 012074. https://doi.org/10.1088/1742-6596/271/1/012074.
28. Howe, R., Christensen-Dalsgaard, J., Hill, F., Komm, R., Larson, T. P., Rempel, M., et al. (2013). *The Astrophysical Journal Letters, 767*(1), L20. https://doi.org/10.1088/2041-8205/767/1/L20
29. Howe, R., Hill, F., Komm, R., Chaplin, W. J., Elsworth, Y., Davies, G. R., et al. (2018). *The Astrophysical Journal Letters, 862*(1), L5. https://doi.org/10.3847/2041-8213/aad1ed
30. Hunter, J. D. (2007). *Computing in Science & Engineering , 9*(3), 90. https://doi.org/10.1109/MCSE.2007.55
31. Jones, E., Oliphant, T., & Peterson, P., (2001). SciPy: Open source scientific tools for Python. Retrieved January 13 2019. http://www.scipy.org/
32. Kosovichev, A. G., Schou, J., Scherrer, P. H., Bogart, R. S., Bush, R. I., Hoeksema, et al. (1997). *Solar Physics, 170*(1), 43. https://doi.org/10.1023/A:1004949311268
33. Larson, T. P., & Schou, J. (2015). *Solar Physics, 290*(11), 3221. https://doi.org/10.1007/s11207-015-0792-y
34. Lekshmi, B., Nandy, D., & Antia, H. M. (2018). *The Astrophysical Journal, 861*(2), 121. https://doi.org/10.3847/1538-4357/aacbd5
35. Miesch, M. S. (2005). *Living Reviews in Solar Physics, 2*(1), 1. https://doi.org/10.12942/lrsp-2005-1
36. Oliphant, T. E. (2006). *A guide to NumPy*. USA: Trelgol Publishing (2nd edition published in 2015 by CreateSpace Independent Publishing Platform, with ISBN-10: 151730007X; ISBN-13: 978-1517300074).
37. Paularena, K. I., Szabo, A., & Richardson, J. D. (1995). *Geophysical Research Letters, 22*(21), 3001. https://doi.org/10.1029/95GL02802
38. Pijpers, F. P., & Thompson, M. J. (1994). *Astronomy & Astrophysics, 281*(1), 231.
39. Price-Whelan, A. M., Sipőcz, B. M., Günther, H. M., Lim, P. L., Crawford, S. M., Conseil, S., et al. (2018). *Astronomy Journal, 156*, 123. https://doi.org/10.3847/1538-3881/aabc4f
40. Rempel, M. (2012). *The Astrophysical Journal Letters, 750*(1), L8. https://doi.org/10.1088/2041-8205/750/1/L8
41. Richardson, J. D., Paularena, K. I., Belcher, J. W., & Lazarus, A. J. (1994). *Geophysical Research Letters, 21*(14), 1559. https://doi.org/10.1029/94GL01076
42. Scherrer, P. H., Bogart, R. S., Bush, R. I., Hoeksema, J. T., Kosovichev, A. G., Schou, J. et al. (1995). *Solar Physics, 162*(1–2), 129. https://doi.org/10.1007/BF00733429
43. Schou, J., Christensen-Dalsgaard, J., & Thompson, M. (1992). *The Astrophysical Journal Letters, 385*, L59. https://doi.org/10.1086/186277
44. Schou, J., Christensen-Dalsgaard, J., & Thompson, M. (1994). *The Astrophysical Journal, 433*, 389. https://doi.org/10.1086/174653
45. Schou, J., Antia, H. M., Basu, S., Bogart, R. S., Bush, R. I., Chitre, S. M., et al. (1998). *The Astrophysical Journal, 505*(1), 390 (1998). https://doi.org/10.1086/306146
46. Schou, J., Scherrer, P. H., Bush, R. I., Wachter, R., Couvidat, S., Rabello-Soares, M. C., et al. (2012). *Solar Physics, 275*(1–2), 229. https://doi.org/10.1007/s11207-011-9842-2
47. Schunker, H., Schou, J., Gaulme, P., & Gizon, L. (2018). *Solar Physics, 293*(6), 95. https://doi.org/10.1007/s11207-018-1313-6
48. Silverman, S. M., & Shapiro, R. (1983). *Journal of Geophysical Research, 88*(A8), 6310. https://doi.org/10.1029/JA088iA08p06310
49. Thompson, M. J. (1990). *Solar Physics, 125*(1), 1. https://doi.org/10.1007/BF00154773

50. Thompson, M. J., Toomre, J., Anderson, E. R., Antia, H. M., Berthomieu, G., Burtonclay, D., et al., *Science, 272*(5266), 1300. https://doi.org/10.1126/science.272.5266.1300
51. Vorontsov, S. V., Christensen-Dalsgaard, J., Schou, J., Strakhov, V. N., & Thompson, M. J. (2002). *Science, 296*(5565), 101. https://doi.org/10.1126/science.1069190
52. Wilson, P. R. (1992). *The Astrophysical Journal, 399*, 294. https://doi.org/10.1086/171924

On Solar and Solar-Like Stars Convection, Rotation and Magnetism

Allan Sacha Brun

Abstract We honor Mike J. Thompson's legacy on solar and stellar convection, rotation and magnetism and their seismic probing by discussing how his major contributions have impacted or challenged the current state of our understanding and guided the development of advanced numerical simulations of the magnetohydrodynamics (MHD) of the Sun and Sun-like stars.

1 Magnetohydrodynamics of the Sun and Sun-Like Stars

The Sun and sun-like stars possess a large variety of dynamical phenomena ranging from turbulence surface convection to large scale flows (differential rotation, meridional circulation) and intense magnetic activity. Of key interest is to characterize how stellar rotation and magnetism vary along the secular evolution of sun-like stars.

As shown in Fig. 1, there is subtle and complex interplay between stellar rotation, stellar dynamo and stellar wind. Indeed dynamo of sun-like stars have a clear dependency on rotation rate, faster the stars is rotating, more active it is (up to a saturation level at low Rossby number [78]). This higher level of activity impacts the stellar wind and the associated mass and angular momentum losses, which in turn modify the rotation rate and hence the dynamo [41, 15]. Understanding and characterizing this nonlinear feedback loop over stellar evolution requires sustained observational, theoretical and modelling efforts. During the many years of scientific adventure that I was lucky to share with Mike through conferences or wonderful dinners at his house, it was always a real pleasure to exchange with him on these

A. S. Brun (✉)
Département d'Astrophysique/AIM, CEA/IRFU, CNRS/INSU, Université de Paris-Saclay, Université de Paris, Gif-sur-Yvette, France
e-mail: sacha.brun@cea.fr

M. J. P. F. G. Monteiro et al. (eds.), *Dynamics of the Sun and Stars*, Astrophysics and Space Science Proceedings 57,
https://doi.org/10.1007/978-3-030-55336-4_9

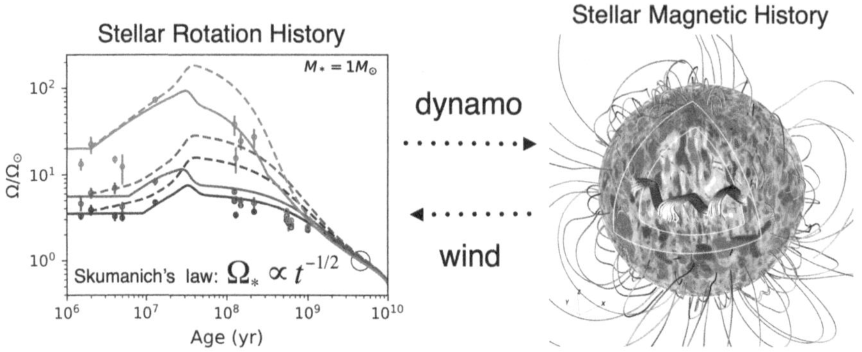

Fig. 1 Fundamental link between stellar rotation and magnetic history through magnetized stellar wind braking. Left: evolution of a one solar mass star's rotation rate over secular age (from [9]). We indicate the Skumanich's law [87] that main sequence stars follow at least up to the solar age (Rossby number). Right: 3-D rendering of radial velocity and magnetic field lines in a nonlinear solar convective dynamo (from [89])

difficult topics and their nonlinear interplay, see for instance the review he lead on the topic [93]. In this contribution, I will discuss high performance numerical simulations of convection, rotation and magnetism of sun-like stars performed with global convection codes such as ASH [29, 17] or Eulag-MHD [88], and how Mike's work influenced my research work and those of my close colleagues on these topics.

2 Large Scale Flows and Fields of the Sun

2.1 A Key Ingredient: Solar and Stellar Differential Rotation

One of the first key paper of Mike that I read is [92], the science paper on solar differential rotation. For the first time, space probing of solar differential rotation thanks to helioseismic inversions was provided. It confirmed what ground observations had inferred: The solar differential rotation was not cylindrical as early numerical simulations were predicting.

Instead it was clearly demonstrated in [92] that the differential rotation had little radial variations over the convective envelope except for a strong radial shear layer, called the tachocline, at its base. The monotonic latitudinal variations was confirmed as well. Such a differential rotation profile had a major impact on the type of solar dynamo models shifting the dominant paradigm from an α-Ω like dynamo to a flux transport Babcock-Leighton type [24, 15].

In a series of papers with Juri Toomre, Mark Miesch and younger colleagues, we have developed specific advanced global 3D numerical simulations with the ASH code [29, 17] to address the source of this very peculiar conical differential

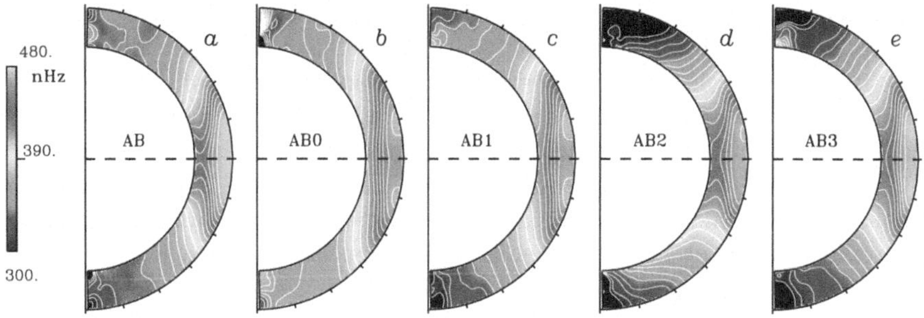

Fig. 2 Several realisations of differential rotation in ASH simulations of the solar convection as published in [70], shown as color contour plots in a meridional plane. All models are prograde (fast equator (reddish colors)—slow poles (blueish colors), but each possess a different level of tilting of iso-contours of Ω, from cylindrical (panel b) to almost horizontal (panel d). Most solar-like solution is obtained in panel e) for a latitudinal temperature contrast of 10 K at the base of the convection zone (bottom of the numerical domain)

rotation of the Sun. In particular I wish here to emphasize the results on the influence of thermal effects on the profile of differential rotation published in [70]. Here a key equation is the so-called *thermal wind balance*, which describes how angular velocity spatial variations are linked to that of entropy, e.g.:

$$\frac{\partial v_\phi}{\partial z} = \frac{g}{2\Omega_\odot r Cp}\frac{\partial S}{\partial \theta},$$

with g the gravitational acceleration, S the entropy fluctuations, Cp the pressure-specific heat and $\Omega_\odot$ the solar rotation rate.

To illustrate this balance, we display in Fig. 2 the angular velocity profile for 5 different cases as color contour plots in the meridional plane. These models differ only by one aspect, their bottom thermal boundary layer. In panel (a), the model is let free to impose the latitudinal variation of the entropy S (or temperature T easier to interpret), as we impose the thermal flux. The angular velocity profile is solar-like, with fast equator and slow poles but its iso-contours are not tilted (conical) enough at mid latitude. The temperature contrast $\Delta_\theta T$ is about 9 K in that case. For cases in panels (b)–(e), we instead imposed the latitudinal variations. In panel (b), we set it to zero, this results in a cylindrical (aligned with rotation axis) differential rotation profile and a weak temperature contrast of 3 K. Increasing the thermal latitudinal contrast from panels (b)–(d), we reach a value of 13 K and an almost horizontal differential rotation profile in panel (d), which is too extreme. Finally in panel (e) the solution is the closest to helioseismic inference, with the right conical differential rotation profile for a temperature contrast of about 10 K. Recall that at the base of the convection zone, the mean temperature $\bar{T}_{bcz}$ is around 2.2 MK. So we are speaking here of very small latitudinal perturbations, e.g. $\Delta_\theta T/\bar{T}_{bcz} \sim 10/2.210^6 \sim 4.510^{-6}$. Thus it is a real challenge to infer the exact latitudinal temperature contrast even with current helioseismic tools. Some attempts have been done to infer the variations of temperature and entropy but were inconclusive at validating the thermal wind balance found in numerical simulation [18]. Rather than using artificial control of the boundary conditions, other studies have shown that adding a coupling to a stable radiative layer below yields more realistic thermal boundary conditions in simulation of the solar convection zone [80, 19]. Note also that the thermal wind balance is a diagnostic equation not a prognostic equation. By that we mean that it does not drive the dynamics, it is the end result of complex heat and angular momentum redistribution in the rotating convective shell. Hence there can exist a geostrophic invariant mean flow [37] that this equation can't "control". Hence the numerical experiments describe in Fig. 2 works well because the simulation are not too turbulent (such that they have an effective Rossby number $Ro = v/\Omega_{\odot}D$ less than 1) and their Péclet number $Pe = vL/\kappa$ is small and the thermal bottom boundary conditions can efficiently be imposed through thermal diffusion in most of the convective envelope and tilt the iso-contours of Ω. For highly turbulent simulations at high Pe number, the tilting of iso-contours (from cylindrical to conical) does not work as efficiently [16]. Nevertheless it is very instructive to see that in order to understand the peculiar angular velocity profile of the Sun one must also understand its thermal latitudinal profile.

A final remark on the differential rotation profile of the Sun and our ability or not to recover its peculiar conical profile by mean of global numerical simulation. Recently more turbulent numerical simulations of solar convection (both local or global) have hit what is now commonly called in the community the *convective conundrum* [52, 71, 64, 53, 38]. Large scale convective flows seem to have in numerical simulation a higher amplitude than inferred from some helioseismic inference (see [48] for a counter example). These large scales are larger than the local Rossby radius of deformation L_{Ro} (a typical expression is $L_{Ro} = \sqrt{gD}/\Omega_{\odot}$, with D the depth of the convective zone). This has for consequences that their effective Rossby number Ro is larger than 1, which is found to typically lead to anti-solar differential rotation [66, 44, 58, 21], see also Fig. 4 for an illustration of this rotation state. Hence the convective conundrum is potentially at the origin of the difficulty for highly turbulent global convective simulations to reproduce the inverted solar differential rotation profile (at least a prograde state of differential rotation (fast equator/slow poles). Simple work around exist, such as computing a numerical simulation at higher rotation rate, lower energy (luminosity) input, or assuming higher viscosity due for instance to the influence of magnetic field [66, 36, 55, 60], all reducing the effective Rossby number of the angular momentum carrying scales. Nevertheless a convincing explanation is still lacking and intense research are currently going on, both on the theoretical/numerical point of view and in validating the complex helioseismic inversion procedure.

2.2 *Meridional Circulation: How Many Cells?*

The change of paradigm for the solar dynamo model due to the new conical differential rotation profile inverted by helioseismology had another consequence: it gave a prominent role to the meridional circulation. In Babcock-Leighton flux transport dynamo, the regeneration of the poloidal magnetic field component is due to the emergence of titled bipolar region at the solar surface. Since most of the toroidal field is generated deep in the convective envelope via stretching by the large scale shear (see Sect. 3 below), the poloidal field must be transported from the surface to the bottom of the convective envelope, some 200 mm below. Three candidates can be chosen for that: magnetic diffusion, turbulent magnetic pumping or meridional circulation acting like a "conveyor belt" [99, 32].

It soon become obvious that a helioseismic inversion of the meridional circulation was key to obtain. Giles et al. [45] published a study using MDI data that indicated the likelihood that there is only one cell per hemisphere. It was marginal because helioseismic inversions below 0.9 solar radius are difficult, but assuming mass conservation and a one cell flow, they showed that such a profile was compatible with the data down to 0.7 solar radius. Inspired by these first helioseismic inversions of the meridional circulation and by its simplicity, many authors started to adopt in their mean field dynamo model a single cell per hemisphere [25, 30, 24]. The meridional flow is assumed poleward at the solar surface and equatorward at the base of the convective zone. While the surface poleward flow is a robust feature, in observations and numerical simulations, the return flow is still largely unconstrained. For instance, 3-D rotating global simulations performed to reproduce the Sun, are finding multi-cellular meridional cells as a function of depth. We illustrate in Fig. 3 such meridional circulation patterns [39, 21]. At solar rotation rate (middle panel), we see that the meridional flow is quite complex with at least 3 cells per hemisphere (we defer for now the discussion of the 2 other panels). So for about at least 10 years there was a heated debate between theory and helioseismic inversion about the numbers of meridional cells, one or more per hemisphere? Here again comes Mike, with a paper in 2007 about the meridional circulation and the possible existence of more than one cell per hemisphere [72], confirming an early finding by Haber et al. [50], of the possible existence of such a second cell at the maximum of activity of cycle 23. Since Mike's pioneering publication, several papers using helioseismic techniques have also indicated the possible existence of at least 2 cells stacked in radius [100, 65, 11]. This is encouraging, it means that theory, models and observations converge toward a consensus. Note that Babcock-Leighton dynamo models using more complex meridional circulations have been tested. They all show that if the meridional circulation is the main transport mechanism of the magnetic flux from the surface to the base of the solar convective envelope, such multi cells profiles must either be temporary or an odd number of radial cells must be considered [56, 54, 31]. As we will briefly discuss in the next subsection, in 3D convective dynamo simulations, the meridional circulation plays very little role in transporting magnetic flux in the tachocline region, turbulent pumping seems

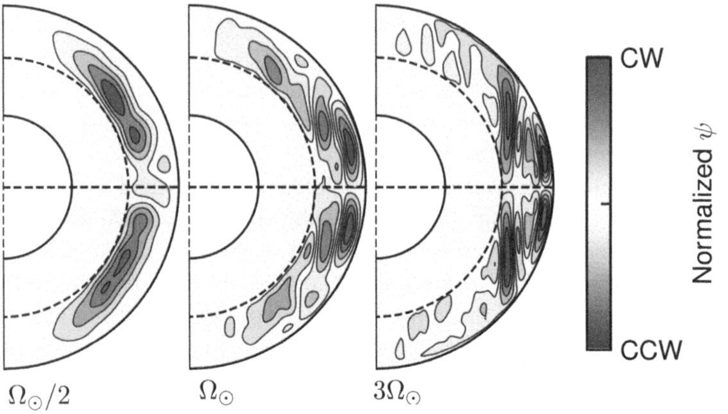

Fig. 3 Several realisations of meridional circulation profiles in ASH simulations of solar convection (from [21]). From left to right the rotation rate goes from half to 3 times the solar rotation rate (of a Rossby number larger than 1 to about 0.2). Such profiles of meridional circulations are associated in turn to: a anti-solar like differential rotation (fast pole, slow equator; such as in the bottom right panel of Fig. 4), a solar-like profile (fast equator, slow poles; such as in the middle right panel of Fig. 4) and to a highly rotationally constrained solution

more efficient, when magnetic diffusion has been lowered enough. However note that [26], favor a diffusive solution to the magnetic flux transport problem. Today the debate is still open, and the reality is likely that all contributes more or less depending on global stellar parameters.

2.3 Magnetic Field in Solar Interior

Another key element of the solar dynamo current paradigm, is what is called the large scale poloidal and toroidal field sources segregation. It means that it is more effective to have the regeneration of the toroidal and poloidal fields performed at different spatial locations, the former at the base of the convection zone and in the tachocline and the latter near the surface. It is traditionally called the interface dynamo model [76]. Being able to provide a constraint on the maximum field strength in the tachocline is thus key to constraint dynamo models. Using advanced helioseismic inversion Mike and colleagues [2] published a detailed study on the magnetic field amplitude in the solar interior, being able to set an upper limit of respectively 20 kg around 30 mm deep and 300 kg inside the tachocline region. Such constraints on the amplitude of the inner solar magnetic field was novel and unique and reinforced the need in order to interpret these findings for theoretical work and renewed 3D MHD numerical simulations of the solar interior, in the spirit of what was started in the 1980s by P. Gilman and G. Glatzmaier [46, 47]. Aware of these challenges, Juri Toomre and many colleagues (including the author) started the active development of the ASH code and its extension to MHD, work that lead to several publications in late 1990s early 2000s [69, 17]. Later we included in our simulations a stable layer below the turbulent convective envelope [14]. We were able to confirm the key role of this zone in the Sun to organize the large scale toroidal field. We found that it takes the form of a large scale magnetic wreath straddling the base of the convection zone. Other research groups joined the effort and nonlinear cycle dynamo solution started to become more common as we entered 2010s.

Nowadays we have a whole suite of interesting nonlinear dynamo models of the Sun obtained by various numerical techniques [79, 36, 98, 55, 49, 89, 90, 97]. We illustrate in Fig. 4, such variety of nonlinear dynamo solutions, by displaying

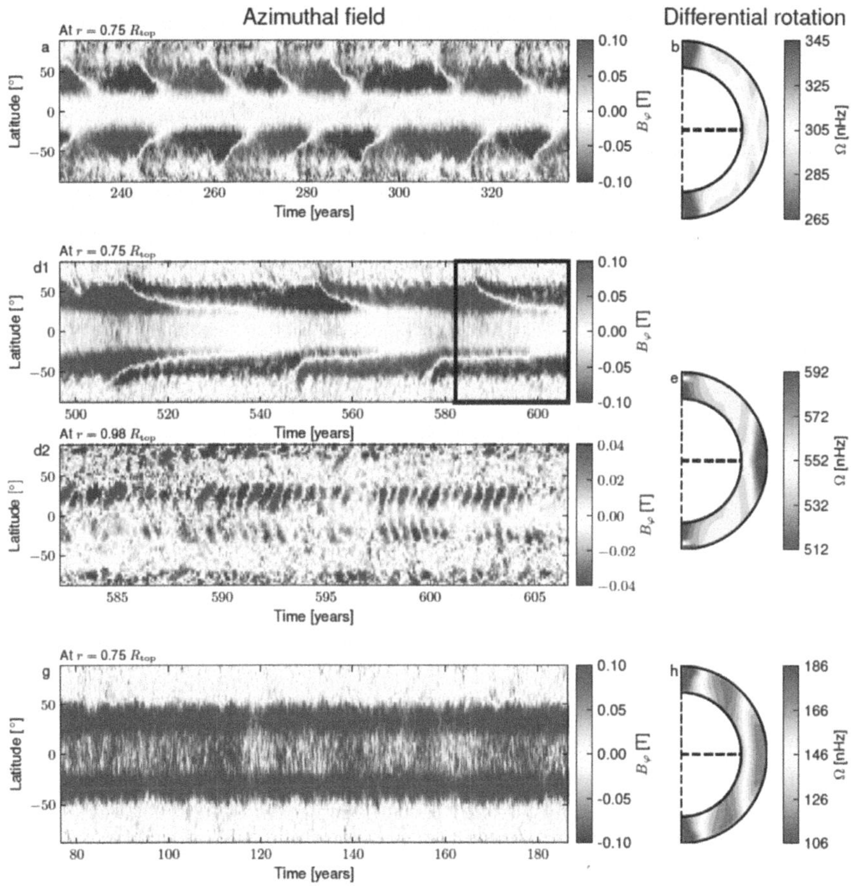

Fig. 4 Typical dynamo runs of a solar-like stars as obtained by Strugarek et al. [90]. Shown on the left column are butterfly (time-latitude) diagrams of the toroidal magnetic field at the base of the convection zone. Contours range between +/-1000G, with red representing positive polarity. On the right column, we display the associated differential rotation state in a meridional plane (as in Fig. 2). Top right panel shows an angular velocity profile with little contrast, middle panel a solar-like profile (fast equator, slow poles) and bottom panel an anti-solar differential profile (slow equator, fast poles)

butterfly (time-latitude) diagrams of the toroidal magnetic field at the base of the convective envelope. We can notice 3 main behavior: long cycle (top panel), short cycle (middle panel, here accompanying a longer cycle), and steady state (bottom panel). We also display the associated differential rotation state in a meridional plane (as in Fig. 2). Top right panel shows an angular velocity profile with little contrast due to the strong feedback of the magnetic field (Ω-quenching effect), middle panel a solar-like profile and bottom panel an anti-solar differential profile (slow equator, fast poles).

Theory and numerical simulations can also be used to derive scaling laws of the magnetic energy as a function of global stellar parameters (rotation, energy input/mass, plasma properties (Prandtl numbers)) [27]. These are useful laws if one wishes to compute trends on long time scales as 3-D MHD simulations can hardly be computed for longer than few millennia. In a recent study [5] we confirmed that dynamo regime change as a function of the Rossby number from equipartition state to a magnetostrophic state dominated by a balance between Lorentz and Coriolis forces. We also find that the magnetic Prandtl number Pm plays an important role. By analyzing many recent dynamo simulations we showed that the ratio of magnetic to kinetic energy, adjusted by Pm follows:

$$\frac{1}{Pm}\frac{ME}{KE} = a0 + b0 * Ro^{-1},$$

with $a0 = 0.053 \pm 0.004$ and $b0 = 0.062 \pm 0.009$, and Ro the Rossby number.

Hence it is expected that Sun-like stars will possess properties close to that of the Sun in a limited range of parameters such as the Rossby number. We now turn to discuss in more detail the connection between the Sun and Sun-like stars, a topic that was of keen interest also to Mike.

3 Connecting Solar and Stellar Magnetism

Mike very early with his long time friend Jørgen Christensen-Dalsgaard, were among the scientists who realized the importance of using *Kepler* satellite photometric data to characterize the thousands of stars that the satellite would monitor. They of course envisioned when possible to perform asteroseismic analysis to invert their dynamical properties [23]. So more recently he was working on characterizing solar-type stars extending his deep understanding of the Sun to them as exposed in [91]. Likewise, over the years, 3D MHD numerical simulations started to consider various type of stars, with a special attention to solar-like stars from 0.5 to 1.2 Msol [12, 13, 3, 4, 59, 95, 89, 90, 97] as more and more data was being accumulated on Sun-like stars and even solar analogues through photometric, asteroseismic and spectropolarimetric techniques [85, 8, 51, 75, 33, 42, 62, 57, 10] (and references therein), adding new constraints on both stellar dynamo and rotation trends, continuing the pioneering work of [73, 74, 6, 34]. In the previous section we

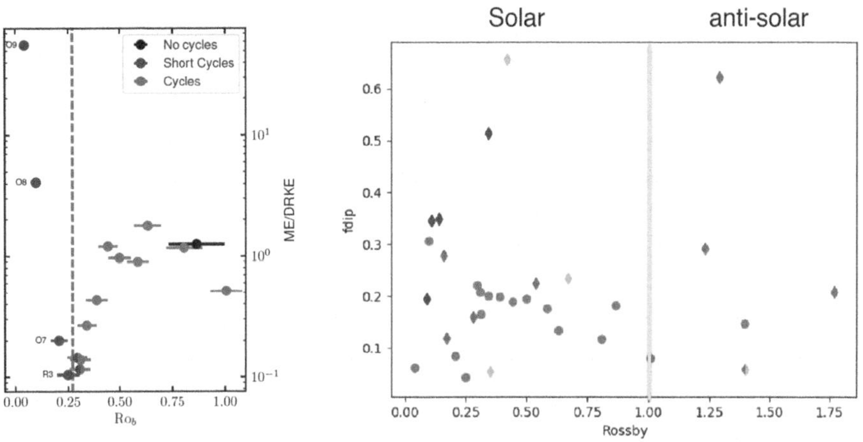

Fig. 5 Left panel: various types of dynamos (long cycle, mix long and short, steady) classified depending on their Rossby number and magnetic energy over differential rotation kinetic energy ratio. Right panel: Amplitude of f_{dip} versus the Rossby number (from [90, 22])

have discussed the properties of key ingredients (differential rotation, meridional circulation, dynamo action) thought to be at the origin and organisation of the large scale solar magnetic field. Given what we know today about solar-like stars it is clear that both their differential rotation (mean flows) and the associated magnetic activity vary with global stellar parameters such as stellar rotation, mass and chemical composition (for the later see [61]). Here we wish to come back to Fig. 1, where the subtle feedback loop between rotation, magnetic activity and secular evolution of stars is discussed putting the Sun in an astrophysical context [67, 1]. I do not have the space to discuss all aspects here, I refer to these recent reviews for more in depth discussion on stellar magnetism and rotation (including surface differential rotation) [20, 15].

In Fig. 5 we display key quantities of several solar-type dynamos computed with respectively Eulag-MHD [89, 90] and ASH codes [95, 22]. In particular we are interested to characterize the magnetic geometry of stars as they cross the Rossby equals 1 limit. Indeed [68], following [94], advocate that the Sun may be currently undergoing a transition in rotation and hence dynamo property, that would have for consequences to make wind braking inefficient and hence stop stellar spin down. This would indicate that the famous Skumanich's law of gyrochronology, e.g. $\Omega(t) \propto t^{-0.5}$ [87, 7] would not apply for solar-like stars older than the Sun. So it is essential to be able to characterize if such a change of rotation and dynamo states really occurs.

On the left panel of Fig. 5, we display how the ratio of the magnetic energy (ME) over the kinetic energy in the differential rotation (DRKE) is changing as a function of the Rossby number. This Figure does clearly show that the transition from low Rossby number ($Ro < 0.25$) to large Rossby number ($Ro > 1$) has an impact on both the type of dynamo (when considering also Fig. 4) and the magnetic energy content with respect to the energy contained in the shear (DRKE). One possibility to explain the drop of braking efficiency of stellar magnetized wind at a given rotation rate (that corresponding to the transition $Ro < 0.25$ to $Ro > 1$), could be a change of magnetic geometry (from large to small scales) on top of the magnetic field amplitude variation and change of cyclic behavior.

To this end, we display in the right panel of Fig. 5, a quantity called f_{dip}, which measures the amplitude of the dipole with respect to the first 12 magnetic multipoles [28, 35]. We focus here on the dipole because it is well known that the wind braking efficiency is mostly dependent on the first low order magnetic multipoles: dipole, quadrupole and octupole [81, 82, 83, 40, 43, 86]. If a drastic change in magnetic geometry was occurring at the Rossby equals 1 limit (for solar-like stars older than the Sun), this should be clearly visible in the quantity f_{dip}. As can be seen in Fig. 5 (right panel), this is not the case, the dipole is not found to collapse for anti-solar rotating dynamos. What is lost is possibly the cyclic behavior (see Fig. 4 above), but steady dynamo do maintain strong dipolar field and hence the associated magnetized wind should also be efficient at braking the star. This results is also found in observations of cool star magnetism [96]. So if the break of the gyrochronology advocated by recent asteroseismic inversion using Kepler satellite data [94] is confirmed, then the origin of this less efficient wind braking is more likely due to a change of the heating mechanism accelerating the wind rather than due to a change of the field geometry. Work is in progress to study this issue by developing more advanced wind models [63, 84] and by coupling them dynamically to dynamo simulations [77].

4 Perspectives

Understanding the convection, rotation and magnetism of the Sun and Sun-like stars and how they evolve in an astrophysical context is a fascinating research adventure, which Mike really enjoyed and I of course share his enthusiasm. Yet, even considering the huge steps forwards in collecting ever improving observations, developing high performance simulations and new theories that we collectively did and for which Mike did major contributions, there is still much to do. I list few routes or new constraints (non exhaustive) that hopefully will help us moving forward even more.

- Infer latitudinal ΔT at the base of the convective envelope and T' vs r to help elucidate the convective conundrum
- Confirm profile of MC flow and the number of cells per hemisphere
- Confirm the dependency of the angular velocity contrast vs rotation and mass
- Validate $\Omega(t)$ laws for stars older than the Sun, is Skumanich's law (e.g. $\Omega(t) \propto t^{-0.5}$) still valid? As a corollary, improve stellar wind models.
- Resolve the spot-dynamo paradox, i.e. the lack of magnetic spots in 3D nonlinear dynamo.

Fig. 6 Mike, Juri and in the back Juri's wife Linda, at a BBQ party that Kyle Augustson and I have organized in our 2017 rental house in Boulder, the summer of the recent US total solar eclipse

I would like to end this paper with a picture of Mike (here with Juri) that I took during the summer 2017 at a BBQ party (Fig. 6). We all miss you very much Mike.

Acknowledgments I would like to dedicate this paper to Kate and Robin. I am grateful to many colleagues over the years for many great discussions about stars, their rotation and magnetism: J. Toomre, M. Miesch, A. Strugarek, P. Charbonneau, S. Matt, M. Browning, K. Augustson, B. Brown, N. Featherstone, V. Réville and of course Mike J. Thompson and Jean-Paul Zahn. I acknowledge funding by European Research Council with grants ERS-Stg 207430 STARS2 and ERC-Syg 810218 WHOLE SUN, CNES for Solar Orbiter and PLATO support, INSU/PNST and IRS SPACEOBS.

References

1. Ahuir, J., Brun, A. S., Strugarek, A. (2020). arXiv:2002.00696.
2. Antia, H. M., Chitre, S. M., & Thompson, M. J. (2000). *Astronomy & Astrophysics, 360*, 335.
3. Augustson, K. C., Brun, A. S., & Toomre, J. (2013). *Astrophysical Journal, 777*, 153. https://doi.org/10.1088/0004-637X/777/2/153
4. Augustson, K., Brun, A. S., Miesch, M., & Toomre, J. (2015). *Astrophysical Journal, 809*, 149 (2015). https://doi.org/10.1088/0004-637X/809/2/149
5. Augustson, K. C., Brun, A. S., & Toomre, J. (2019). *Astrophysical Journal, 876*(1), 83. https://doi.org/10.3847/1538-4357/ab14ea
6. Baliunas, S. L., Donahue, R. A., Soon, W. H., Horne, J. H., Frazer, J., Woodard-Eklund, L., et al. (1995). *Astrophysical Journal, 438*, 269. https://doi.org/10.1086/175072
7. Barnes, S. A. (2003). *Astrophysical Journal, 586*(1), 464. https://doi.org/10.1086/367639
8. Barnes, J. R., Collier Cameron, A., Donati, J. F. James, D. J., Marsden, S. C., & Petit, P. (2005). *Monthly Notices of the Royal Astronomical Society, 357*, L1. https://doi.org/10.1111/j.1745-3933.2005.08587.x

9. Benbakoura, M., Réville, V., Brun, A. S., Le Poncin-Lafitte, C., & Mathis, S. (2019). *Astronomy & Astrophysics, 621*, A124. https://doi.org/10.1051/0004-6361/201833314
10. Benomar, O., Bazot, M., Nielsen, M. B., Gizon, L., Sekii, T., Takata, M. et al. (2018). *Science, 361*(6408), 1231. https://doi.org/10.1126/science.aao6571.
11. Böning, V. G. A., Roth, M., Jackiewicz, J., & Kholikov, S. (2017). *Astrophysical Journal, 845*(1), 2. https://doi.org/10.3847/1538-4357/aa7af0
12. Brown, B. P., Browning, M. K., Brun, A. S., Miesch, M. S., & Toomre, J. (2010). *Astrophysical Journal, 711*, 424. https://doi.org/10.1088/0004-637X/711/1/424.
13. Brown, B. P., Miesch, M. S., Browning, M. K., Brun, A. S., & Toomre, J. (2011). *Astrophysical Journal, 731*, 69. https://doi.org/10.1088/0004-637X/731/1/69
14. Browning, M. K., Miesch, M. S., Brun, A. S., & Toomre, J. (2006). *Astrophysical Journal, 648*, L157. https://doi.org/10.1086/507869
15. Brun, A. S., & Browning, M. K. (2017). *Living Reviews in Solar Physics, 14*(1), 4. https://doi.org/10.1007/s41116-017-0007-8
16. Brun, A. S., & Rempel, M. (2009). *Space Science Reviews, 144*, 151. https://doi.org/10.1007/s11214-008-9454-9
17. Brun, A. S., Miesch, M. S., & Toomre, J. (2004). *Astrophysical Journal, 614*, 1073. https://doi.org/10.1086/423835
18. Brun, A. S., Antia, H. M., & Chitre, S. M. (2010). *Astronomy & Astrophysics, 510*, A33. https://doi.org/10.1051/0004-6361/200913166
19. Brun, A. S., Miesch, M. S., & Toomre, J. (2011). *Astrophysical Journal, 742*, 79. https://doi.org/10.1088/0004-637X/742/2/79
20. Brun, A.S., Browning, M. K., Dikpati, M., Hotta, H., & Strugarek, A. (2015). *Space Science Reviews, 196*, 101. https://doi.org/10.1007/s11214-013-0028-0
21. Brun, A. S., Strugarek, A., Varela, J., Matt, S. P., Augustson, K. C., Emeriau, C., et al. (2017). *Astrophysical Journal, 836*(2), 192. https://doi.org/10.3847/1538-4357/aa5c40
22. Brun, A.S. Strugarek, A., Perri, B., Varela, J., Noraz, Q., Augustson, K. C., et al. *Astrophysical Journal* (2020, submitted).
23. Chaplin, W. J., Appourchaux, T., Elsworth, Y., García, R. A., Houdek, G., Karoff, C., et al. (2010). *Astrophysical Journal, 713*, L169. https://doi.org/10.1088/2041-8205/713/2/L169
24. Charbonneau, P. (2010). *Living Reviews in Solar Physics, 7*(1), 3. https://doi.org/10.12942/lrsp-2010-3
25. Choudhuri, A. R., Schussler, M., & Dikpati, M. (1995). *Astronomy & Astrophysics, 303*, L29.
26. Choudhuri, A. R., Chatterjee, P., & Jiang, J. (2007). *Phys. Rev. Letter, 98*(13), 131103 (2007). https://doi.org/10.1103/PhysRevLett.98.131103
27. Christensen, U. R. (2020). *Space Science Reviews, 152*(1-4), 565. https://doi.org/10.1007/s11214-009-9553-2
28. Christensen, U. R., & Aubert, J. (2006). *Geophysical Journal International, 166*(1), 97. https://doi.org/10.1111/j.1365-246X.2006.03009.x
29. Clune, T. C., Elliot, J. R., Miesch, M. S., Toomre, J., & Glatzmaier, G. A. (1999). *Parallel Computing, 25*, 361. http://dx.doi.org/10.1016/S0167-8191(99)00009-5
30. Dikpati, M., & Charbonneau, P. (1999). *Astrophysical Journal, 518*(1), 508. https://doi.org/10.1086/307269
31. Dikpati, M., Mitra, D., & Anderson, J. L. (2016). *Advances in Space Research, 58*(8), 1589. https://doi.org/10.1016/j.asr.2016.08.004
32. Do Cao, O., Brun, A. S. (2011). *Astronomische Nachrichten, 332*, 907. https://doi.org/10.1002/asna.201111623
33. do Nascimento, J. J.-D., García, R. A., Mathur, S., Anthony, F., Barnes, S. A., Meibom, S., et al. (2014). *Astrophysical Journal, 790*(2), L23. https://doi.org/10.1088/2041-8205/790/2/L23. http://stacks.iop.org/2041-8205/790/i=2/a=L23
34. Donahue, R. A., Saar, S. H., & Baliunas, S. L. (1996). *Astrophysical Journal, 466*, 384. https://doi.org/10.1086/177517
35. Emeriau-Viard, C., & Brun, A. S. (2017). *Astrophysical Journal, 846*(1), 8. https://doi.org/10.3847/1538-4357/aa7b33

36. Fan, Y., & Fang, F. (2014). *Astrophysical Journal, 789*(1), 35. https://doi.org/10.1088/0004-637X/789/1/35
37. Fearn, D. R. (1998). *Reports on Progress in Physics, 61*(3), 175. https://doi.org/10.1088/0034-4885/61/3/001
38. Featherstone, N. A., & Hindman, B. W. (2016). *The Astrophysical Journal Letters, 830*(1), L15. https://doi.org/10.3847/2041-8205/830/1/L15
39. Featherstone, N. A., & Miesch, M. S. (2015). *Astrophysical Journal, 804*, 67. https://doi.org/10.1088/0004-637X/804/1/67
40. Finley, A. J., & Matt, S. P. (2018). *Astrophysical Journal, 854*(2), 78. https://doi.org/10.3847/1538-4357/aaaab5
41. Gallet, F., & Bouvier, J. (2015). *Astronomy & Astrophysics, 577*, A98. https://doi.org/10.1051/0004-6361/201525660
42. García, R. A., Ceillier, T., Salabert, D., Mathur, S., van Saders, J. L., Pinsonneault, M., et al. (2014). *Astronomy & Astrophysics, 572*, A34. https://doi.org/10.1051/0004-6361/201423888
43. Garraffo, C., Drake, J. J., Dotter, A., Choi, J., Burke, D. J., Moschou, et al. (2018). *Astrophysical Journal, 862*(1), 90. https://doi.org/10.3847/1538-4357/aace5d
44. Gastine, T., Yadav, R. K., Morin, J., Reiners, A., & Wicht, J. (2014). *Monthly Notices of the Royal Astronomical Society, 438*, L76. https://doi.org/10.1093/mnrasl/slt162
45. Giles, P. M., Duvall, T. L., Scherrer, P. H., Bogart, R. S. (1997). *Nature, 390*(6655), 52. https://doi.org/10.1038/36294
46. Gilman, P. A. (1983). *Astrophysical Journal Supplement, 53*, 243. https://doi.org/10.1086/190891
47. Glatzmaier, G. A. (1985). *Astrophysical Journal, 291*, 300. https://doi.org/10.1086/163069
48. Greer, B. J., Hindman, B. W., Featherstone, N. A., & Toomre, J. (2015). *The Astrophysical Journal Letters, 803*(2), L17. https://doi.org/10.1088/2041-8205/803/2/L17
49. Guerrero, G., Smolarkiewicz, P. K., de Gouveia Dal Pino, E. M., Kosovichev, A. G., & Mansour, N. N. (2016). *Astrophysical Journal, 819*(2), 104. https://doi.org/10.3847/0004-637X/819/2/104
50. Haber, D. A., Hindman, B. W., Toomre, J., Bogart, R. S., Larsen, R. M., & Hill, F. (2002). *Astrophysical Journal, 570*(2), 855.https://doi.org/10.1086/339631
51. Hall, J. C., Lockwood, G. W., & Skiff, B. A. (2007). *The Astronomical Journal, 133*(3), 862. http://stacks.iop.org/1538-3881/133/i=3/a=862
52. Hanasoge, S. M., Duvall, T. L., & Sreenivasan, K. R. (2012). *Proceedings of the National Academy of Science, 109*(30), 11928. https://doi.org/10.1073/pnas.1206570109
53. Hanasoge, S.,, Gizon, L., & Sreenivasan, K. R. (2016). *Annual Review of Fluid Mechanics, 48*(1), 191. https://doi.org/10.1146/annurev-fluid-122414-034534
54. Hazra, G., Karak, B. B., & Choudhuri, A. R. (2014). *Astrophysical Journal, 782*(2), 93. https://doi.org/10.1088/0004-637X/782/2/93
55. Hotta, H., Rempel, M., & Yokoyama, T. (2016). *Science, 351*(6280), 1427. https://doi.org/10.1126/science.aad1893
56. Jouve, L., & Brun, A. S. (2007). *Astronomy & Astrophysics, 474*(1), 239. https://doi.org/10.1051/0004-6361:20077070
57. Judge, P. G., Egeland, R., Metcalfe, T. S., Guinan, E., & Engle, S. (2017). *Astrophysical Journal, 848*(1), 43. https://doi.org/10.3847/1538-4357/aa8d6a
58. Käpylä, P. J., Käpylä, M. J., & Brandenburg, A. (2014). *Astronomy & Astrophysics, 570*, A43. https://doi.org/10.1051/0004-6361/201423412
59. Karak, B. B., Käpylä, P. J., Käpylä, M. J., Brandenburg, A., Olspert, N., & Pelt, J. (2015). *Astronomy & Astrophysics, 576*, A26. https://doi.org/10.1051/0004-6361/201424521
60. Karak, B. B., Miesch, M., & Bekki, Y. (2018). *Physics of Fluids, 30*(4), 046602. https://doi.org/10.1063/1.5022034
61. Karoff, C., Metcalfe, T. S., Santos, Â. R. G., Montet, B. T., Isaacson, H., Witzke, V., et al. (2018). *Astrophysical Journal, 852*(1), 46. https://doi.org/10.3847/1538-4357/aaa026
62. Kochukhov, O., Petit, P., Strassmeier, K. G., Carroll, T. A., Fares, R., Folsom, C. P. et al. (2017). *Astronomische Nachrichten, 338*(4), 428. https://doi.org/10.1002/asna.201713310

63. Linker, J. A., Caplan, R. M., Downs, C., Riley, P., Mikic, Z., Lionello, R., et al. (2017). *Astrophysical Journal, 848*(1), 70. https://doi.org/10.3847/1538-4357/aa8a70
64. Lord, J. W., Cameron, R. H., Rast, M. P., Rempel, M., & Roudier, T. (2014). *Astrophysical Journal, 793*(1), 24. https://doi.org/10.1088/0004-637X/793/1/24
65. Mandal, K., Hanasoge, S. M., Rajaguru, S. P., & Antia, H. M. (2018). *Astrophysical Journal, 863*(1), 39. https://doi.org/10.3847/1538-4357/aacea2
66. Matt, S. P., Do Cao, O., Brown, B. P., & Brun, A. S. (2011). *AN, 332*, 897. https://doi.org/10.1002/asna.201111624
67. Matt, S. P., Brun, A. S., Baraffe, I., Bouvier, J., & Chabrier, G. (2015). *Astrophysical Journal Letter, 799*(2), L23. https://doi.org/10.1088/2041-8205/799/2/L23
68. Metcalfe, T. S., & van Saders, J. (2017). *Solar Physics, 292*(9), 126. https://doi.org/10.1007/s11207-017-1157-5
69. Miesch, M. S., Elliott, J. R. Toomre, J., Clune, T. L., Glatzmaier, G. A., & Gilman, P. A. (2000). *Astrophysical Journal, 532*, 593. https://doi.org/10.1086/308555.
70. Miesch, M. S., Brun, A. S., & Toomre, J. (2006). *Astrophysical Journal, 641*, 618. https://doi.org/10.1086/499621
71. Miesch, M. S., Featherstone, N. A., Rempel, M., & Trampedach, R. (2012). *Astrophysical Journal, 757*(2), 128. https://doi.org/10.1088/0004-637X/757/2/128
72. Mitra-Kraev, U., & Thompson, M. J. (2007). *Astronomische Nachrichten, 328*(10), 1009. https://doi.org/10.1002/asna.200710873.
73. Noyes, R. W., Hartmann, L. W., Baliunas, S. L., Duncan, D. K., & Vaughan, A. H. (1984). *Astrophysical Journal, 279*, 763. https://doi.org/10.1086/161945
74. Noyes, R. W., Weiss, N. O., & Vaughan, A. H. (1984). *Astrophysical Journal, 287*, 769 (1984). https://doi.org/10.1086/162735
75. Oláh, K., Kolláth, Z., Granzer, T., Strassmeier, K. G., Lanza, A. F., Järvinen, S. et al. (2009). *Astronomy & Astrophysics, 501*, 703. https://doi.org/10.1051/0004-6361/200811304
76. Parker, E. N. (1993). *Astrophysical Journal, 408*, 707. https://doi.org/10.1086/172631
77. Perri, B., Brun, A. S., Strugarek, A., & Réville, V. (2019). arXiv:1912.01271.
78. Pizzolato, N., Maggio, A., Micela, G., Sciortino, S., & Ventura, P. (2003). *Astronomy & Astrophysics, 397*, 147. https://doi.org/10.1051/0004-6361:20021560
79. Racine, E., Charbonneau, P., Ghizaru,M., Bouchat, A., & Smolarkiewicz, P. K. (2011). *Astrophysical Journal, 735*(1), 46. 10.1088/0004-637X/735/1/46. http://stacks.iop.org/0004-637X/735/i=1/a=46
80. Rempel, M. (2005). *Astrophysical Journal, 622*(2), 1320. https://doi.org/10.1086/428282
81. Réville, V., Brun, A. S., Matt, S. P., Strugarek, A., & Pinto, R.F. (2015). *Astrophysical Journal, 798*(2), 116. https://doi.org/10.1088/0004-637X/798/2/116
82. Réville, V., Brun, A. S., Strugarek, A., Matt, S. P., Bouvier, J., Folsom, C. P. et al. (2015). *Astrophysical Journal, 814*(2), 99. https://doi.org/10.1088/0004-637X/814/2/99
83. Réville, V., Folsom, C. P., Strugarek, A., & Brun, A. S. (2016). *Astrophysical Journal, 832*(2), 145. https://doi.org/10.3847/0004-637X/832/2/145.
84. Réville, V., Velli, M., Panasenco, O., Tenerani, A., Shi, C., Badman, S. T., Bale, et al. (2020). *Astrophysical Journal Supplement, 246*(2), 24. https://doi.org/10.3847/1538-4365/ab4fef
85. Saar, S. H., & Brandenburg, A. (1999). *Astrophysical Journal, 524*, 295. https://doi.org/10.1086/307794
86. See, V., Matt, S. P., Finley, A. J., Folsom, C. P., Boro Saikia, S., Donati, J. F. et al. (2019). *Astrophysical Journal, 886*(2), 120. https://doi.org/10.3847/1538-4357/ab46b2
87. Skumanich, A. (1972). *Astrophysical Journal, 171*, 565. https://doi.org/10.1086/151310
88. Smolarkiewicz, P. K., & Charbonneau, P. (2013). *Journal of Computational Physics, 236*, 608 (2013). https://doi.org/10.1016/j.jcp.2012.11.008
89. Strugarek, A., Beaudoin, P., Charbonneau, P., Brun, A. S. & do Nascimento, J. D. (2017). *Science, 357*(6347), 185. https://doi.org/10.1126/science.aal3999
90. Strugarek, A., Beaudoin, P., Charbonneau, P., & Brun, A. S. (2018). *Astrophysical Journal, 863*(1), 35. https://doi.org/10.3847/1538-4357/aacf9e

91. Thompson, M. (2014). *Frontiers in Astronomy and Space Sciences, 1*, 1. https://doi.org/10.3389/fspas.2014.00001
92. Thompson, M. J., Toomre, J., Anderson, E. R. Antia, H. M., Berthomieu, G., Burtonclay, D., et al. (1996). *Science, 272*(5266), 1300. https://doi.org/10.1126/science.272.5266.1300
93. Thompson, M. J., Christensen-Dalsgaard, J., Miesch, M. S., & Toomre, J. (2003). *Annual Review of Astronomy and Astrophysics, 41*, 599. https://doi.org/10.1146/annurev.astro.41.011802.094848
94. van Saders, J. L., Ceillier, T., Metcalfe, T. S., Silva Aguirre, V., Pinsonneault, M. H., García, R. A., et al. (2016). *Nature, 529*(7585), 181. https://doi.org/10.1038/nature16168
95. Varela, J., Strugarek, A., & Brun, A. S. (2016). *Advances in Space Research, 58*(8), 1507. https://doi.org/10.1016/j.asr.2016.06.032
96. Vidotto, A.A., Donati, J. F., Jardine, M., See, V., Petit, P., Boisse, I., et al. (2016). *Monthly Notices of the Royal Astronomical Society, 455*(1), L52. https://doi.org/10.1093/mnrasl/slv147.
97. Viviani, M., Käpylä, M. J., Warnecke, J., Käpylä, P. J., & Rheinhardt, M. (2019). *Astrophysical Journal, 886*(1), 21. https://doi.org/10.3847/1538-4357/ab3e07
98. Warnecke, J., Käpylä, P. J., Käpylä, M. J., & Brandenburg, A. (2014). *The Astrophysical Journal Letters, 796*(1), L12. https://doi.org/10.1088/2041-8205/796/1/L12
99. Yeates, A. R., Nandy, D., & Mackay, D. H. (2008). *Astrophysical Journal , 673*(1), 544. https://doi.org/10.1086/524352.
100. Zhao, J., Bogart, R. S., Kosovichev, A. G., Duvall, T. L., Jr., & Hartlep, T. (2013). *The Astrophysical Journal Letters, 774*(2), L29. https://doi.org/10.1088/2041-8205/774/2/L29

Recent Progress in Local Helioseismology

A. C. Birch

Abstract Local helioseismology is a collection of methods for using observations of solar oscillations at the surface to estimate three-dimensional mass flows and variations in wave speed in the solar interior. Michael Thompson contributed to the development of the techniques used to solve inverse problems in local helioseismology. A few selected important targets for current local helioseismic inferences are discussed: Rossby waves and flows around active regions.

1 Introduction

The Sun is currently the only star that we can study at high spatial resolution. Observations of the Sun are important tests of our ideas about astrophysical fluid dynamics (a topic of great interest to Michael Thompson; [60]). Local helioseismology, based on correlations in the pattern of solar oscillations, can be used to map time-dependent three-dimensional mass flows and wave-speed variations in the solar interior. Local helioseismology allows access to characteristics of the solar interior that cannot be inferred solely from the frequencies of the solar global oscillations, as these frequencies are only sensitive to longitude and north-south averages of the solar structure and dynamics. Some important targets for local helioseismology are the deep meridional circulation, convection in the solar interior, global-scale Rossby waves (Sect. 4.1), and flows around active regions (Sect. 4.2).

Gizon and Birch [14] and Gizon et al. [16] provide comprehensive reviews of the methods of local helioseismology and of the main results. In between these two larger reviews, Thompson and Zharkov [61] described the status of local helioseismology at that time, with a focus on open questions about data analysis for sunspots and active regions. More recently, Zhao and Chen [63] reviewed

A. C. Birch (✉)
Max-Planck-Institut für Sonnensystemforschung, Göttingen, Germany
e-mail: birch@mps.mpg.de

M. J. P. F. G. Monteiro et al. (eds.), *Dynamics of the Sun and Stars*, Astrophysics and Space Science Proceedings 57,
https://doi.org/10.1007/978-3-030-55336-4_10

recent inferences about subsurface flows from time-distance and ring-diagram helioseismology. Hanasoge et al. [24] reviewed helioseismic measurements of solar convection. Nagashima [46] reviewed results from local helioseismology from the Solar Optical Telescope onboard the Hinode spacecraft.

Here, for the sake of readability, I provide a minimal introduction to local helioseismology but mostly focus on a few recent developments. In addition, I will highlight two important contributions of Michael Thompson to the topic of inversions for local helioseismology (Sect. 3.1.2).

2 Data for Local Helioseismology

Most local helioseismology is currently carried out with observations of the Doppler velocity (line-of-sight velocity) from SOHO/MDI (1996–2011, [56]), GONG (from 1995, [26, 28]), and/or SDO/HMI (from 2010, [57, 58]). The cadence of the SOHO/MDI and GONG observations is 1 min, while the Doppler images from SDO/HMI are made at a cadence of 45 s. These cadences are sufficient to capture the solar oscillations, which have a dominant period of about 5 min (though oscillation power can be seen at much higher frequency, see Fig. 1). The spatial resolutions of the different instruments are different, but all are sufficient to capture the acoustic and surface gravity waves that are important for local helioseismology.

Figure 1 shows an example power spectrum for SDO/HMI Doppler observations near the center of the disk. The ridges of large power correspond to the resonance frequencies of the acoustic modes (called p modes, pressure modes) and the surface gravity mode (called the f mode). The resonance frequencies and eigenfunctions of these modes depend on conditions in the solar interior and thus surface observations of the waves can be used to study the physics of the solar interior.

3 Methods of Local Helioseismology

The traditional methods of local helioseismology are ring-diagram analysis, time-distance helioseismology, and helioseismic holography. As described in Sect. 2, the input data for these methods are time series of, typically, Doppler images. Here I focus on specific topics related to ring-diagram analysis and time-distance helioseismology where there have been important recent developments. I will also describe some areas where the work of Michael Thompson has had a particular impact. I will not cover the topic of mode-coupling even though this topic is rapidly developing ([22] provides a recent summary of the formalism). In addition, I will not cover the Legendre-Fourier Decomposition method for measuring the subsurface meridional circulation [7], but see the conference contribution by Braun et al. [8].

Measurements of surface flows using non-helioseismic methods have played an important role in validating helioseismic methods. As one example, granules

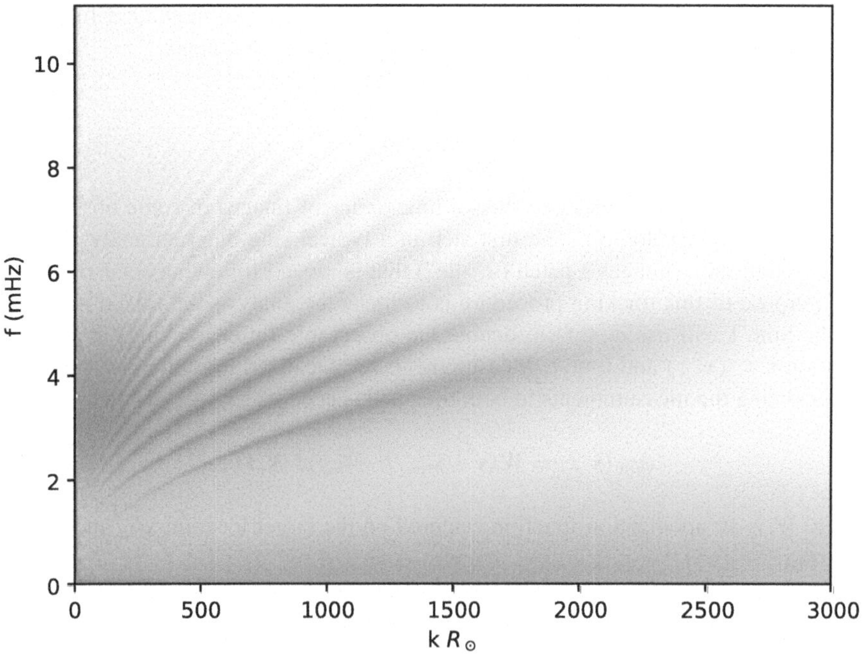

Fig. 1 Example power spectrum of Doppler velocity observations from SDO/HMI. The dark ridges have large power and are due to the surface gravity wave and the acoustic waves. For $kR_{\odot}$ less than about 1200 these ridges visibly extend up to the Nyquist frequency of the observations (11.1 mHz). The high power at low temporal frequency is due to convection. The lighter gray show regions of lower power. The f mode is the ridge with the lowest frequency, the other ridges are the p modes. The lowest frequency p mode at each k is the mode with a single radial node (the p_1 mode). Figure adapted from Nagashima et al. [48]

(small-scale convective features) act as tracers of larger-scale flows. Measuring the movement of the granulation pattern from one image to the next is therefore a measure of the larger-scale flow field. Granulation tracking and local helioseismology measurements of surface flows are in good agreement for intermediate spatial scales (e.g. [39]).

3.1 Ring-Diagram Analysis

Ring-diagram analysis was introduced by Hill [27] and involves measuring the local dispersion relation of acoustic and surface-gravity waves. The local dispersion relation depends on the strength and direction of subsurface (and surface) flows as well as the isotropic wave speed and the magnetic field. Section 3.1.1 summarizes the method of ring-diagram analysis, Sect. 3.1.2 describes the SOLA inversion method (Michael Thompson was involved in developing this method with Frank Pijpers), and Sect. 3.1.3 reviews a few selected recent developments.

3.1.1 Ring-Diagram Analysis: Background

The input for ring-diagram analysis is a time-series of images of some observable (e.g. line-of-sight velocity). The first step in a typical ring-diagram analysis is to extract from these images a patch (a "tile") that is moved following solar rotation. The purpose of this tracking procedure is to focus the analysis on a fixed position on the Sun. I will use $\phi(\mathbf{x}, t)$ to denote the observable (after tracking) at surface location $\mathbf{x} = (x, y)$ and time t. Measurements are localized to a time and position by apodizing the measurements in both time and space:

$$\phi_{\text{tile}}(\mathbf{x}, t) = W(\mathbf{x} - \mathbf{x}_{\text{tile}}, t - t_{\text{tile}})\phi(\mathbf{x}, t) \, , \tag{1}$$

where W is an apodization function centered on the target location $\mathbf{x}_{\text{tile}}$ and target time t_{tile}.

The three-dimensional power spectrum ("ring diagram") of the apodized data is

$$P(\mathbf{k}, \omega) = |\phi_{\text{tile}}(\mathbf{k}, \omega)|^2 \, , \tag{2}$$

where $\mathbf{k} = (k_x, k_y)$ is the horizontal wavevector and $\phi_{\text{tile}}(\mathbf{k}, \omega)$ is the three-dimensional Fourier transform of $\phi_{\text{tile}}(\mathbf{x}, t)$.

The next step in ring-diagram analysis is to determine the positions of the resonances in the 3D power spectrum P. Bogart et al. [4] briefly describes the commonly used fitting procedures in the context of the HMI ring-diagram pipeline. The results of the fitting procedure are resonance frequencies, line widths, and flow parameters u_x and u_y for different modes. Subsurface flows alter the power spectrum through the Doppler effect; the u_x and u_y are measures of this effect in the two horizontal directions. Each of these parameters depends on the radial order and angular degree of the mode (n and $\ell \approx kR_\odot$, respectively, the solar radius is $R_\odot$).

The u_x and u_y parameters are connected to flows in the solar interior by sensitivity functions that satisfy, e.g.,

$$u_{x,n\ell} = \int K_{n\ell}(z) v_x(z) \, \mathrm{d}z + N_{n\ell} \, , \tag{3}$$

where z is depth, v_x is the true flow in the x direction, $N_{n\ell}$ denotes the noise (see [46] for a detailed discussion), and the integral (and all following integrals) is taken over all depths. The simple form of the above equation relies on the assumptions that the true flow v_x is weak (non-linear terms are neglected), horizontally uniform over the ring tile (see [3] for the case of horizontally varying flows), and steady during the observing time. This simple form also assumes that v_x doesn't depend on v_y (see [62] for discussion of cross-talk).

Equation (3) is helpful for clarifying the inversion problem of ring diagrams: given a set of measured flow parameters $u_{x,n\ell}$ for various values of n and ℓ and a set of theoretically computed kernel functions $K_{n\ell}$, estimate the depth-dependent flow field $v_x(z)$. Thompson [59] used this simple one-dimensional system, though for the case of 1D rotation inversions from global-mode frequency splittings, to illustrate concepts of inverse theory for helioseismology.

3.1.2 Ring-Diagram Analysis: The SOLA Method and the Noise Correlation

Pijpers and Thompson [51, 52] introduced the "subtractive optimally localized averaging" (SOLA) method for inversions in the context of inversions of global-mode helioseismology. It was later used in local helioseismology (e.g. [32, 34]. In the spirit of the conference, I will here use the ring-diagram inversion problem described in the previous section to demonstrate the SOLA method (the mathematics below is essentially the same as the original demonstration of [51], in the context of 1D rotation inversions).

For any linear 1D ring inversion, the estimated flow $\tilde{v}_x$ at a depth z is a linear combination of the u_x parameters. For the sake of simplicity I will drop the x subscripts and use a collective mode index $i = (n, \ell)$. For a target depth z_t any particular inversion will produce a vector of coefficients $\mathbf{c}$ so that

$$\tilde{v}(z_\mathrm{t}) = \sum_i c_i(z_\mathrm{t}) u_i \ , \tag{4}$$

where the sum over i is over all modes for which measurements u_i are available. Equation (3) implies that

$$\tilde{v}(z_\mathrm{t}) = \int \mathcal{K}(z; z_\mathrm{t}) v(z) \, \mathrm{d}z + \text{noise} \ , \tag{5}$$

where the averaging kernel $\mathcal{K}$ shows how the inferred flow is related to the true flow. Generically, a "good" inversion has a narrow averaging kernel that is localized at the target location and a noise level that is acceptable. The averaging kernel is a linear combination of the original kernels

$$\mathcal{K}(z; z_\mathrm{t}) = \sum_i c_i(z_\mathrm{t}) K_i(z) \ . \tag{6}$$

The key idea of SOLA is to minimize the difference between the averaging kernel and some target function while also controlling the noise level. In particular, Pijpers and Thompson [51] proposed to minimize, for a specific target location,

$$X(\mathbf{c}) = \int [\mathcal{K}(z) - \mathcal{T}(z)]^2 \, \mathrm{d}z + \mu \mathbf{c}^{\mathrm{T}} \Lambda \mathbf{c} \tag{7}$$

where $\mathcal{T}$ is a target function, μ is a control parameter, Λ is the covariance matrix of the input measurements, and the superscript T indicates the transpose. The minimization of X with respect to the coefficients $\mathbf{c}$ (the averaging kernel $\mathcal{K}$ depends on $\mathbf{c}$, Eq. (6)) is carried out with the linear constraint that the averaging kernel integrates to one. The target function is typically a function that is localized around the target depth z_t with a width matched to the expected spatial resolution of the inversion (expected to be connected to the wavelength). The control parameter, or regularization parameter, μ allows for a trade-off between matching the target function and reducing the noise in the inversion result.

SOLA was an improvement over the existing OLA methods as it allows inversions for multiple choices of target functions $\mathcal{T}$ for almost the same numerical cost as a single inversion. To see this schematically, consider the minimization of X in Eq. (7). The linear constraint that the averaging kernel integrates to one can be introducing by using a Lagrange multiplier λ. The minimization results in the linear system

$$M\mathbf{c} + \lambda \mathbf{w} = \mathbf{t} \tag{8}$$
$$\mathbf{w} \cdot \mathbf{c} = 1$$

where the matrix M contains the pair-wise overlap integrals of the kernels, the vector $\mathbf{t}$ contains the projection of the target function on the kernels, and the vector $\mathbf{w}$ contains the integrals of the kernels

$$M_{ij} = \mu \Lambda_{ij} + \int K_i(z) K_j(z) \, \mathrm{d}z \, , \tag{9}$$

$$t_i = \int K_i(z) \mathcal{T}(z) \, \mathrm{d}z \, , \tag{10}$$

$$w_i = \int K_i(z) \, \mathrm{d}z \, . \tag{11}$$

It is important that the matrix M does not depend on the target depth z_t: solving the linear system (8) for multiple choices of the target function doesn't involve recomputing or refactoring the overlap matrix. This makes SOLA much faster than traditional OLA where the target depth appears in the analog of the M matrix (see [51] for more details). Also notice that the input data (the u_i) don't appear in the inversion problem—inversions for multiple data sets, as long they involve a common set of modes, are essentially free. For the case of 3D inversions where the kernels are translation invariant in the horizontal directions, the full inversion problem (Eq. (7)) can be split into independent (small) inversion problems at each horizontal wavevector and thus solved very efficiently (e.g. [36, 35]).

Howe and Thompson [30] emphasized that the error correlation is an additional quantity to consider when assessing the result of an inversion. Suppose that two sets of coefficients, $\mathbf{c}_1$ and $\mathbf{c}_2$, correspond to inversions for two different target depths. The error covariance for the flow estimates, $\tilde{v}_1$ and $\tilde{v}_2$, at these two target depths is

$$\mathrm{Cov}\left[\tilde{v}_1, \tilde{v}_2\right] = \mathbf{c}_1^{\mathrm{T}} \Lambda \mathbf{c}_2 \,. \tag{12}$$

Howe and Thompson [30] showed that for reasonable inversions of solar rotation, the width of the averaging kernel and the width of the error correlation are often comparable. They emphasized that for OLA methods the error covariance between far-away target points can be large, and that this effect can be reduced (without substantially altering the averaging kernels) by careful choice of regularization parameters.

3.1.3 Ring-Diagram Analysis: Recent Developments

Motivated by the difference between the strength of subsurface convection inferred by Hanasoge et al. [23] and Greer et al. [20], Nagashima et al. [49] carried out a detailed study of the multi-ridge fitting method used by Greer et al. [20]. Multi-ridge fitting is a class of ring-fitting methods where instead of fitting a single ridge (i.e., a single radial order) at a time, the fitting function takes into account all of the significant ridges inside some fitting interval (in the case of [20], all ridges at fixed horizontal wavenumber). Nagashima et al. [49] found and corrected a number of issues with the fitting code used by Greer et al. [20]: the main issue was that the cost function was based on an inaccurate model for the distribution of noise in the power spectrum. Nagashima et al. [49] used Monte Carlo tests to show that the updated code produced unbiased (u_x, u_y) flow parameters (within the noise level of the Monte Carlo); this work was partially shown in a conference poster [47]. This updated fitting method didn't, however, explain the orders-of-magnitude difference between the results of Hanasoge et al. [23] and Greer et al. [20].

In another recent development in methods for ring-diagram analysis, Alshehhi et al. [1] presented a machine-learning approach to fast inversions. This work trained a neural network on a large set of existing ring-diagram inversions. The trained network takes ring fit parameters (u_x, u_y) as input and estimates the flow field at a fixed depth. Alshehhi et al. [1] show that this machine-learning approach allows very fast inversions: all of the 15° ring diagram measurements for a single Carrington rotation can be inverted in seconds on a single core (the pipeline code requires about 30 CPU hours).

3.2 Time-Distance Helioseismology

Time-distance helioseismology [12] is a set of methods for measuring the times taken for acoustic or surface-gravity wave packets to travel between two distinct points on the solar surface. In the case of acoustic waves, these wave packets can be thought of as propagating along ray paths through the interior. Thus, surface measurements provide constraints on wave propagation speeds, and thus physical conditions, in the solar interior. Section 3.2.1 provides, for the sake of readability, a very condensed introduction to time-distance helioseismology. Section 3.2.2 briefly outlines two recent developments in theory for time-distance helioseismology: computational helioseismology and a comparison of the diagnostic power of amplitudes and travel-times.

3.2.1 Time-Distance Helioseismology: Background

The time-distance cross covariance computed from an observable ϕ is

$$C(\mathbf{x}_1, \mathbf{x}_2, t) = A(T, t) \int W(t')W(t' + t)\phi(\mathbf{x}_1, t')\phi(\mathbf{x}_2, t' + t)\, \mathrm{d}t' \,, \tag{13}$$

where $\mathbf{x}_1$ and $\mathbf{x}_2$ are two points on the solar surface, t is the time lag, $W(t)$ is the temporal apodization function (i.e. $W(t)$ is one for $0 \le t \le T$ and zero otherwise, T is the duration of the observations), and $A(T, t)$ is a normalization constant that depends only on the duration of the observations T and the time lag. For example, Thompson and Zharkov [61] used $A(T, t) = (T - |t|)^{-1}$. In practice, the cross-covariance is often computed in the Fourier domain.

The typical next step in time-distance helioseismology is to extract wave travel times from the measured cross-covariances (or spatial averages of cross-covariances). There are a handful of methods for achieving this. Thompson and Zharkov [61] reviewed the standard methods.

Hughes et al. [32] studied two approaches to producing time-distance measurements that are sensitive to flows near a particular target point in the interior. In one case, they considered the problem of determining optimum combinations of travel times. This case is directly analogous to the ring-diagram inversion shown in Sect. 3.1.2, with travel times playing the role of ring-fit parameters. In a second case, they looked for the optimum weights for averaging the Doppler signal before computing the cross-covariance (i.e., replacing the $\phi(\mathbf{x}_1, t)$ and $\phi(\mathbf{x}_2, t)$ above with weighted averages of ϕ at different spatial locations). This leads to a non-linear inversion for the weights. In the examples they considered, the SOLA approach (linear inversion of travel times) outperformed the non-linear inversion to find optimal weights for averaging the Doppler signal. I discuss a possible generalization of the weighted Doppler approach in the outlook (Sect. 5).

3.2.2 Time-Distance Helioseismology: Recent Developments

As described above (Sect. 3.2.1), traditional time-distance helioseismology is based on measuring and interpreting wave travel times obtained from cross-covariances. The amplitude of the cross-covariance, group time (envelope time), and parameters describing the frequency content (the mean frequency and the bandwidth) are sometimes by-products of fitting cross-covariances to extract travel times (e.g. [14, 61]). These parameters are not often used to provide quantitative constraints on subsurface conditions.

Pourabdian et al. [53] used a toy model to compare the sensitivity of time-distance travel times and amplitudes to local changes in sound speed. They showed that amplitude measurements are sensitive to changes in the sound speed located along the ray path connecting the two observation points. This is in contrast with travel times, which are more sensitive to changes in the sound speed located at the border of the first Fresnel zone. This result is known in the geophysics literature (e.g. [9]). Pourabdian et al. [53] showed that a consequence of these different spatial sensitivity patterns is that "deep focusing" averaging of amplitudes produces well-localized kernels, while the same kind of averaging of travel times produces sensitivity functions that are not maximum at the nominal focus position. They showed that in the simple model, a localized sound-speed perturbation, and with a straightforward deep-focusing scheme, amplitude measurements have better signal-to-noise ratios than travel-time measurements. This work demonstrates the importance of continued efforts on the theoretical background for local helioseismology, a point I will come back to in Sect. 5.

In another recent development, Gizon et al. [17] introduced and demonstrated a framework for computing the local helioseismic measurements (e.g., wave travel times) that would be expected to result from solar flows. Their approach is to compute frequency-domain Green's functions (impulse responses) for a simplified scalar wave equation describing the propagation of acoustic waves. The cross-covariance function and travel-time sensitivity kernels are then computed in terms of the Green's functions. Their method is an improvement over previous semi-analytical methods as it can be easily generalized to include observational effects that are cumbersome to include in the semi-analytical calculations (e.g., line-of-sight projection and foreshortening). Fournier et al. [13] extended this work by demonstrating an efficient method for computing sensitivity functions in the case of spherically symmetric reference solar models; sensitivity functions in spherical geometry were discussed also in a conference poster [2].

4 Selected Recent Observational Results

Some important recent results obtained using local helioseismology are the discovery and characterization of global-scale internal Rossby waves (Sect. 4.1) and measurements of near-surface flows associated with active regions (Sect. 4.2). It is notable that these results don't depend on inversions, but rather on the direct interpretation of helioseismic parameters (e.g. travel times or ring-fit parameters) as being proportional to flows. Another common theme is that the comparison between helioseismic measurements of surface flows and granulation tracking measurements of the same flows are important for demonstrating the robustness of the results.

4.1 Rossby Waves

Löptien et al. [43] unambiguously detected classical Rossby waves in a six-year time series of maps of horizontal surface flows measured using granulation tracking and ring-diagram analysis applied to SDO/HMI observations. These waves are global scale (azimuthal wavenumbers $m \leq 15$) and are concentrated around the equator. The measured dispersion relation, in a frame rotating at the equatorial surface rotation rate Ω, is close to that of classical sectoral ($m = \ell$) Rossby waves

$$\omega_m = -\frac{2\Omega}{m+1} . \tag{14}$$

These modes have a retrograde phase speed and prograde group speed.

Figure 2 shows theoretical eigenfunctions for the $m = 8$ sectoral Rossby mode on a uniformly rotating sphere (e.g. [55]). The north-south velocity is symmetric about the equator and maximum at the equator. The east-west velocity is anti-symmetric across the equator and has an overall smaller amplitude than the north-south velocity. The radial vorticity (radial component of the curl of velocity), like the north-south velocity, is symmetric about the equator and has its maximum there. Löptien et al. [43] and Proxauf et al. [54] showed that the horizontal eigenfunctions of the Rossby waves on the Sun are more concentrated near the equator than the theoretical expectations for uniformly rotating stars. Furthermore, they showed that

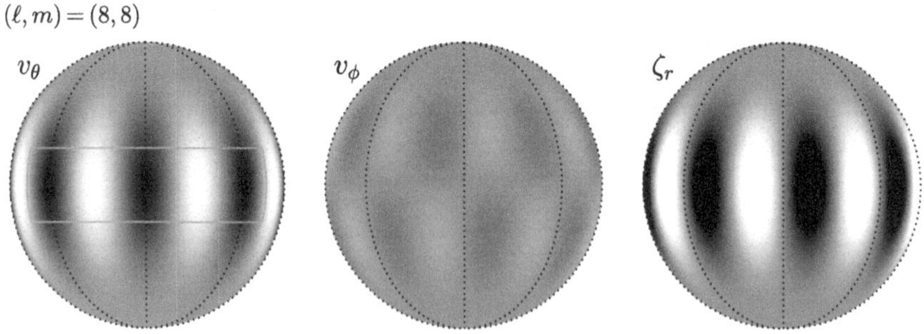

Fig. 2 Patterns of north-south velocity (left), east-west velocity (middle), and radial vorticity (right) for a model of the $m = 8$ sectoral ($m = \ell$) Rossby mode on a uniformly rotating sphere. Figure adapted from Liang et al. [40]

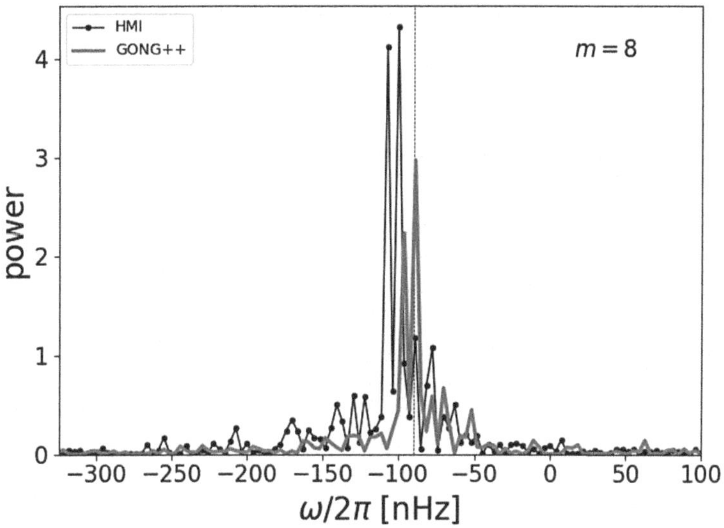

Fig. 3 Power spectra associated with the $m = 8$ Rossby wave. The power spectrum from SDO/HMI observations is shown in black and the GONG++ result is shown in the grey curve. The dashed vertical line shows the frequency determined by Löptien et al. [43] from surface flows measured using granulation tracking. Hanson et al. [25] compare the mode parameters measured from these data sets. Figure adapted from Hanson et al. [25]

the eigenfunctions of radial vorticity have zero crossings in latitude. Proxauf et al. [54] show that the zero crossings are at roughly 25° to 30° latitude. A physical mechanism leading to these zero crossings which is based the interaction of Rossby waves with latitudinal differential rotation has been proposed by Gizon et al. [18].

Liang et al. [40] applied time-distance helioseismology to 21 years of data (SOHO/MDI and also SDO/HMI) to measure the north-south velocities associated with the Rossby waves studied by Löptien et al. [43]. These time-distance measurements confirmed the detection of Löptien et al. [43]. The detection of Rossby waves was also confirmed by Alshehhi et al. [1] and Hanson et al. [25] using ring-diagram analysis and Mandal and Hanasoge [44] using mode coupling. Figure 3 shows the power spectrum of the $m = 8$ Rossby wave as seen in the near-surface layers using ring-diagram analysis of SDO/HMI and GONG++ observations, both for the time period of May 2010 to December 2018, from Hanson et al. [25]. The cause for the difference between the GONG++ and SDO/HMI results is not known (see [25] for more discussion). Some of these results were described in a conference poster [19]. In another conference poster, Duvall [11] used time-distance helioseismology to make preliminary estimates of the depth dependence of the Rossby waves.

4.2 Flows Associated with Active Regions

Flows around active regions have been observed with helioseismology (e.g. [15, 21, 29]) and are a potentially important ingredient for understanding the solar dynamo (e.g. [10, 45]). Löptien et al. [42] confirmed that these flows can be also be seen in the surface flows measured using granulation tracking. In a conference poster, Korzennik [38] used time-distance helioseismology to search for signatures of active regions before their emergence at the photosphere. The conference poster of Komm and Gosain [37] looked for connections between kinetic helicity in the flows around active regions and flaring activity.

One of the open questions is how, or if, flows around active regions are connected to properties of the active regions (e.g., age, total flux, number of sunspots). Braun [6], also shown in a conference poster [5], applied helioseismic holography to SDO/HMI observations to infer the near-surface horizontal flows associated with 336 active regions. By computing average flow fields for active regions binned by field strength, this work showed that the strength of the large-scale converging flows are similar for active regions with total magnetic flux above about 10^{21} Mx. Figure 4 shows the average flow fields for the lowest-flux regions in the sample (left panel) and highest-flux regions (right panel). Braun [6] also detected a retrograde flow on the poleward side of the average active region and suggested that this retrograde flow may be an important component of the observed torsional oscillations (time-varying zonal flows associated with the solar cycle, see [31], for a recent overview).

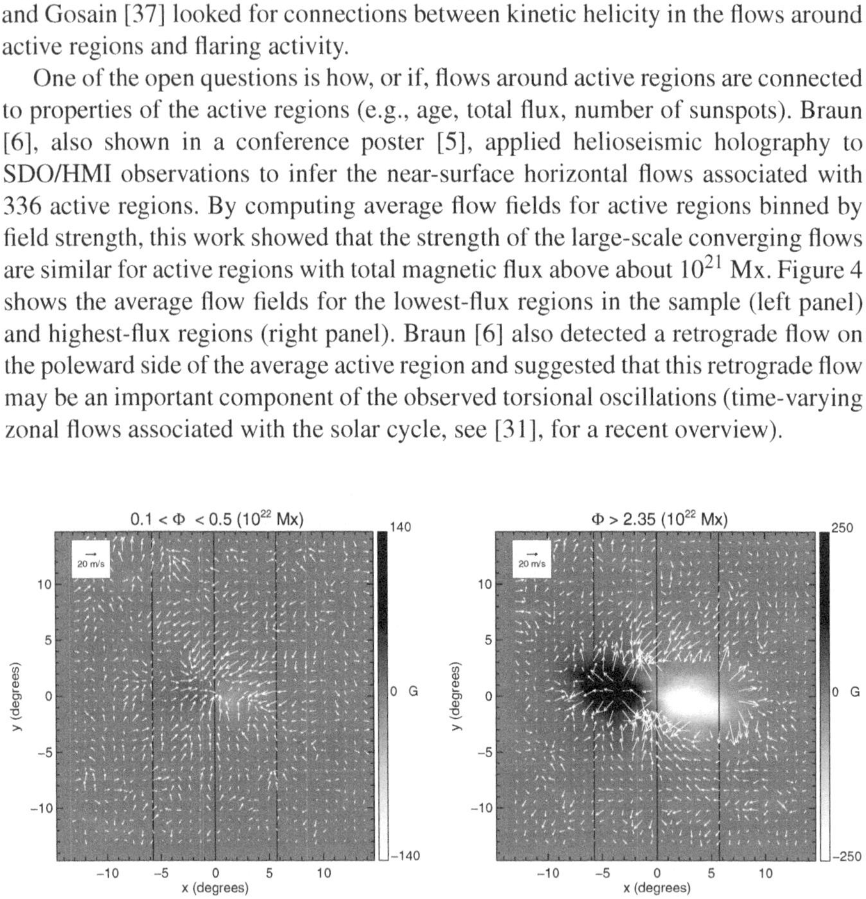

Fig. 4 Near-surface horizontal flow fields (white arrows) from helioseismic holography after averaging over active regions grouped by total magnetic flux. The background grey scale shows the average line-of-sight field. The left panel shows the average over the active regions with magnetic flux between 10^{21} Mx and 5×10^{21} Mx (773 individual maps) in the sample of Braun [6] and the right panel shows the average over the largest (flux larger than 2.35×10^{22} Mx, 1097 individual maps). The $\hat{\mathbf{x}}$ direction is prograde and the $\hat{\mathbf{y}}$ direction is poleward. Figure adapted from Braun [6]

5 Outlook

Two particular areas where I think that there is potential for improvement in inversions for local helioseismic methods are the use of time-distance parameters other than travel times and optimal weighting of time-distance measurements.

Most recent time-distance helioseismology work has been done with single-skip (also called first-bounce) travel times. This procedure could be generalized to employ other parameters measured from the cross-covariance (e.g. the amplitude, Sect. 3.2.2) and to use measurements for other skips. The work of Pourabdian et al. [53] is a first step in this direction. Measurements from other skips are central to far-side imaging [41, 64] but are not commonly used in near-side imaging.

Another promising area is to consider the methods used for averaging two-point time-distance measurements. Hughes et al. [32] detailed an approach for computing optimal weightings. With only a slight generalization to include frequency-dependent and complex-valued weights, this optimization procedure would be a generalization of helioseismic holography. This approach may help unify time-distance helioseismology and helioseismic holography.

Acknowledgments The Solar and Stellar Interiors Department at the MPS acknowledges support by the ERC Synergy Grant WHOLESUN 81021 and support from the German Aerospace Center (DLR) for the German Data Center for SDO (GDC-SDO). The HMI data used are courtesy of NASA/SDO and the HMI science team. ACB thanks Doug Braun, Laurent Gizon, Bastian Proxauf, Jesper Schou, and Hannah Schunker for helpful comments and Doug Braun, Chris Hanson, Zhi-Chao Liang, and Kaori Nagashima for providing figures or input data for figures. This work used the NumPy [50] and Matplotlib [33] Python packages.

References

1. Alshehhi, R., Hanson, C. S., Gizon, L., & Hanasoge, S. (2019). *Astronomy & Astrophysics, 622*, A124.
2. Bhattacharya, J. (2019). *Helioseismic Sensitivity Kernels to Probe Solar Subsurface Convection in Spherical Geometry*. Conference poster. https://www2.hao.ucar.edu/MJTWorkshop2019/Agenda
3. Birch, A. C., Gizon, L., Hindman, B. W., & Haber, D. A. (2007). *The Astrophysical Journal, 662*, 730.
4. Bogart, R. S., Baldner, C., Basu, S., Haber, D. A., & Rabello-Soares, M. C. (2011). *Journal of Physics Conference Series, 271*, 012008. GONG-SoHO 24: A New Era of Seismology of the Sun and Solar-Like Stars.
5. Braun, D. (2019). *Active Region Flows and Their Contribution to Varying Global Dynamics*. Conference poster. https://www2.hao.ucar.edu/MJTWorkshop2019/Agenda
6. Braun, D. C. (2019). *The Astrophysical Journal, 873*, 94.
7. Braun, D. C., & Fan, Y. (1998). *The Astrophysical Journal Letters, 508*, L105.
8. Braun, D., Birch, A., & Fan, Y. (2019). *Probing the Variation with Depth of the Solar Meridional Circulation Using Legendre Function Decomposition*. Conference poster. https://www2.hao.ucar.edu/MJTWorkshop2019/Agenda
9. Dahlen, F. A., & Baig, A. M. (2002). *Geophysical Journal International, 150*, 440.

10. De Rosa, M. L., & Schrijver, C. J. (2006). *Proceedings of SOHO 18/GONG 2006/HELAS I, Beyond the Spherical Sun* (Vol. 624), 12. ESA Special Publication.
11. Duvall, T. (2019). *Measurement of Rossby Wave Radial Eigenfunctions Using Deep-Focus Time-Distance Helioseismology*. Conference poster. https://www2.hao.ucar.edu/MJTWorkshop2019/Agenda
12. Duvall, T. L., Jefferies, S. M., Harvey, J. W., & Pomerantz, M. A. (1993). *Nature, 362*, 430.
13. Fournier, D., Hanson, C. S., Gizon, L., & Barucq, H. (2018). *Astronomy & Astrophysics, 616*, A156.
14. Gizon, L., & Birch, A. C. (2005). *Living Reviews in Solar Physics, 2*, 6.
15. Gizon, L., Duvall, T. L. J., & Larsen, R. M. (2001). In P. Brekke, B. Fleck, & J. B. Gurman (Eds.), Recent Insights into the Physics of the Sun and Heliosphere: Highlights from SOHO and Other Space Missions. *IAU Symposium* (Vol. 203, p. 189).
16. Gizon, L., Birch, A. C., & Spruit, H. C. (2010). *Annual Review of Astronomy and Astrophysics, 48*, 289.
17. Gizon, L., Barucq, H., Duruflé, M., Hanson, C. S., Leguèbe, M., Birch, A. C., et al. (2017). *Astronomy & Astrophysics, 600*, A35.
18. Gizon, L., Fournier, D., & Albekioni, M. (2020). *Astronomy & Astrophysics* (accepted).
19. Gizon, L., & The MPS Team. (2019). *Global-Scale Rossby Waves in the Solar Interior*. Conference poster. https://www2.hao.ucar.edu/MJTWorkshop2019/Agenda
20. Greer, B. J., Hindman, B. W., Featherstone, N. A., & Toomre, J. (2015). *The Astrophysical Journal Letters, 803*, L17.
21. Haber, D. A., Hindman, B. W., Toomre, J., Bogart, R. S., & Hill, F. (2001). In A. Wilson & P. L. Pallé (Eds.), SOHO 10/GONG 2000 Workshop: Helio- and Asteroseismology at the Dawn of the Millennium (Vol. 464, pp. 209–212). ESA Special Publication.
22. Hanasoge, S. (2018). *The Astrophysical Journal, 861*, 46.
23. Hanasoge, S. M., Duvall, T. L., & Sreenivasan, K. R. (2012). *Proceedings of the National Academy of Science, 109*, 11928.
24. Hanasoge, S., Gizon, L., & Sreenivasan, K. R. (2016). *Annual Review of Fluid Mechanics, 48*, 191.
25. Hanson, C. S., Gizon, L., & Liang, Z.-C. (2020). arXiv e-prints, arXiv:2002.01194.
26. Harvey, J. W., Hill, F., Hubbard, R. P., Kennedy, J. R., Leibacher, J. W., Pintar, J. A., et al. (1996). *Science, 272*, 1284.
27. Hill, F. (1988). *The Astrophysical Journal, 333*, 996.
28. Hill, F., Bolding, J., Toner, C., Corbard, T., Wampler, S., Goodrich, B., et al. (2003). In H. Sawaya-Lacoste (Eds.), GONG+ 2002. Local and Global Helioseismology: The Present and Future (Vol. 517, pp. 295–298). ESA Special Publication.
29. Hindman, B. W., Haber, D. A., & Toomre, J. (2009). *The Astrophysical Journal, 698*, 1749.
30. Howe, R., & Thompson, M. J. (1996). *Monthly Notices of the Royal Astronomical Society, 281*, 1385.
31. Howe, R., Hill, F., Komm, R., Chaplin, W. J., Elsworth, Y., Davies, G. R., et al. (2018). *The Astrophysical Journal Letters, 862*, L5.
32. Hughes, S. J., Pijpers, F. P., & Thompson, M. J. (2007). *Astronomy & Astrophysics, 468*, 341.
33. Hunter, J. D. (2007). *Computing in Science & Engineering, 9*, 90.
34. Jackiewicz, J., Gizon, L., Birch, A. C., & Thompson, M. J. (2007). *Astronomische Nachrichten, 328*, 234.
35. Jackiewicz, J., Birch, A. C., Gizon, L., Hanasoge, S. M., Hohage, T., Ruffio, J.-B., et al. (2012). *Solar Physics, 276*, 19.
36. Jacobsen, B., Moller, I., Jensen, J., & Efferso, F. (1999). *Physics and Chemistry of the Earth A, 24*, 215.
37. Komm, R., & Gosain, S. (2019). *Long-Lived Activity Complexes, Their Kinetic Helicity, Lifetime, and Flare Activity*. Conference poster. https://www2.hao.ucar.edu/MJTWorkshop2019/Agenda
38. Korzennik, S. (2019). *On Active Region Emergence Precursors*. Conference poster. https://www2.hao.ucar.edu/MJTWorkshop2019/Agenda

39. Langfellner, J., Gizon, L., & Birch, A. C. (2015). *Astronomy & Astrophysics, 581*, A67.
40. Liang, Z.-C., Gizon, L., Birch, A. C., & Duvall, T. L. (2019). *Astronomy & Astrophysics, 626*, A3.
41. Lindsey, C., & Braun, D. C. (2000). *Solar Physics, 192*, 261.
42. Löptien, B., Birch, A. C., Duvall, T. L., Gizon, L., Proxauf, B., & Schou J. (2017). *Astronomy & Astrophysics, 606*, A28.
43. Löptien, B., Gizon, L., Birch, A. C., Schou, J., Proxauf, B., Duvall, T. L., et al. (2018). *Nature Astronomy, 2*, 568.
44. Mandal, K., & Hanasoge, S. (2019). arXiv e-prints, arXiv:1908.05890.
45. Martin-Belda, D., & Cameron, R. H. (2016). *Astronomy & Astrophysics, 586*, A73.
46. Nagashima, K. (2018). In T. Shimizu, S. Imada, & M. Kubo (Eds.), *Local helioseismology analyses with Hinode/SOT datasets* (Vol. 449, p. 183). Astrophysics and Space Science Library. Singapore: Springer
47. Nagashima, K. (2019). *Monte-Carlo Test of an Improved Multi-Ridge Fitting Code for Ring-Diagram Analysis*. Conference poster. https://www2.hao.ucar.edu/MJTWorkshop2019/Agenda
48. Nagashima, K., Löptien, B., Gizon, L., Birch, A. C., Cameron, R., Couvidat, S., et al. (2014). *Solar Physics, 289*, 3457.
49. Nagashima, K., Birch, A. C., Schou, J., Hindman, B. W., & Gizon, L. (2020). *Astronomy & Astrophysics, 633*, A109.
50. Oliphant, T. E. (2006). *A guide to NumPy* (Vol. 1) USA: Trelgol Publishing.
51. Pijpers, F. P., & Thompson, M. J. (1992). *Astronomy & Astrophysics, 262*, L33.
52. Pijpers, F. P., & Thompson, M. J. (1994). *Astronomy & Astrophysics, 281*, 231.
53. Pourabdian, M., Fournier, D., & Gizon, L. (2018). *Solar Physics, 293*, 66.
54. Proxauf, B., Gizon, L., Löptien, B., Schou, J., Birch, A. C., & Bogart, R. S. (2020). *Astronomy & Astrophysics, 634*, A44.
55. Saio, H. (1982). *The Astrophysical Journal, 256*, 717.
56. Scherrer, P. H., Bogart, R. S., Bush, R. I., Hoeksema, J. T., Kosovichev, A. G., Schou, J., et al. (1995). *Solar Physics, 162*, 129.
57. Scherrer, P. H., Schou, J., Bush, R. I., Kosovichev, A. G., Bogart, R. S., Hoeksema, J. T., et al. (2012). *Solar Physics, 275*, 207.
58. Schou, J., Scherrer, P. H., Bush, R. I., Wachter, R., Couvidat, S., Rabello-Soares, M. C., et al. (2012). *Solar Physics, 275*, 229.
59. Thompson, M. J. (1991). In D. Gough & J. Toomre (Eds.), *Helioseismic inversion* (Vol. 388, p. 61). Berlin: Springer.
60. Thompson, M. J. (2006). *An introduction to astrophysical fluid dynamics*. London: Imperial College Press.
61. Thompson, M. J., & Zharkov, S. (2008). *Solar Physics, 251*, 225.
62. Švanda, M., Gizon, L., Hanasoge, S. M., & Ustyugov, S. D. (2011). *Astronomy & Astrophysics, 530*, A148.
63. Zhao, J., & Chen, R. (2016). *Asian Journal of Physics, 25*, 325.
64. Zhao, J., Hing, D., Chen, R., & Hess Webber, S. (2019). *The Astrophysical Journal, 887*, 216.

Time-Distance Helioseismology of Deep Meridional Circulation

S. P. Rajaguru and H. M. Antia

Abstract A key component of solar interior dynamics is the meridional circulation (MC), whose poleward component in the surface layers has been well observed. Time-distance helioseismic studies of the deep structure of MC, however, have yielded conflicting inferences. Here, following a summary of existing results we show how a large center-to-limb systematics (CLS) in the measured travel times of acoustic waves affects the inferences through an analysis of frequency dependence of CLS, using data from the Helioseismic and Doppler Imager (HMI) onboard Solar Dynamics Observatory (SDO). Our results point to the residual systematics in travel times as a major cause of differing inferences on the deep structure of MC.

1 Introduction

Large-scale organisation of plasma flows in the convection zones of the Sun and sun-like stars is central to a host of problems related to stellar interior dynamics and magnetic dynamos. Poleward meridional flow, well observed on the solar surface through a variety of techniques, is recognised as a surface component of deep meridional circulation (MC), which traces back to a nearly century old prediction [5]. There have been a good number of theoretical studies of MC with current numerical approaches recognising well that its understanding requires solving the complex fluid dynamical problem involving exchanges of energy and momentum between convection, rotation, thermal stratification and magnetic fields (see [6] and references therein). Clearly, reliable helioseismic inferences of the

S. P. Rajaguru (✉)
Indian Institute of Astrophysics, Bangalore, India
e-mail: rajaguru@iap.res.in

H. M. Antia
Tata Institute of Fundamental Research, Colaba, Mumbai, India
e-mail: antia@tifr.res.in

M. J. P. F. G. Monteiro et al. (eds.), *Dynamics of the Sun and Stars*, Astrophysics and Space Science Proceedings 57,
https://doi.org/10.1007/978-3-030-55336-4_11

deep structure of MC is crucial to make progress in this field [10, 13]. Recently, helioseismology, especially time-distance helioseismology, has made significant progress in this direction, however with unsettling differences between the published results [15, 7, 11, 2, 9]. A large part of these differences are thought to be related to the identification and accounting of a large systematics in travel-time measurements [14]. Here, in this article, we summarise these developments and show that further progress in this field depends heavily on understanding the origin of this systematics, fully characterising it and removing it reliably.

2 Time-Distance Helioseismology of Deep MC: Current Results

Most of the inferences on the deep structure of MC have largely been from time-distance helioseismology [4]. Travel times of acoustic waves propagating in meridional planes are measured in deep-focus geometry for a range of travel distances Δ covering depths from the surface down to the base of the convection zone to capture the meridional flows and inverted, commonly, in ray theory approximation. Although different studies have followed the above basic method, we refer readers to respective publications for finer details of the measurement and inversion procedures adopted. We point out that Rajaguru and Antia [11] implemented an in-built mass conservation constraint in terms of the stream function to invert travel times thereby determining both the meridional (u_θ) and radial (u_r) components of the flow. The data used by different authors are as follows: [15], [11] and [2] used first 2, 4 and 6 years of SDO/HMI Doppler data respectively, [7] used 2 years (2010–2012) of Global Oscillation Network Group (GONG) data, and [8] have used SOHO/MDI data.

All the different results from the above studies are summarised in Fig. 1. Here, meridional flow profiles obtained by different authors have been hemispherically symmetrized and plotted. A prominent feature in the results of [15] and [2] is a double-cell MC in depth covering most of the latitudes with the outer cell having a rather shallow return flow at about $0.9R_\odot$. Results of [7] agree with this shallow return flow but fail to reproduce the deeper second cell of MC. Distinct from the above inferences is a single cell deep MC derived by Rajaguru and Antia [11], with a depth of $\approx 0.77R_\odot$ for a large-scale reversal of flow. The results of [8] cover only the low latitudes ($<30°$) and show a double-cell structure. It is also worthwhile to note that all the profiles in Fig. 1 indicate a complicated flow structure at low latitudes ($<20°$) with two or more reversals of flows over depth, corresponding to two or more cells.

The above differences among the different studies, employing basically the same method and in some cases even the same data set, obviously demand a thorough relook at the analysis procedures, signals and systematics in the measurement. The major difference pertaining to the large-scale flow profile, viz. single- or double-

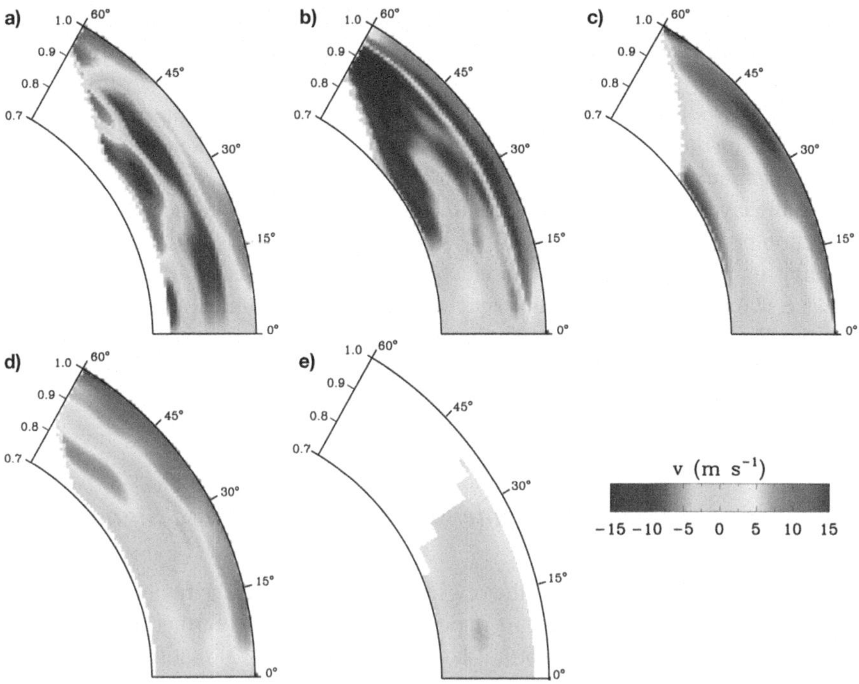

Fig. 1 Hemispherically symmetrized meridional flow profiles obtained from time-distance helioseismic inversions by (**a**) Zhao et al. [15], (**b**) Jackiewicz et al. [7], (**c**) Rajaguru and Antia [11], (**d**) Chen and Zhao [2], and (**e**) Lin and Chou [8]. Poleward flows are positive and equatorward flows are negative. Figure courtesy of Junwei Zhao

cell profiles of MC, may, at a basic level, be taken as due to the implementation or not of mass conservation constraint in the inversion. However, inaccuracies in the identification and separation of signals from systematics could have large impacts in the inferences and we focus on these in the rest of this article.

3 Systematics and Signals in Travel Times

A major development that led to the above presented time-distance helioseismic studies of MC, in the first place, has been the identification and removal of a large systematic center-to-limb effect in the measurements [14, 15], especially improving the identification of flow signals in the deeper layers [15, 11, 7]. As shown by Zhao et al. [14], the above effect is a large systematic increase in travel time differences against angular distance from the solar disk center mimicking a radial outflow from the centre towards the limb, and which increases as Δ increases. This large center-to-limb systematics (CLS) in travel times is still of unknown origin, although the analyses of [14, 15] involving comparisons of travel times from different observables (corresponding to different heights of formation in the solar atmosphere) from the SDO/HMI as well as with that from another instrument (SOHO/MDI) pointed to possible physical causes in the solar atmosphere related to observation height differences. A study by Baldner and Schou [1] showed that the near-surface granular convection could affect the wave-propagation in the observable layers leading to a similar effect as the CLS in travel times. An empirical correction procedure was suggested by Zhao et al. [14]: estimate CLS from waves traveling in W-E direction over an equatorial belt as a function of center-to-limb distance (*i.e. longitude*) and subtract it from N-S travel times measured against latitude. This prescription was tested using different observables and also validated by comparing with direct surface measurements of meridional flows by Zhao et al. [14].

3.1 Wave-Frequency Dependence of Systematics and Signals

Given that we do not understand the origin of CLS, hence lack a model to remove it from the measurements reliably it is imperative that we devise further diagnostics to characterise it better. In this regard, it has been suggested [12] that the dependence of CLS on wave frequencies is examined: filter acoustic waves forming resonant p modes in frequency domain over several narrow frequency bands, and measure travel times against these frequency bands. Here, we implement the above by fitting frequency filtered cross-correlations: a total of 15 different frequencies spaced every 0.2 mHz interval starting from 2.0 mHz with a FWHM of 1 mHz. Both the N-S and W-E (CLS) cross-correlations are frequency filtered, and the meridional flow travel times at each frequency are estimated by subtracting the CLS from the N-S travel times. We have used 7 years (2010–2017) of SDO/HMI data in this study. Figure 2 displays the CLS as estimated from W-E travel times and the

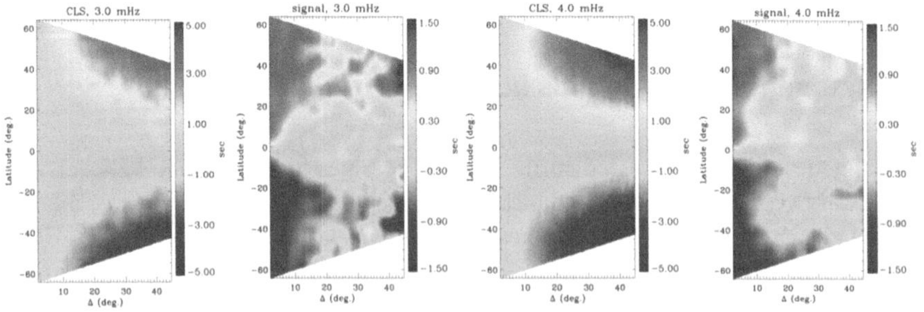

Fig. 2 Estimates of CLS and meridional flow signals at two representative frequencies, 3.0 and 4.0 mHz. Flow signals are obtained after subtracting the CLS from the N-S travel times at the same frequencies

meridional flow signals, which are obtained after subtracting the CLS from the N-S travel times, for two representative frequencies, 3.0 and 4.0 mHz. The CLS has a strong frequency dependence with rapid increase up to about 4 mHz before levelling off and decreasing slightly at higher frequencies. A detailed look at the frequency dependence of CLS and travel times is deferred to a separate publication, except showing the main impact of this on flow inferences (see next Section). We note that Chen and Zhao [3] have also performed a detailed analysis of frequency dependence of CLS via a direct Fourier phase analysis of cross-correlations. Since the magnitudes of CLS is several times that of signals due to meridional flows, any slight inaccuracies in the estimation of CLS would cause significant errors in the signals due to flows. Comparison of signals and CLS in Fig. 2 already indicates a correlation between them, viz. the sign change in signals coincides with the rapid rise of CLS against frequency. It is unclear as to at which frequency the estimation of CLS is robust and hence the signals too.

4 Differing Meridional Flow Solutions over Wave-Frequency

The travel times at different frequency bands estimated as above are inverted in ray theory approximation using the same method as in [11] with proper accounting of the frequency dependent ray paths, and the results are shown in Fig. 3. Surprisingly, the solutions differ, with the lower frequency (3.0 mHz) travel times yielding a single-cell profile while those from the higher frequency (4.0 mHz), where the CLS is maximum, yielding a double-cell profile. This property of solutions obtained clearly implicate the impact of the procedure used to estimate CLS and the strong frequency dependence of CLS. On the other hand if we assume that the frequency dependence of travel times seen in Fig. 2 is due to the finite wavelength effects and not due to leakage of CLS, then it cannot be modeled by ray-theory approximation and require wave modeling such as first Born approximation. We defer a detailed analysis of the nature of solutions against frequency, comparisons with Born approximation treatments of wave-propagation, including the temporal evolution of MC solutions, to a forthcoming publication.

5 Conclusions

The difficult proposition of measuring the deep MC, down to the base of the convection zone, is further compounded by the large systematics, CLS, described and analysed in this work. We have shown that the very different inferences drawn by different studies in the field are attributable to the CLS: travel time signals estimated at frequency bands ($\approx$4 mHz and above) where the CLS is large yield double-cell MC, while those at lower frequency bands yield single-cell MC. We however point out that we have not established the robustness of our estimates of

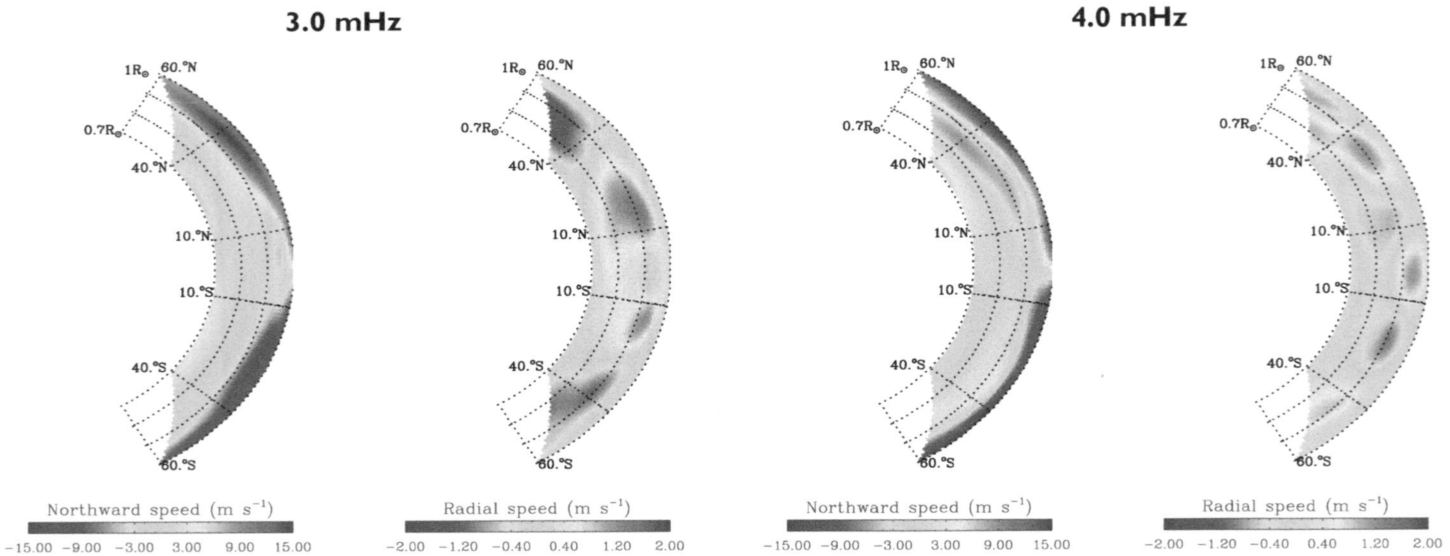

Fig. 3 Meridional circulation profiles obtained from ray-theory inversion of travel time signals shown in Fig. 2. While travel times measured around 3.0 mHz give a single-cell profile, those around 4.0 mHz, where the CLS is maximum, yield a double-cell profile

CLS and that it is unclear as to at which frequency the estimation of CLS and the signals are robust and hence which of the obtained solutions for deep MC are closer to reality. We therefore caution that more detailed analyses are necessary to draw firmer conclusions on the above.

Acknowledgments The authors are greatly appreciative of their long professional association with Michael Thompson. The first author SPR is much grateful to Michael Thompson for research mentorship, which initiated him in the field of time-distance helioseismology.

References

1. Baldner, C. S., & Schou, J. (2012). *The Astrophysical Journal Letters, 760*, L1.
2. Chen, R., & Zhao, J. (2017). *The Astrophysical Journal, 849*, 144.
3. Chen, R., & Zhao, J. (2018). *The Astrophysical Journal, 853*, 161.
4. Duvall, T. L., et al. (1993). *Nature, 362*, 430.
5. Eddington, A.S. (1925). *The Observatory, 48*, 73.
6. Featherstone, N. A., & Miesch, M. S. (2015). *The Astrophysical Journal, 804*, 67.
7. Jackiewicz, J., Serebryanskiy, A., & Kholikov, S. (2015). *The Astrophysical Journal, 805*, 133.
8. Lin, C.-H., & Chou, D.-Y. (2018). *The Astrophysical Journal, 860*, 48.
9. Mandal, K., Hanasoge, S. M., Rajaguru, S. P., & Antia, H. M. (2018). *The Astrophysical Journal, 863*, 39.
10. Miesch, M. S., & Toomre, J. (2009). *Annual Review of Fluid Mechanics, 41*, 317.
11. Rajaguru, S. P., & Antia, H. M. (2015). *The Astrophysical Journal, 813*, 114.
12. Rajaguru, S. P., & Antia, H. M. (2017). ISSI Team Meeting Proceedings on "Studies of Deep Meridional Flow". http://science-media.org/presentation/241
13. Toomre, J., & Thompson, M. J. (2015). *Space Science Reviews, 196*(1–4), 1–14.
14. Zhao, J., Nagashima, K., Bogart, R. S., Kosovichev, A. G., & Duvall, T. L. Jr. (2012). *The Astrophysical Journal Letters, 749*, L5.
15. Zhao, J., Bogart, R. S., Kosovichev, A. G., Duvall, T. L. Jr., & Hartlep, T. (2013). *The Astrophysical Journal Letters, 774*, L29.

Surface Rotation and Magnetic Activity of Solar-Like Stars: Impact on Seismic Detections

S. Mathur, A. R. G. Santos, R. A. García, P. G. Beck, S. N. Breton, L. Bugnet, T. S. Metcalfe, M. H. Pinsonneault, N. Santiago, G. Simonian, and J. van Saders

Abstract Rotation plays an important role in the evolution of stars. For decades rotation-age relationships have been used as surface rotation can be measured rather easily. Here we show the result of the measurement of rotation periods for a large number of *Kepler* solar-like stars along with their level of magnetic activity. We also study solar-like stars for which oscillations have not been detected using 1 month of *Kepler* data in order to understand what can be the reason behind these non-detections. Surprisingly the conclusion of our analysis is that magnetic activity is not the main reason for not detecting acoustic modes in those stars.

S. Mathur (✉)
Instituto de Astrofísica de Canarias, La Laguna, Tenerife, Spain

Departamento de Astrofísica, Universidad de La Laguna, La Laguna, Tenerife, Spain
e-mail: smathur@iac.es

A. R. G. Santos · T. S. Metcalfe
Space Science Institute, Boulder, CO, USA

R. A. García · S. Breton · L. Bugnet
IRFU, CEA, Université Paris-Saclay, Gif-sur-Yvette Cedex, France

Université Paris Diderot, AIM, Sorbonne Paris Cité, CEA, CNRS, Gif-sur-Yvette Cedex, France

P. G. Beck
Institute for Physics, Karl-Franzens University of Gray, NAWI Graz, Graz, Austria

Instituto de Astrofísica de Canarias, La Laguna, Tenerife, Spain

Departamento de Astrofísica, Universidad de La Laguna, La Laguna, Tenerife, Spain

M. H. Pinsonneault · G. Simonian
Department of Astronomy, The Ohio State University, Columbus, OH, USA

N. Santiago
University of Puerto Rico at Cayey, Cayey, Puerto Rico

J. van Saders
Institute for Astronomy, University of Hawai'i, Honolulu, HI, USA

M. J. P. F. G. Monteiro et al. (eds.), *Dynamics of the Sun and Stars*, Astrophysics and Space Science Proceedings 57,
https://doi.org/10.1007/978-3-030-55336-4_12

1 Introduction

Stars with an external convective envelope have rotation periods evolving with time, roughly as the square root of their ages [27, 13, 16] during an important part of their lives. Thus, by measuring the surface rotation period of a solar-like star, we can have an estimate of its age using these empirical gyrochronology relations that also depend on the mass and the colors of the stars [14, 2, 3, 6]. While it was shown using *Kepler* data that these relations seem to be valid only up to a given Rossby number (ratio of the rotation period and the convective turnover time of about 2) (see [29]), they still hold for K and M dwarfs. This is great news as measuring rotation is somewhat easier by studying the modulation in the light curves due to the passage of spots/active regions on the stellar discs as done for thousands of stars for example by [22, 10, 21] and see [7] for a review. Moreover, as these measurements depend on the presence of spots, we can also derive a proxy of the magnetic activity based on the photometric observations that we call $S_{\rm ph}$ (photometric S-index in analogy with the spectroscopic one from CaHK observations). In this work, we present the analysis of the rotation periods of more than 26,000 M and K dwarfs observed in long cadence for 4 years with the *Kepler* mission [25]. First, we removed all kind of pollution (e.g. classical pulsators, known binaries, etc.). For this analysis we used our improved rotation pipeline where we combine three different methods (time-frequency analysis, auto-correlation function and composite spectrum) with different ways of calibrating the photometric signal. We obtained reliable rotation periods for more than 15,000 stars, including 4000 stars without rotation periods reported up to now. We also derived the magnetic activity levels for those stars. Finally, as we know for the Sun and many solar-like stars now, magnetic activity has an impact on the amplitude of the acoustic modes by suppressing them [8]. We will present the analysis of a sample of 1000 solar-like stars observed in short-cadence during the survey phase of the mission and for which no modes where detected.

2 Studying the Surface Rotation and Magnetic Activity of Main-Sequence Stars

2.1 Data and Methodology

Different methods can be used to measure the surface rotation from the analysis of photometric time series.

The classical method to detect a periodicity is to compute the periodogram, which has been done in many of the *Kepler* targets [22]. However one issue that arises from this analysis is it can happen that we only measure half of the real rotation period when we take the highest peak in the periodogram. Indeed this is due to the fact that there can be spots diametrically opposite to the visible disk.

Another way is to do a time-frequency analysis, which can help to overcome that problem. We used the wavelet tools [28] that we adapted for our photometric data [15, 17]. This method also allows us to follow the change of magnetic activity and of rotation periods in some cases. In addition, we can also see if there is an instrumental problem in the data that biases the result.

Finally another method that is widely used is the auto-correlation function [21]. This method is more robust and likely to recover the real rotation periods (and not half of it) in many cases. This method has been applied on several tens of thousands of stars observed by *Kepler*.

We decided to use a combination of the two last methods in order to better retrieve a reliable rotation period. Having a rotation period obtained with two independent methods suggests that the value obtained is more likely to be due to real signal. Our methodology was tested on a sample of simulated data validating it and showing that it did indeed provide reliable results in more than 90% of the cases [1].

Since we measure the rotation through the presence of starspots, it means that the star is active and hence we can measure the level of magnetic activity with a proxy based on the standard deviation of the time series. We developed the S_{ph} that is obtained by computing the standard deviation of subseries of length $5 \times P_{\mathrm{rot}}$ [18, 19]. Comparison with classical magnetic proxies for the Sun showed a good agreement and validated the S_{ph} as a good proxy for stellar magnetic activity [23].

2.2 Results

We first applied this methodology to solar-like stars observed by the *Kepler* mission and for which acoustic modes had been detected [10]. Out of the 540 stars, reliable rotations were obtained for 310 stars by using the long-cadence data (~30 min) of the mission. This analysis showed that the Sun is similar to many solar-like stars in terms of its level of magnetic activity. Furthermore, these rotation periods were used to show that the classical gyrochronology relation does not reproduce the rotation-age relation when computing the age from seismic analyses [29]. We also point out that those rotation periods are used for the inversion analysis (Santos et al., this volume).

Recently, we analyzed a sample of 26,521 K and M dwarfs. After removing polluters such as classical pulsators, red giants, close-binary candidates, we did a similar anlaysis as the previous set. We analyzed 4 years of long-cadence data using lightcurves calibrated with the KADACS software [9] and filtered with a 20, 55, and 80-day filter. We also used the PDC-MAP lightcurves [12] to check for possible pollution as the KADACS apertures are slightly larger than the NASA original ones. After comparison of the results from different methods and datasets, we obtained reliable rotation periods for 15,290 stars, leading to more than 60% yield. In particular, compared to previously published catalogs, we report more than 4000 rotation periods (for these spectral types) for fainter and cooler stars.

We are now in the process extending this analysis to the rest of G, F, and subgiants stars ($\sim$100,000 stars) (Santos et al. in prep.). The automatic selection for 70,000 stars led to rotation periods for around 20,000 stars. For the next step, instead of doing a visual check of the stars without agreement, we are developing a machine learning algorithm [4].

3 Stars Without Detection of Oscillations

Among the sample of $\sim$2000 solar-like stars observed in short cadence ($\sim$1 min) by the *Kepler* mission during the survey phase, modes were detected for only $\sim$540 stars. While the magnetic activity of those stars was studied in [5], it was found that they were in general more active than the Sun. This could explained the non detection of the modes. Recently we analyzed a subsample of 1017 stars after removing some red giants, classical pulsators and new detection of modes using the DR25 of the *Kepler* data [20]. We computed the predicted amplitude of the modes from scaling relations and compared it to the high-frequency noise in the power spectrum. Focusing on the ones where we were expecting to detect the modes, we computed the rotation period and the $S_{\rm ph}$ that we obtained reliably for 323 stars.

In Fig. 1, we show the stars without oscillations detected with black crosses and the ones with oscillations detected with black squares. While we clearly see that the stars with detections barely have a magnetic activity level larger than the Sun at maximum activity, only 32% of the stars without oscillations are above the magnetic activity of the Sun. We looked at other reasons for not detecting the modes such as the metallicity that can have an effect on the excitation and thus on the amplitude of the modes [24], binarity also known to affect oscillations [11, 26], and inclination angle that can bias the $S_{\rm ph}$ measurement. However the sample with additional observations, such as spectroscopic ones, are too sparse, making it difficult to conclude.

Some other explanations for not detecting the modes is that the prediction of the amplitudes based on scaling relations is not perfect and should be improved (using the non detections) or that the magnetic activity levels of the stars were particularly high during the short-cadence observations.

4 Conclusions and Summary

We have analyzed the rotation and magnetic activity of more than 25,000 solar-like stars and are now analyzing 100,000 additional stars observed by the *Kepler* mission. The analysis of stars without detection of oscillations suggests that magnetic activity is not the main reason and more work with additional spectroscopic observations as well as improving the amplitude prediction should be done.

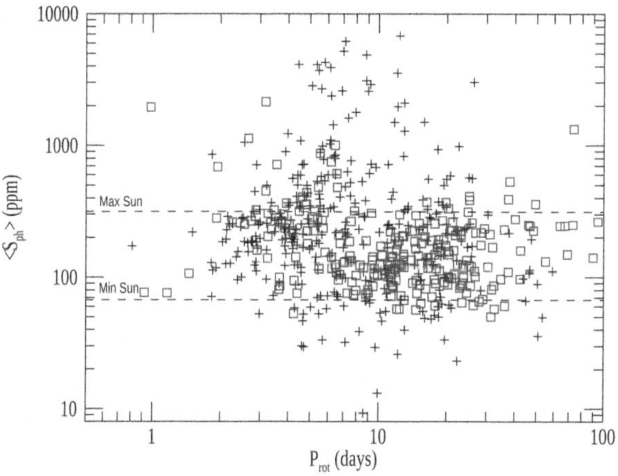

Fig. 1 Average magnetic activity proxy S_{ph} as a function of surface rotation period P_{rot} for 323 stars selected as explained in Sect. 3. The black squares represent the stars with detected oscillations while the black crosses represent the stars without detection of oscillations. The two dashed lines show the levels of magnetic activity of the Sun at minimum and maximum activity. Based on [20]

Acknowledgments S.M. acknowledges support by the National Aeronautics and Space Administration under Grant NNX15AF13G, by the National Science Foundation grant AST-1411685 and the Ramon y Cajal fellowship number RYC-2015-17697. R.A.G. acknowledges the support from PLATO and GOLF CNES grants. A.R.G.S. acknowledges the support from National Aeronautics and Space Administration under Grant NNX17AF27G. P.G.B. acknowledges the support of the MINECO under the fellowship program 'Juan de la Cierva incorporacion' (IJCI-2015-26034).

References

1. Aigrain, S., Llama, J., Ceillier, T., Chagas, M. L. d., Davenport, J. R. A., García, R. A., et al. (2015). *Monthly Notices of the Royal Astronomical Society, 450*, 3211. https://doi.org/10.1093/mnras/stv853
2. Barnes, S. A. (2003). *The Astrophysical Journal, 586*, 464. https://doi.org/10.1086/367639
3. Barnes, S. A. (2007). *The Astrophysical Journal, 669*, 1167. https://doi.org/10.1086/519295
4. Breton, S. N., Bugnet, L., Santos, A. R. G., Le Saux, A., Mathur, S., Palle, P. L., et al. (2019). arXiv e-prints, arXiv:1906.09609.
5. Chaplin, W. J., Bedding, T. R., Bonanno, A., Broomhall, A. M., García, R. A., Hekker, S., et al. (2011). *The Astrophysical Journal Letters, 732*, L5. https://doi.org/10.1088/2041-8205/732/1/L5
6. Epstein, C. R., & Pinsonneault, M. H. (2014). *The Astrophysical Journal, 780*, 159. https://doi.org/10.1088/0004-637X/780/2/159
7. García, R. A., & Ballot, J. (2019). *Living Reviews in Solar Physics, 16*(1), 4. https://doi.org/10.1007/s41116-019-0020-1
8. García, R. A., Mathur, S., Salabert, D., Ballot, J., Régulo, C., Metcalfe, T. S., et al. (2010). *Science, 329*, 1032. https://doi.org/10.1126/science.1191064

9. García, R. A., Hekker, S., Stello, D., Gutiérrez-Soto, J., Handberg, R., Huber, D., et al. (2011). *Monthly Notices of the Royal Astronomical Society, 414*, L6. https://doi.org/10.1111/j.1745-3933.2011.01042.x
10. García, R. A., Ceillier, T., Salabert, D., Mathur, S., van Saders, J. L., Pinsonneault, M., et al. (2014). *Astronomy & Astrophysics, 572*, A34. https://doi.org/10.1051/0004-6361/201423888
11. Gaulme, P., Jackiewicz, J., Appourchaux, T., & Mosser, B. (2014). *The Astrophysical Journal, 785*, 5. https://doi.org/10.1088/0004-637X/785/1/5
12. Jenkins, J. M., Caldwell, D. A., Chandrasekaran, H., Twicken, J. D., Bryson, S. T., Quintana, E. V., et al. (2010). *The Astrophysical Journal Letters, 713*, L87. https://doi.org/10.1088/2041-8205/713/2/L87
13. Kawaler, S. D. (1988). *The Astrophysical Journal, 333*, 236. https://doi.org/10.1086/166740
14. Kraft, R. P. (1967). *The Astrophysical Journal, 150*, 551. https://doi.org/10.1086/149359
15. Liu, Y., Liang, X., & Weisberg, R. (2007). *Journal of Atmospheric and Oceanic Technology, 24*, 2093.
16. MacGregor, K. B., & Brenner, M. (1991). *The Astrophysical Journal, 376*, 204. https://doi.org/10.1086/170269
17. Mathur, S., García, R. A., Régulo, C., Creevey, O. L., Ballot, J., Salabert, D., et al. (2010). *Astronomy & Astrophysics, 511*, A46. https://doi.org/10.1051/0004-6361/200913266
18. Mathur, S., García, R. A., Ballot, J., Ceillier, T., Salabert, D., Metcalfe, T. S., et al. (2014). *Astronomy & Astrophysics, 562*, A124. https://doi.org/10.1051/0004-6361/201322707
19. Mathur, S., Salabert, D., García, R. A., & Ceillier, T. (2014). *Journal of Space Weather and Space Climate, 4*(27), A15. https://doi.org/10.1051/swsc/2014011
20. Mathur, S., García, R. A., Bugnet, L., Santos, Â. R. G., Santiago, N., & Beck, P. G. (2019). *Frontiers in Astronomy and Space Sciences, 6*, 46. https://doi.org/10.3389/fspas.2019.00046
21. McQuillan, A., Mazeh, T., & Aigrain, S. (2014). *The Astrophysical Journal Supplement Series, 211*, 24. https://doi.org/10.1088/0067-0049/211/2/24
22. Nielsen, M. B., Gizon, L., Schunker, H., & Karoff, C. (2013). *Astronomy & Astrophysics, 557*, L10. https://doi.org/10.1051/0004-6361/201321912
23. Salabert, D., García, R. A., Jiménez, A., Bertello, L., Corsaro, E., & Pallé, P. L. (2017). *Astronomy & Astrophysics, 608*, A87. https://doi.org/10.1051/0004-6361/201731560
24. Samadi, R., Ludwig, H. G., Belkacem, K., Goupil, M. J., & Dupret, M. A. (2010). *Astronomy & Astrophysics, 509*, A15. https://doi.org/10.1051/0004-6361/200911867
25. Santos, A. R. G., García, R. A., Mathur, S., Bugnet, L., van Saders, J. L., Metcalfe, T. S., et al. (2019). *The Astrophysical Journal Supplement Series, 244*(1), 21. https://doi.org/10.3847/1538-4365/ab3b56
26. Sekaran, S., Johnston, C., Tkachenko, A., Beck, P. G., Prša, A., & Hambleton, K. M. (2019). *Astronomy & Astrophysics, 624*, A140. https://doi.org/10.1051/0004-6361/201834095
27. Skumanich, A. (1972). *The Astrophysical Journal, 171*, 565. https://doi.org/10.1086/151310
28. Torrence, C., & Compo, G. P. (1998). *Bulletin of the American Meteorological Society, 79*, 61. https://doi.org/10.1175/1520-0477(1998)079
29. van Saders, J. L., Ceillier, T., Metcalfe, T. S., Aguirre, V. S., Pinsonneault, M. H., García, R. A., et al. (2016). *Nature, 529*, 181. https://doi.org/10.1038/nature16168

Study of Acoustic Halos in NOAA Active Region 12683

S. C. Tripathy, K. Jain, S. Kholikov, F. Hill, and P. Cally

Abstract We characterize the spatio-temporal power distribution around the active region 12683 as a function of height in the solar atmosphere, wave frequencies, magnetic field strength and inclination of the magnetic field.

1 Data and Results

Acoustic halos were first observed in Dopplergrams as a power enhancement around 6 mHz and are believed to be signatures of wave interaction with the magnetic field ([1], and references therein). In this context, we analyze the NOAA active region (AR) 12683 which consists of two sunspots surrounded by plage of both polarities when it was near the disk center on September 30, 2017. Since our goal is to investigate the variation in the acoustic power as a function of height in the solar atmosphere, we analyze Doppler (V) data from the Helioseismic and Magnetic Imager (HMI) and 1700 Å intensity data (I_{1700}) from the Atmospheric Imaging Assembly (AIA) instruments on board the Solar Dynamics Observatory (SDO). In order to characterize the power as a function of the magnetic field strength and field inclination, we also use vector magnetic field data from SDO. Details of the data reduction procedure is described in [2].

Figure 1 displays the acoustic power as a function of the frequency, magnetic field and field inclination. In order to compare the results with previously analyzed AR 11330 [2], we average the magnetic field strength over 10 G bins with an upper limit of 850 G while the field inclination is averaged over 4° bins. We find that

S. C. Tripathy (✉) · K. Jain · S. Kholikov · F. Hill
National Solar Observatory, Boulder, CO, USA
e-mail: stripathy@nso.edu

P. Cally
Monash Center for Astrophysics, School of Mathematical Sciences, Monash University, Clayton, VIC, Australia

M. J. P. F. G. Monteiro et al. (eds.), *Dynamics of the Sun and Stars*, Astrophysics and Space Science Proceedings 57,
https://doi.org/10.1007/978-3-030-55336-4_13

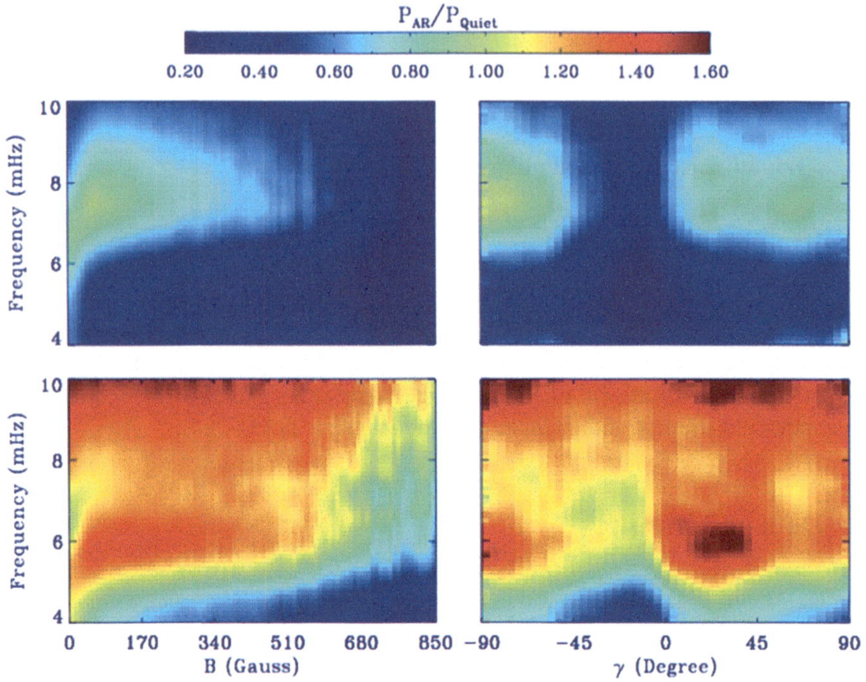

Fig. 1 Normalized power as a function of the total magnetic field strength (left) and inclination (right). Panels from bottom to top represent HMI Doppler and AIA 1700 Å intensity observations

the power enhancement in V begins about 4.5 mHz and extends up to 10 mHz. In contrast, the halo in I_{1700} is confined to a very narrow region around 8 mHz and does not extend to higher frequencies as was observed in AR 11330. This could be due to the different field strengths associated with the active regions. We also observe that the distribution of power as a function of the field inclination angle is different as compared to AR 11330. In summary, we note that the power halos seen in different active regions behave differently and warrants a statistical analysis of many active regions.

Acknowledgments SDO data courtesy of SDO (NASA) and the HMI and AIA consortium.

References

1. Schunker, H., & Braun, D. C. (2011). Newly identified properties of surface acoustic power. *Solar Physics, 268*, 349–362. https://doi.org/10.1007/s11207-010-9550-3
2. Tripathy, S. C., Jain, K., Kholikov, S., Hill, F., Rajaguru, S. P., & Cally, P. S. (2018). A study of acoustic halos in active region NOAA 11330 using multi-height SDO observations. *Advances in Space Research, 61*(2), 691–704. https://doi.org/10.1016/j.asr.2017.10.033

Helioseismic Center-to-Limb Effect and Measured Travel-Time Asymmetries Around Sunspots

Junwei Zhao and Ruizhu Chen

Abstract Center-to-limb effect, an asymmetry in the measured travel times for waves propagating away from and toward the solar disk center, was previously reported in time–distance helioseismic measurements. Meanwhile, another travel-time asymmetry between waves traveling away from and into sunspots is well known. It is possible that both of the travel-time asymmetries are partially, albeit not entirely, due to different atmospheric heights where oscillatory signals are taken for computing the acoustic travel times. Here, we take simultaneous observations of two atmospheric heights in a quiet-Sun region, and measure acoustic travel times between the lower and higher atmosphere. The results demonstrate that the measured travel-time asymmetries resemble those measured in the center-to-limb effect and in sunspot regions.

Helioseismic center-to-limb (CtoL) effect [3] is a systematic effect that exhibits as the measured travel-time asymmetries for acoustic waves traveling away and toward the solar disk center. For a helioseismic inference of the Sun's internal meridional circulation, the CtoL effect must be removed from raw travel-time measurements [4]. Depending on what observables are used to measure the travel times, the sign of this effect may vary; e.g., if Doppler velocities or line intensities are used for measurements, the travel times for waves traveling away from the disk center are usually shorter than those traveling toward the disk center. Due to the limb-darkening effect, the observed atmosphere is often higher near the limb than near the disk center. More recently, Chen and Zhao [1] found that the CtoL effect is highly frequency (frequency is for acoustic waves, not for electromagnetic waves) dependent, and the CtoL effect shows opposite signs for waves above and below the cut-off frequency. There is another well-known travel-time asymmetry: travel

J. Zhao (✉) · R. Chen
W. W. Hansen Experimental Physics Laboratory, Stanford University, Stanford, CA, USA
e-mail: junwei@sun.stanford.edu

M. J. P. F. G. Monteiro et al. (eds.), *Dynamics of the Sun and Stars*, Astrophysics and Space Science Proceedings 57,
https://doi.org/10.1007/978-3-030-55336-4_14

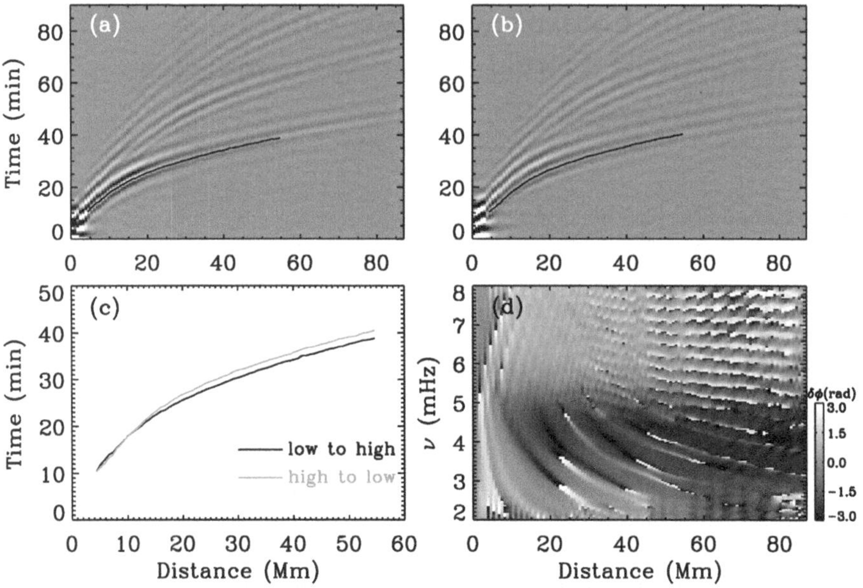

Fig. 1 Time–distance diagrams measured for waves traveling from the lower to the higher atmosphere (**a**), and from the higher to the lower atmosphere (**b**). (**c**) Comparison of the two travel times fitted from panels (**a**) and (**b**). (**d**) Frequency-dependent relative phase shifts between the time–distance diagrams in panels (**a**) and (**b**)

times measured in waves traveling away from sunspots are shorter than those toward sunspots [2]. Both of the above-mentioned effects show travel-time asymmetries in a same sense that the waves traveling from the lower to higher atmosphere are measured with shorter travel times than the other way around.

To examine whether these two observational effects are caused partly by differences in atmospheric heights, we select a quiet-Sun region observed in *SDO*/HMI line-core intensity (higher atmosphere) and continuum intensity (lower) for a test. As shown in Fig. 1a–c, the measured travel times, which are fitted from the cross-correlation functions using Gabor wavelet function, indeed show a travel-time asymmetry, with waves traveling from the lower to higher atmosphere faster. An analysis in the Fourier domain (Fig. 1d) also shows that the travel-time asymmetry shows a clear frequency dependence, which highly resembles the CtoL effect [1].

References

1. Chen, R., & Zhao, J. (2018). *The Astrophysical Journal, 853*, 161.
2. Lindsey, C., & Braun, D. C. (2005). *The Astrophysical Journal Supplement Series, 155*, 209.
3. Zhao, J., Nagashima, K., Bogart, R. S., Kosovichev, A. G., & Duvall, T. L. (2012). *The Astrophysical Journal Letters, 749*, L5.
4. Zhao, J., Bogart, R. S., Kosovichev, A. G., Duvall, T. L., & Hartlep, T. (2013). *The Astrophysical Journal Letters, 774*, L29.

Probing the Variation with Depth of the Solar Meridional Circulation Using Legendre Function Decomposition

D. C. Braun, A. Birch, and Y. Fan

Abstract We apply the helioseismic methodology of Legendre Function Decomposition to 7.5 years of Dopplergrams obtained by the Helioseismic and Magnetic Imager (HMI) as the basis of inferring the depth variation of the meridional flow between 20° and 60° latitude in both hemispheres. We assess and remove center-to-limb artifacts using measurements obtained by applying the procedure to pseudo poles at the east and west limbs. Forward modeling is carried out to evaluate the consistency of the corrected frequency shifts with models of the depth variation of the meridional circulation in the top half of the convection zone. The observations are consistent with a return flow in the northern hemisphere below about 40 Mm depth, but no obvious return flow in the south.

1 Summary of Procedure and Results

The solar meridional circulation is a crucial component of magnetic flux transport and dynamo models, but no consensus exists regarding the variation of its properties with depth [6, 3, 5]. Here we use the method of Legendre Function Decomposition [1, 4, 2] to measure the meridional flow. Our analysis is based on [1] with the following modifications: (1) In addition to the measurements described by Braun and Fan [1] we assess frequency shifts due to the center-to-limb effect using

D. C. Braun (✉)
NWRA, Boulder, CO, USA
e-mail: dbraun@nwra.com

A. Birch
MPS, Göttingen, Germany
e-mail: birch@mps.mpg.de

Y. Fan
HAO/NCAR, Boulder, CO, USA
e-mail: yfam@ucar.edu

M. J. P. F. G. Monteiro et al. (eds.), *Dynamics of the Sun and Stars*, Astrophysics and Space Science Proceedings 57,
https://doi.org/10.1007/978-3-030-55336-4_15

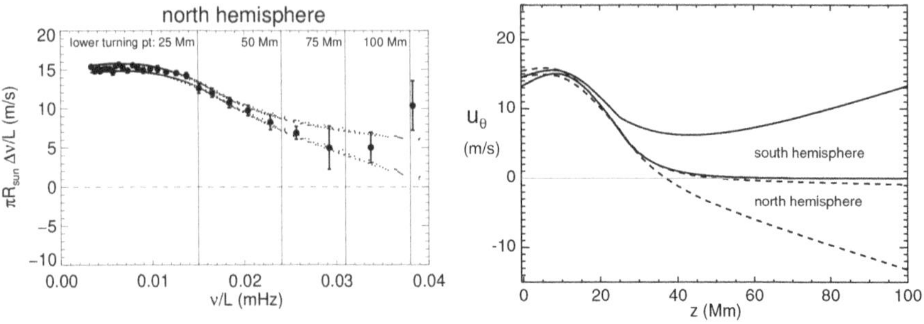

Fig. 1 Left panel: small points indicate predictions of forward models designed to straddle the error bars of the binned, averaged, corrected scaled frequency shifts measured in the north hemisphere. Right panel: the pair of flows for both hemispheres (north = dashed lines, south = solid lines) derived from the modeling

spherical coordinate systems with poles centered on the east and west limbs (summing the power spectra to remove rotation), and (2) instead of the centroid method, we fit multiple Gaussian peaks with a polynomial term to account for background power. We use 90 months of HMI data, sampled in 8-h intervals, between 2010 and 2017 for the nominal measurements and 12 months (2011) for the east-west measurements. Forward modeling (Fig. 1) is performed as in [1]. The observations are consistent with a return flow in the northern hemisphere but no obvious return flow in the south. Details will follow in a forthcoming publication.

Acknowledgments DCB is supported by NASA Heliophysics Div. (80NSSC18K0066, 80NSSC18K0068) and by the National Science Foundation (AGS-1623844). ACB is partially supported from the European Research Council Synergy Grant WHOLE SUN #810218. This presentation is dedicated to Michael Thompson.

References

1. Braun, D. C., & Fan, Y. (1997). Helioseismic measurements of the subsurface meridional flow. *The Astrophysical Journal Letters, 508*, L105–L108.
2. Doerr, H.-P., Roth, M., Zaatri, A., Krieger, L., & Thomspon, M. J. (2010). A new code for Fourier-Legendre analysis of large datasets: First results and a comparison with ring-diagram analysis. *Astronomische Nachrichten, 331*, 911–914.
3. Jackiewicz, J., Serebryanskiy, A., & Kholikov, S. (2015). Meridional flow in the solar convection zone. II. Helioseismic inversions of GONG data. *The Astrophysical Journal, 805*, 133–141.
4. Mitra-Kraev, U., & Thompson, M. J. (2007). Meridional flow profile measurements with SOHO/MDI. *Astronomische Nachrichten: Astronomical Notes, 328*, 1009–1012.
5. Rajaguru, S. P., & Antia, H. M. (2015). Meridional circulation in the solar convection zone: time-distance helioseismic inferences from four years of HMI/SDO observations. *The Astrophysical Journal, 813*, 114.
6. Zhao, J., Bogart, R. S., Kosovichev, A. G., Duvall, T. L. Jr., & Hartlep, T. (2013). Detection of equatorward meridional flow and evidence of double-cell meridional circulation inside the Sun. *The Astrophysical Journal Letters, 774*, L29–L34.

New Inversion Scheme for Time-Distance Helioseismology

Jason Jackiewicz

Abstract Helioseismic inversions typically rely on some variant of least-squares methods, such as SOLA (Pijpers and Thompson, Astron Astrophys 281:231–240, 1994). Optimization procedures that rely on matrix inversions may suffer from numerical instabilities caused by ill-conditioned matrices. Markov Chain Monte Carlo (MCMC) methods sample the parameter space, independent of its complexity, while Bayesian methods are powerful ways of of understanding the posterior distribution using prior information known about a model. Here we demonstrate an idealized example of these techniques applied in the context of meridional circulation and time-distance helioseismology.

1 The Toy Problem and a Probabilistic Solution

One outstanding problem in local helioseismology is the determination of meridional circulation from travel-time measurements. We start with the meridional flow model developed in [4]. It parametrizes the radial profile of the latitudinal flow, and then uses mass conservation to derive the radial profile of the radial flow. The latitudinal variation of the flows are modulated by a sine function.

The model is modified slightly here to include more radial points, and our test profile is a double cell which reverses at $\sim 0.9\ R_{\odot}$ and then again at $\sim 0.8\ R_{\odot}$ (Fig. 1). Constructed this way, the model has 7 free parameters: one that controls the location of the peak velocity in latitude, and the velocity at 6 interior points. We then compute ray kernels for a set of latitudes and distances that sample the model relatively well. From the kernels and the flow model, forward travel times are computed, and artificial random noise is added to the travel times at a rather high level (about 20% of the rms of the measurements, or about 0.1 s).

J. Jackiewicz (✉)
Department of Astronomy, New Mexico State University, Las Cruces, NM, USA
e-mail: jasonj@nmsu.edu

M. J. P. F. G. Monteiro et al. (eds.), *Dynamics of the Sun and Stars*, Astrophysics and Space Science Proceedings 57,
https://doi.org/10.1007/978-3-030-55336-4_16

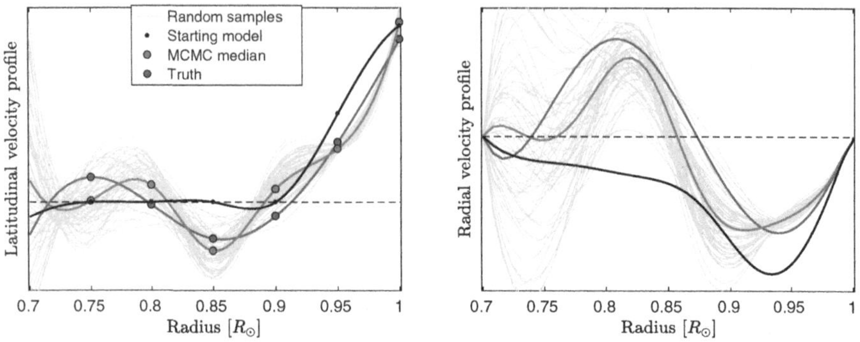

Fig. 1 A solution to the problem in model space. The left panel is the function that determines the depth dependence of the latitudinal flow ($F(r)$ in [4]), which is determined by 6 free parameters (filled circles). The gray lines are models drawn from 50 random samples of the posterior distribution. The right panel is the function that controls the radial velocity

A Bayesian MCMC method [2] has been developed and written in Matlab [3] based on a popular Python toolkit [1]. Figure 1 shows the median of the posterior distribution of an inversion (of the synthetic travel times) in model space, i.e., the resulting radial profiles of the two flow components. A few random samples are shown, along with the known input "truth" and the starting model, to show how far the models have gone in about 10^5 steps. The random samples give a sense of the distribution of the posteriors, within which the blue curves generally lie.

2 Conclusions and Future Work

MCMC inversions in a Bayesian framework for time-distance helioseismology are quite powerful for problems that allow generative models with a modest number of parameters, as demonstrated here. There is also somewhat less freedom in the interpretation of the results compared to standard inversions [5], and they are a robust way to interpret astronomical observations.

References

1. Foreman-Mackey, D., Hogg, D. W., Lang, D., & Goodman, J. (2013, March). emcee: The MCMC Hammer. *Publications of the Astronomical Society of the Pacific, 125*, 306.
2. Goodman, J., & Weare, J. (2010). Ensemble samplers with affine invariance. *Communications in Applied Mathematics and Computational Science, 5*(1), 65–80.
3. Jackiewicz, J. (2020). Probabilistic inversions for time-distance helioseismology (2020). arXiv e-prints, arXiv:2007.01432.
4. Liang, Z.-C., Birch, A. C., Duvall, T. L. Jr., Gizon, L., & Schou, J. (2017, May). Comparison of acoustic travel-time measurements of solar meridional circulation from SDO/HMI and SOHO/MDI. *Astronomy & Astrophysics, 601*, A46.
5. Pijpers, F. P., & Thompson, M. J. (1994, January). The SOLA method for helioseismic inversion. *Astronomy & Astrophysics, 281*, 231–240.

On Active Region Emergence Precursors

Sylvain G. Korzennik

Abstract I present intriguing results from a systematic look at possible indicators of active region emergence precursors using both acoustic power mapping and time distance travel time anomalies. I analyzed a set of active regions that have emerged at low latitudes and close to the central meridian, and were observed with HMI. Such regions, since they emerged near disk center, could be tracked for days before and after their emergence. Data cubes of tracked surface observable were analyzed, using time-series of various length, using HMI's intensity, velocity and magnetic field data. While each active region displays its own particularities, this systematic approach allowed me to pick some interesting statistical properties that will be summarized. Using different time-series length allowed me to investigate the trade-off between precision and resolution, whether temporal or spatial. It also revealed to which extent one sees a tight correlation between various metric of surface activity and its effect of the measured acoustic signal, whether seen in raw of filtered power map or travel time anomalies.

1 Introduction

I present results from analyzing 25 emerging active regions (EARs) looking for potential emergence signal precursors, using velocity, intensity and magnetic field observations acquired with HMI; the EARs were selected for emerging near disk center.

Tracked data cubes were extracted on a 512×512 pixel grid using JSOC's `mtrack`, with a resolution of $0.0303°$ per pixel to match the HMI resolution at disk center. Since these EARs were selected to emerge near disk center, each region could be tracked for nearly 5 days before and after the time of emergence. I characterized

S. G. Korzennik (✉)
Center for Astrophysics | Harvard & Smithsonian, Cambridge, MA, USA
e-mail: skorzennik@cfa.harvard.edu

M. J. P. F. G. Monteiro et al. (eds.), *Dynamics of the Sun and Stars*, Astrophysics and Space Science Proceedings 57,
https://doi.org/10.1007/978-3-030-55336-4_17

each EAR by computing the acoustic power map, $P_{i,j}$, the time propagation anomalies, $\delta\tau_{i,j}$, the corresponding mean field, $B_{i,j}$, and the corresponding mean continuum intensity, $I_{c,i,j}$.

Raw and filtered acoustic power maps were computed using the velocity observable, after applying a high pass filter ($t < t_o = 20$ m), and various notch filters (i.e. $\nu_c(i) = 1.33, (1), 7.33$ mHz; and $\Delta\nu = 0.5$ mHz as width), or none, and for time-series of various lengths. I also computed time propagation anomalies, $\delta\tau$, from classical time-distance analysis with a few different phase filtering. I reduced the spatial resolution to increase the SNR of the time-distance correlation functions. For each phase filter, I computed a mean anomaly by averaging over a range in Δ. The corresponding mean continuum intensity, I_c, and a mean surface magnetic field, B were also computed. In order to explore the trade-off between SNR and time resolution I used time-series of different lengths ($N_{\rm TMS} = 256, 384,$ or 768, i.e.: $T_{\rm TMS} = 3.2, 4.8,$ or 9.6 h) and for the time-distance, adjusted the spatial averaging accordingly (i.e.: 7×7, 6×6, or 5×5).

2 Results

Figure 1 shows the characteristics of some of the EARs analyzed, showing in one set of panels, as a function of time and radius, the annulus-averaged magnetic field. That same figure also shows the resulting regressions between either intensity, acoustic power and magnetic field, as well as the correlation of the mean time propagation anomaly with the intensity, the magnetic field, or the acoustic power. These are shown for two time-series, one corresponding to a temporal offset with respect to the time of emergence, ΔT, or -14.2 h and one for ΔT of $+14.2$ h. (i.e.: $\pm 4 \times T_{\rm TMS}$). Since each EAR has a different and very specific topology, it is not obvious how to combine them in order to increase the SNR of any precursor signal. They cannot simply be averaged. Instead I opted to combine the resulting regressions for a given time offset with respect to the emergence time, and bin those with respect to either B, I_c, or P.

3 Conclusions

Initial results indicate that because of the variable topology it is not obvious how to best combine EARs, yet the differences in topology could be indicative of the nature of underlying processes. While there is a very clear post-emergence signature, each EAR displays distinct spatio-temporal patterns, and the regression curves show a somewhat complicated pattern. The post-emergence signal is a time anomaly, $\delta\tau$ of -67 s per kG, or -3 s per $\%(\delta I_c)$, 0.33 s per $\%(\delta P)$ and 0.66 s per $\%(\delta P_6)$.

The time evolution of these regression curves show a slow onset of the time anomaly ($\delta\tau$) with respect to B and I_c, but not P (or P_6), especially when using

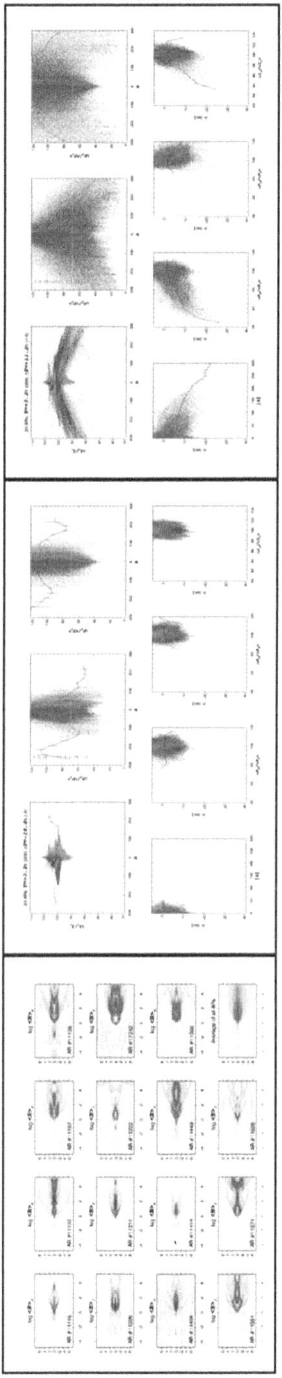

Fig. 1 Left: Surface magnetic field, averaged over annuli, in a time versus radius format for 15 EARs, and the corresponding average for all regions (lower right panel). Middle and right: combined regression, for $\Delta T = -14.2$h and $+14.2$h (i.e.: $\pm 4 \times T_{\mathrm{TMS}}$), with $N_{\mathrm{TMS}} = 256$

long T_{TMS}, raising the questions whether this is an observational artifact or whether B and I_c are pure surface indicators, while P might be a *better* acoustic indicator, since a change in acoustic power is seen shortly before the emergence.

Acknowledgments HMI data are courtesy of NASA/SDO and the HMI consortium. The author is supported by NASA grant NNX17AB72G. Most of the computations done for this work were carried out on the Smithsonian Institution's High Performance Cluster (*SI/HPC*).

Solar Hemispheric Helicity Rules: A New Explanation

Bhishek Manek, Nicholas Brummell, and Dongwook Lee

Abstract Large-scale solar surface magnetic fields are believed to emerge from the deep solar interior due to magnetic buoyancy and are observed as sunspots, embedded in active regions. The geometry and bi-polar nature of the pairwise emergence of sunspots has motivated the consideration of the rising and emerging strong toroidal magnetic flux concentrations in the interior as cylindrical flux tubes. Since magnetic flux is not expected to exist as isolated flux tubes in the deep solar interior, we have investigated the dynamics of more realistic magnetic flux concentrations embedded in a larger-scale diffused field. We find that the presence of a weak large-scale background field, through which the flux tube rises, can significantly alter the dynamics and lead to interesting insights about solar observations of current/magnetic helicity. We further carry out a Monte Carlo study to explore these dynamics quantitatively for a direct comparison with Solar Hemispheric Helicity rules.

Large-scale solar surface magnetic fields are believed to emerge from the deep solar interior due to magnetic buoyancy and are observed as sunspots, embedded in active regions. The observations of these emergences have revealed many patterns, such as the butterfly diagram, Hale's polarity law etc. A relatively new observational tracer is the measured magnetic/current helicity at the solar surface. Helicity measures the twisting, linking and connectivity of the magnetic fields, and is an important quantity since the "undoing" of strong helicity (magnetic reconnection) can release stored energy. Hence, it is suspected to be the ultimate driver for magnetic

B. Manek (✉) · N. Brummell · D. Lee
University of California Santa Cruz, Santa Cruz, CA, USA
e-mail: bmanek@ucsc.edu; brummell@soe.ucsc.edu; dlee79@ucsc.edu

M. J. P. F. G. Monteiro et al. (eds.), *Dynamics of the Sun and Stars*, Astrophysics and Space Science Proceedings 57,
https://doi.org/10.1007/978-3-030-55336-4_18

phenomenon on many scales such as solar flares, CMEs etc. The origin of helicity is therefore a crucial question. Observations of current helicity, $H_C = \int B.(\nabla \times B)dV$, have shown a preferential negative sign in the northern hemisphere and a preferential positive sign in the southern hemisphere. This is known as the "Solar Hemispheric Helicity Rule" (hereby referred to as SHHR), see [3]. The SHHR is however only weakly followed, with around 60–75% of active regions conforming to the rule. We have investigated the origin of the SHHR with 2.5D simulations of flux tubes embedded in a background field. Our paper [1] simulates the evolution of an initially twisted toroidal magnetic flux embedded in a horizontal poloidal magnetic field that drops off exponentially with height, in an adiabatic, polytropic Cartesian domain. In the solar sense, the tube represents the strong toroidal field concentrations and the background field mimics the poloidal field (see Fig. 1a). As these flux tubes are twisted and have both axial and azimuthal field, they possess either a positive or negative helicity depending on the orientation of the fields. We find that even a relatively weak background field can hinder the rise of the flux tubes in a dramatic way. More surprisingly, the orientation of the background field also has a significant effect on the rise (see Fig. 1b). We find that a positively-oriented background field with a strength of around 16% or greater (compared to the initial axial field strength) suppresses the rise of the tube whereas a negatively-oriented background field of strength around 6% or greater is sufficient to suppress the rise. The result of this is to select a particular combination of background field orientation and flux tube helicity (where the 'local' twist at the trailing edge of the tube is enhanced) to preferentially rise. This selection mechanism, when translated into solar context, agrees with the established observations of SHHR. The work in [1] concentrated on the rise of a single embedded magnetic flux structure. However, the SHHR is just a statistical result based on the rise of many such structures. In an attempt to reproduce a result that is more directly comparable with the SHHR, we have carried out a study with 50 Monte Carlo simulations of five randomly twisted and randomly located flux tubes embedded in a large-scale background field of fixed strength. Figure 2a shows a typical case with $B_s = 5\%$ at $t = 7$ and $t = 45$ corresponding to the solar southern hemisphere. For a quantitative indicator of emerging current helicity, we sum over the z-component of the quantity $B.(\nabla \times B)$ over time for any flux structure that successfully rises through 75% of the simulation box. The statistics presented in Fig. 2b show the magnitude and sign of current helicity from these 50 simulations. About 76% of the flux structures that successfully rise show positive helicity signatures, commensurate with SHHR expectation. The 24% that is non-compliant represents the SHHR scatter, and is due to the extremes in the random twist fluctuations, tube interactions and other effects. Simulations of different background field strengths represent different parts of the solar cycle and provide further results in agreement with the SHHR. A detailed description of these simulations will be forthcoming in a future work [2].

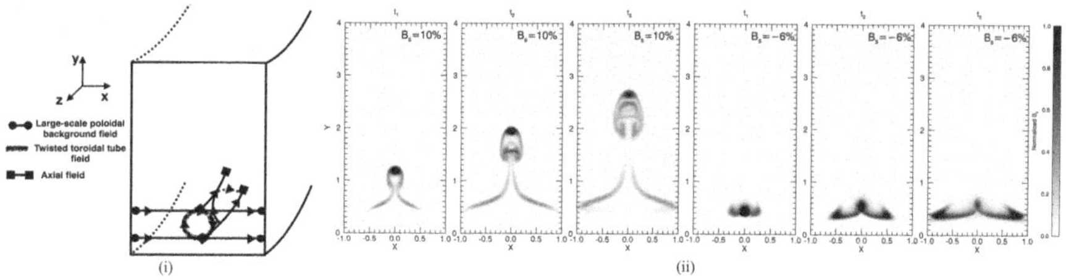

Fig. 1 (**a**) Model Setup for Flux tube rise simulations. (**b**) 2.5 D flux tube dynamics shown for different orientation of background field, $B_s = 10\%$ and $B_s = -6\%$ with 3 time steps during the buoyant rise shown for each

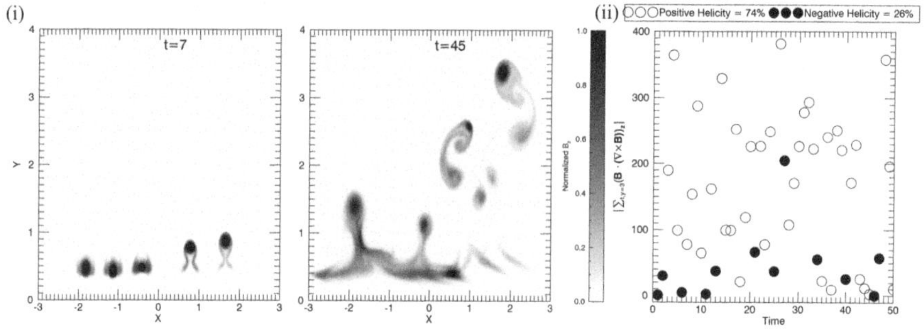

Fig. 2 (**a**) A typical rise of five randomly twisted flux tubes at $t = 7, 45$ for $B_s = 5\%$; (**b**) Sign of current helicity summed over time at $y = 3$ in the simulation box for model southern hemisphere field orientations when $B_s = 5\%$

References

1. Manek, B., Brummell, N., & Lee, D. (2018). *The Astrophysical Journal Letters, 859*, L27.
2. Manek, B. & Brummell, N. (2020). *The Astrophysical Journal* (in preparation).
3. Pevtsov, A. A., et al. (2014). *Space Science Reviews, 186*, 285.

Comparing Solar Activity Minima Using Acoustic Oscillation Frequencies

Kiran Jain, Sushanta C. Tripathy, and Frank Hill

Abstract Using the variability of solar acoustic oscillation mode frequencies, we investigate the width and timing of last three solar minima at various depths in the solar interior. We find that the current minimum is as deep as the previous minimum, however, more data are needed to reveal the true depth of the current minimum.

1 Introduction

Uninterrupted observations of the Sun's acoustic oscillations over more than two decades have established a strong connection between sub-surface layers and the magnetic activity visible above the surface [2]. However, during the minimum between cycle 22 and 23 (MIN 23/24), the deeper layers in solar interior sensed minimum about a year earlier than the outer layers ([1], and references therein). In this paper, we investigate if the finding were unique to MIN 23/24 or similar trend does exist during the current minimum between cycles 24 and 25 (MIN 24/25).

2 Results

We use 247 36-day independent frequency tables for (n, ℓ, m) multiplets from GONG covering a period from 1995 May 7 to 2019 Sept 9. In Fig. 1, we plot temporal variation of frequency shifts for modes with their lower turning point radius in the core (r_t/R_0 = 0.0–0.3), the radiative zone (r_t/R_0 = 0.3–0.7), near the tachocline (r_t/R_0 = 0.71–0.73), and the convection zone (r_t/R_0 = 0.7–1.0) where R_0 is the radius of the Sun. It is evident that MIN 24/25 is deeper than MIN 22/23

K. Jain (✉) · S. C. Tripathy · F. Hill
National Solar Observatory, Boulder, CO, USA
e-mail: kjain@nso.edu

M. J. P. F. G. Monteiro et al. (eds.), *Dynamics of the Sun and Stars*, Astrophysics and Space Science Proceedings 57,
https://doi.org/10.1007/978-3-030-55336-4_19

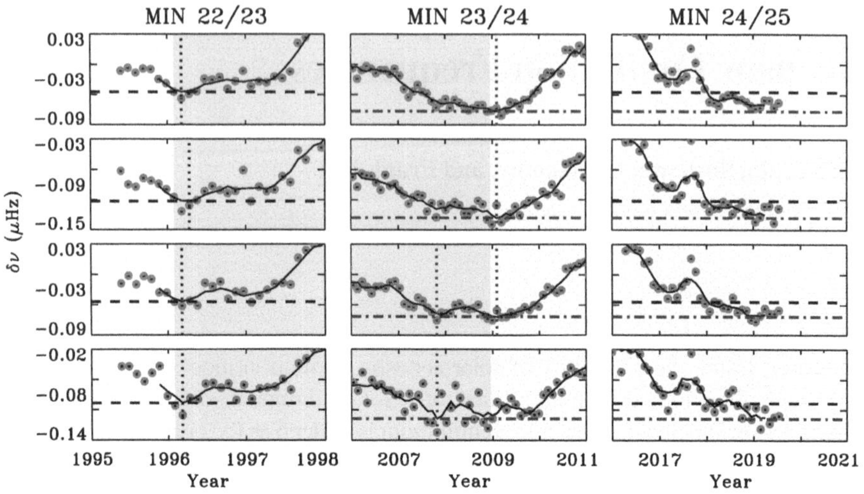

Fig. 1 Variation in frequency shifts (circles) for the modes with lower turning points in four different layers in the interior during the last three minima; convection zone (top row), tachocline region (second row), radiative zone (third row), and the core (bottom row). The solid line represents a 5-point running mean of the frequency shifts while vertical dotted lines are drawn to identify the epochs of lowest values in frequency shifts (i.e., the seismic minima). Horizontal dashed and dashed-dot lines represent lowest frequency shifts during MIN 22/23 and MIN 23/24, respectively

but comparable with MIN 23/24. In addition, all four layers seem to show similar response during MIN 24/25, which is different from MIN 23/24. Since both solar activity and the oscillation frequencies still continue to decrease, it is possible that the solar activity has not reached at the minimum level yet (Sept, 2019). If this trend continues, the MIN 24/25 will be much deeper than MIN 23/24.

Acknowledgments This work utilizes GONG data from NSO, which is operated by AURA under a cooperative agreement with NSF.

References

1. Jain, K., Tripathy, S. C., & Hill, F. (2011). How peculiar was the recent extended minimum: A hint toward double minima. *The Astrophysical Journal, 739*, 6. https://doi.org/10.1088/0004-637X/739/1/6, arXiv:1107.0049.
2. Jain, K., Tripathy, S., Komm, R., Hill, F., & Simoniello, R. (2018). 22 Year solar magnetic cycle and its relation to convection zone dynamics. In D. Banerjee, J. Jiang, K. Kusano, & S. Solanki (Eds.). *IAU Symposium* (Vol. 340, pp. 9–10). https://doi.org/10.1017/S1743921318001242, arXiv:1805.05371.

Solar Cycle Variation of Large Scale Plasma Flows

B. Lekshmi

Abstract Large scale plasma flows like differential rotation, and meridional flow play an important role in driving the solar dynamo process. Our long-term study of the hemispherical asymmetry in torsional oscillation and meridional flow show that there exists a time delayed correlation (anti-correlation) between the asymmetries in torsional oscillation (meridional flow) and the sunspot cycle, with the asymmetry in the flows preceding the cycle asymmetry.

1 Introduction

Solar torsional oscillation and meridional flow show spatio-temporal variation over the solar cycle and are asymmetric across the equator. They are strongly correlated with the solar magnetic cycle and we study the relationship between the hemispherical asymmetry of the torsional oscillation and the residual (time-varying) meridional flow velocities with the asymmetry in solar magnetic activity [2, 3].

2 Data and Analysis

We use $\sim$17 year of zonal and meridional flow velocities obtained from Global Oscillation Network Group (GONG) and Helioseismic and Magnetic Imager (HMI) ring diagram pipelines [1]. A 11 year mean at each latitude is subtracted to obtain the temporal variation in the flows. Sunspot data from SILSO World Data Center, RGO/NOAA sunspot area data and synoptic magnetograms from Michelson Doppler Imager and HMI are used. The hemispherical asymmetry in velocity

B. Lekshmi (✉)
Center of Excellence in Space Sciences India, IISER Kolkata, Mohanpur, India
e-mail: lb15rs016@iiserkol.ac.in

M. J. P. F. G. Monteiro et al. (eds.), *Dynamics of the Sun and Stars*, Astrophysics and Space Science Proceedings 57,
https://doi.org/10.1007/978-3-030-55336-4_20

(V), flux (F) and Sunspot Number (SSN) are given by $V_N - V_S$, $F_N - F_S$ and $SSN_N - SSN_S$, where the subscripts N and S represent northern and southern hemispheres respectively.

The hemispherical asymmetry in near-surface torsional oscillation over the active latitude at 2 Mm depth shows a correlation of 0.72(99.9%) and 0.78(99.9%) with the magnetic flux and SSN asymmetries respectively. Also the torsional oscillation asymmetry precedes the magnetic flux and SSN asymmetries by 1.41 ± 0.38 year and 1.18 ± 0.23 year respectively.

Asymmetry in the residual meridional flow at 2Mm depth obtained from GONG (HMI) over the active latitude shows a correlation of −0.73 (−0.82) with the magnetic flux and −0.74 (−0.83) with the SSN asymmetries with 99.9% confidence. Also the asymmetry in velocity from GONG (HMI) precedes the asymmetry in magnetic flux and SSN by $3.6\pm0.1(3.4\pm0.1)$ and $3.5\pm0.1(3.4\pm0.1)$ respectively.

3 Conclusion

Hemispherical asymmetry in the deep-seated magnetic flux belt can be responsible for the flow asymmetries through non-local (e.g., thermal shadow) processes. The strong correlation (anti-correlation) of the asymmetries in torsional oscillation (meridional flow) with the asymmetry in sunspot cycle and the corresponding time-delays implies that the flow asymmetries can be a precursor of hemispheric magnetic activity. But more accurate deep flow measurements and magneto hydrodynamic simulations are necessary to conclusively establish the physics behind the generation of these asymmetries and their relationship with the sunspot cycle.

Acknowledgments We acknowledge NSO Synoptic Program, NASA/SDO and HMI science teams, WDC-SILSO, CSIR and MHRD - Government of India.

References

1. Komm, R., González Hernández, I., Howe, R., & Hill, F. (2015, April). Subsurface zonal and meridional flow derived from GONG and SDO/HMI: A comparison of systematics. *Solar Physics, 290*, 1081–1104. https://doi.org/10.1007/s11207-015-0663-6
2. Lekshmi, B., Nandy, D., & Antia, H. M. (2018, July). Asymmetry in solar torsional oscillation and the sunspot cycle. *The Astrophysical Journal, 861*(2), 121. https://doi.org/10.3847/1538-4357/aacbd5
3. Lekshmi, B., Nandy, D., & Antia, H. M. (2019, October). Hemispheric asymmetry in meridional flow and the sunspot cycle. *Monthly Notices of the Royal Astronomical Society, 489*(1), 714–722. https://doi.org/10.1093/mnras/stz2168

Measuring the Dispersion Relation of Acoustic-Gravity Waves in the Solar Atmosphere

Bernhard Fleck, Stuart M. Jefferies, Neil Murphy, and Francesco Berrilli

Abstract We use localized measurements of the dispersion relation for acoustic-gravity waves to generate the first maps of the spatial structure of the sound speed, acoustic cut-off frequency, and radiative damping time in the Sun's lower atmosphere. These maps offer a new diagnostic for the solar atmosphere.

1 Background

Observations of acoustic-gravity waves at different heights in the solar atmosphere have long been used to probe its structure and dynamics. A common diagnostic is the phase difference of the waves, $\Delta\phi$, as observed at different heights. $\Delta\phi$ is related to the height difference between the observations, Δz, the phase travel time of the wave between the observations, Δt_{ph}, and the vertical component of the wave vector, k_z, through the relationship

$$\Delta\phi = k_z \Delta z = \omega \Delta t_{ph}. \tag{1}$$

B. Fleck (✉)
ESA Science and Operations Department, c/o NASA/GSFC, Greenbelt, MD, USA
e-mail: bfleck@esa.nascom.nasa.gov

S. M. Jefferies
GSU, Atlanta, GA, USA

University of Hawaii, Honolulu, HI, USA
e-mail: sjefferies@gsu.edu

N. Murphy
NASA/JPL, Pasadena, CA, USA
e-mail: neil.murphy@jpl.nasa.gov

F. Berrilli
Università di Roma Tor Vergata, Rome, Italy
e-mail: francesco.berrilli@roma2.infn.it

M. J. P. F. G. Monteiro et al. (eds.), *Dynamics of the Sun and Stars*, Astrophysics and Space Science Proceedings 57,
https://doi.org/10.1007/978-3-030-55336-4_21

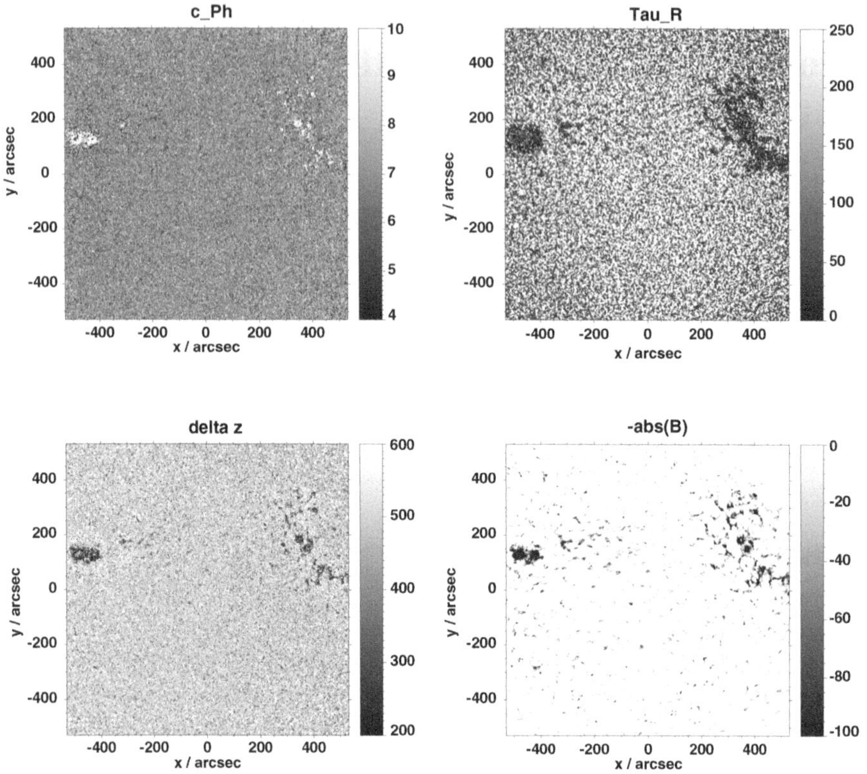

Fig. 1 Results from fitting the measured local phase difference spectra to a model. Top left: sound speed c_s, in km s^{-1}. Top right: The radiative damping time, τ_R, in seconds. Lower left: Δz between the MOTH (Na D) and HMI (Fe 6173) Doppler signals. Lower right: The average unsigned magnetic field measured by HMI over the 11-h observational periods, in Gauss

This equation shows that the observed variation of $\Delta\phi$ with frequency provides a measure of the local magneto-acoustic-gravity wave dispersion relation for upward propagating waves.

We have used Eq. (1) and k_z as defined by the dispersion relation for acoustic-gravity waves in an isothermal stratified atmosphere with constant radiative damping [2], to model the measured phase difference spectra generated from 11-h of Doppler velocity measurements of the full solar disk acquired on January 21, 2017 with the Magneto-Optical Filter at two Heights (MOTH) instrument at the South Pole and the Helioseismic and Magnetic Imager (HMI) on board SDO [1].

The variables in our model are Δz, the sound speed, c_s, the radiative damping in the atmosphere, τ_R, and the modified acoustic cut-off frequency [1], ω_0. We model the phase difference in each spatial pixel in a 1060 arcsec $\times$ 1060 arcsec region near disk center. Figure 1 shows the resulting maps of Δz, c_s and τ_R. The map of the modified acoustic cut-off frequency is presented elsewhere [1]. All of these maps are the first of their kind and offer a new diagnostic for the solar atmosphere.

References

1. Jefferies, S. M., Fleck, B., Murphy, N., & Berrilli, F. (2019). Observed local dispersion relations for magnetoacoustic-gravity waves in the Sun's atmosphere: Mapping the acoustic cutoff frequency. *The Astrophysical Journal Letters, 884*, L8.
2. Souffrin, P. (1972). Radiative relaxation of sound waves in an optically thin isothermal atmosphere. *Astronomy and Astrophysics, 17*, 458.

Non-adiabatic Helioseismology via 3D Convection Simulations

Regner Trampedach

Abstract Eigen-functions of p-modes have been extracted from a 3D simulation of a deep convective solar atmosphere. The part of the eigen-functions that are out of phase with respect to the density wave, is fitted to an analytical expression in depth and frequency and then applied to a non-adiabatic oscillation calculation. The resulting damping widths and frequency shifts are compared with helioseismic observations.

1 Non-adiabatic Modes

Non-adiabatic modes are maybe more accurately described as modes where the eigen-functions have a pressure component that is out of phase with the density eigen-functions, with the result that work is performed on the mode. This can be either mode-dampening in nature (and opposed by stochastic excitation) or mode-exciting, delineating the two distinct regimes of stable and unstable modes, respectively. This contribution is concerned only with the stable regime, also known as *solar-like oscillations*. Being coherently damped and stochastically excited by convective turbulence their amplitudes are small and usually well described as a linear perturbation around a static mean stratification.

The 3D solar simulation is performed on a $6 \times 6 \times 3.5$ Mm Cartesian box with $\sim$6 pressure scale-heights above and $\sim$8 below the photosphere, and open vertical boundaries allowing convective flows in and out of the box. The bottom is constructed to have a modal node, to avoid having the boundary conditions performing work on the modes. Realistic equation of state and opacities are employed, and radiative transfer is performed on long characteristics with full spectral coverage

R. Trampedach (✉)
Space Science Institute, Boulder, CO, USA
e-mail: rtrampedach@SpaceScience.org

M. J. P. F. G. Monteiro et al. (eds.), *Dynamics of the Sun and Stars*, Astrophysics and Space Science Proceedings 57,
https://doi.org/10.1007/978-3-030-55336-4_22

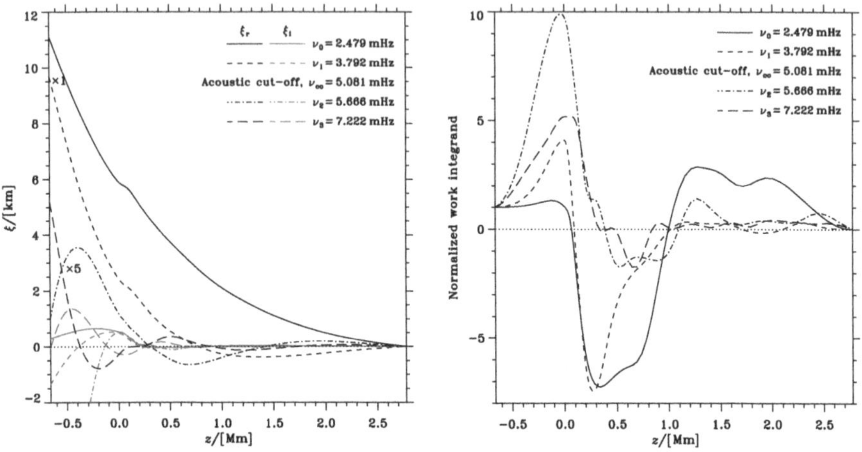

Fig. 1 **Left:** Complex eigen-functions of the four discernible modes excited in the solar simulation. Black lines show the real part, gray thin lines the imaginary part, and line-styles indicate the mode, 0–3. Only the first two modes are trapped by the atmosphere, and the two others are above the acoustic cut-off frequency, $\nu_{\rm co}$, leaking mode-energy into the atmosphere. These two eigenfunctions have been scaled by a factor of 5, in the plot. **Right:** The work integrands, Eq. (1), for the same four modes, normalized at the top. The sign is chosen for damping to be positive

accomplished by an opacity binning scheme, as described in detail by Trampedach et al. [2].

The work integral is (as function time on the left, in Fourier space on the right)

$$W = \int \mathrm{d}t \int \mathrm{d}z\, \delta p \frac{\partial \dot{\xi}}{\partial z} \quad \Leftrightarrow \quad \widetilde{W}_\omega = i\omega \int \mathrm{d}z\, \delta \tilde{p}^*_\omega \frac{\partial \tilde{\xi}_\omega}{\partial z} \tag{1}$$

where ξ is the displacement from the equilibrium model, δp, is the pressure fluctuation (gas + turbulent), and z is the vertical direction. Displacement eigenfunctions and Fourier-components of the work integrand are shown for the four modes found in the simulation, in Fig. 1. This is the first step in evaluating the mode damping properties of the simulation, as currently pursued. A different approach calibrated convective properties in 1D models to the grid of 3D simulations, reproducing the damping in 1D, non-adiabatic, oscillation calculations by Houdek et al. [1].

References

1. Houdek, G., Lund, M. N., Trampedach, R., Christensen-Dalsgaard, J., Handberg, R., & Appourchaux, T. (2019). *Monthly Notices of the Royal Astronomical Society, 487*, 595–608.
2. Trampedach, R., Asplund, M., Collet, R., Nordlund, Å, & Stein, R. F. (2013). *The Astrophysical Journal, 769*, 18, 1–15.

Part III
From the Sun to the Stars

Deciphering Solar Convection

Mark Peter Rast

Abstract Numerical modeling of solar and stellar convection, and by extension modeling of solar and stellar dynamos faces a surprising challenge. No hydrodynamic, magnetohydrodynamic, or radiative magnetohydrodynamic model of solar convection, if conducted in a sufficiently deep domain, achieves the velocity power spectrum implied by observations of the Sun. The horizontal velocity at low wavenumbers in the upper layers of the simulation domains is much too high, monotonically increasing to low wavenumber rather than rolling over at supergranular scales, as on the Sun. This reflects convective amplitudes at depth that are similarly too large, and results in equatorial differential rotation profiles in simulations of rotating spherical shells of opposite sign to those observed. The problem worsens in models with decreasing diffusivities, as the amplitudes of the convective motions increase. This has come to be known as the convective conundrum. Solving it is critical to understanding dynamo behavior on stars, which in turn is central to the assessment of the structure of the asterospheres in which their planetary companions are embedded. This paper examines what is known about solar convection in light of one possible underlying cause of the convective conundrum, that the deep interior of the Sun is even more nearly adiabatically stratified than our models suggest or can achieve. Correcting this in models will likely be difficult, but we point in some potentially fruitful directions.

1 Introduction

Somewhat surprisingly, after nearly four decades of compelling first-principles modeling of solar and stellar convection using three-dimensional radiative hydrodynamic and magnetohydrodynamic solvers (spanning the period since the pioneering

M. P. Rast (✉)
Department of Astrophysical and Planetary Sciences, University of Colorado, Boulder, CO, USA
e-mail: mark.rast@lasp.colorado.edu

M. J. P. F. G. Monteiro et al. (eds.), *Dynamics of the Sun and Stars*, Astrophysics and Space Science Proceedings 57,
https://doi.org/10.1007/978-3-030-55336-4_23

work of [28]), we have arrived at the generally accepted conclusion that there is something fundamentally wrong with our numerical solutions. All numerical models of sufficient depth, whether computed in local area or global spherical shell geometry, produce large scale convective flows with significantly higher amplitudes than those deduced from observations. This mismatch has come to be known as the convective conundrum, and its origin is not likely to be in the subtleties of observations. In its simplest form, the convective conundrum is found in the fact that the Sun's photospheric horizontal velocity spectrum peaks at supergranular scales, while the power in all simulations of adequate depth increases monotonically toward low wavenumbers beyond the supergranular scale. If simulations were correct, giant-cell scale, not supergranular scale, motions would be most evident in the solar photosphere.

Perhaps more critically, this mismatch between the scales of motions evident in observations of the photosphere and those realized in numerical simulations reflects a difference in convective amplitudes at depth. The amplitudes of the deep large scale motions in simulations are too high, thus the Rossby number achieved at mean solar rotation rates is also too high (e.g., [12], and references therein), and global spherical shell models fail to recover the solar differential rotation profile at solar rotation rates. This failure introduces a critical uncertainty in our ability to model solar and stellar dynamos from first principles. Moreover, lowering model diffusivities, worsens the problem because convective amplitudes increase. The importance of this failure is glaring. How can we model differential rotation, and by extension dynamo action, on distant stars (a process central to understanding the asterospheres in which their planetary companions are embedded) if we cannot do so successfully for the Sun for which detailed observations can serve as a guide? In this paper we will briefly review what is known about solar convection, focusing attention primarily on one possibility, that the convective conundrum results because the deep interior of the Sun is even more nearly adiabatically stratified than our models suggest or can achieve. If this is indeed the case, correcting the numerical simulations will likely prove difficult, but we suggest some possible directions forward.

2 Convective Scales

The spectrum of convective motions on the Sun is continuous, but over the years solar physicists have pointed to possible characteristic scales: granulation, meso-granulation, supergranulation and giant cells. Granulation [17] is readily apparent in high resolution images of the solar photosphere and is characterized by $\sim$1000 km diameter cells with $\sim$1 km/s vertical flows and $\sim$0.5 - 0.25 h lifetimes. It reflects the dominant dynamics of the strongly cooled radiative boundary layer, is well observed and robustly modeled (e.g., [29]), even in quite shallow domains (8 Mm), by codes that capture the rapid change in radiative opacity in the photospheric layers and implement open lower boundary conditions to minimize bottom-up dynamical influences. Mesogranulation [31], on the other hand, is observationally elusive, with a reported length scale of ~ 5000 km, ~ 60 m/s vertical flows and ~ 3 h lifetime. It appears to result from the self-organization of granular flows, a process first proposed in the context of supergranulation [40, 34] but likely more relevant on mesogranular scales [6, 3, 23, 9]. The existence of mesogranulation as a real convective feature is still actively debated [30, 4].

At somewhat larger scales supergranulation [14, 22] is well observed either directly in spectral Doppler shifts away from disk center or as traced by the solar magnetic network prominent in low chromospheric lines such as Ca II K. Supergranular cells have a range of diameters centered around 30,000 km, horizontal flow velocities of $\sim$100 m/s and lifetimes of $\sim$20 h. The origin of this characteristic scale of motion has been debated for some time, with recent evidence pointing to it being defined not as a selected scale, but as the scale above which power in the horizontal photospheric flows declines [25, 8]. This interpretation ties its origin to the solution of the convective conundrum. Note that the early suggestion that the second ionization of helium plays an important role [22, 42, 31] is not supported by numerical simulations [36] or simplified models base on them [25]. Finally, convective motions on the scale of the depth of the solar convection zone ($\sim$ 200 Mm in size, with $\sim$ 10 m/s horizontal flow amplitudes and $\sim$ 1 month lifetimes), called giant cells, are very prominent in numerical simulations but very difficult to find and only fairly recently detected in solar observations [15]. It is these scales that dominate global spherical shell simulations and which, because of their significant amplitudes, place simulations with mean rotation rates matching the Sun in the incorrect differential rotation regime.

Only two of the scales above are apparent as features in the spectrum of observed convective motions in the solar photosphere (Fig. 1a, from [16]): granulation, which is responsible for a prominent peak at high spherical harmonic degree, and supergranulation, with power peaking near $l = 120$. Otherwise the spectrum shows no evidence for discrete scales of motion. It is important to note that the spectrum of solar photospheric motions, as that shown in Fig. 1a is composite. Vertical velocities dominate the spectrum at high wave numbers, while horizontal motions are most important at supergranular scales. The amplitude of the vertical velocity contribution monotonically decreases from the granulation peak toward low wavenumbers. The supergranular peak results from an increase in the horizontal velocity contribution up to spherical harmonic degree of $\sim$120 before rolling over beyond that.

The spectra of the convective motions found in simulations of solar convection (e.g., Fig. 1b, from [18]) are fundamentally different. The granular peak may or may not be present depending on the resolution of the model and how well it captures the upper radiative boundary layer, and the horizontal velocity power at low wavenumbers, measured in the photosphere, continues to increase beyond the supergranular scale up to the scale of giant cells if the deep convection zone is included in the calculation. As illustrated by the *blue dotted* line added to Fig. 1a, the convective conundrum is captured by the enhanced power (by a factor of

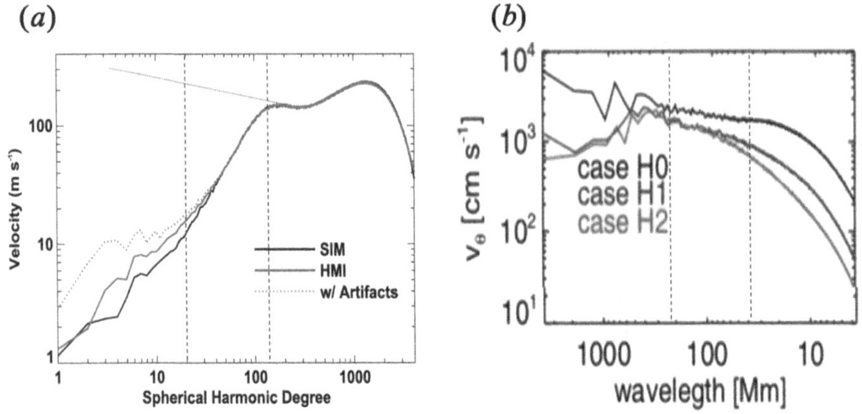

Fig. 1 Solar convective velocity spectra as determined from Helioseismic and Magnetic Imager (HMI) observations (*red* curve in (**a**) from [16]) and from high resolution global simulations (in (**b**) from [18]). Vertical *blue dashed* fiducial lines have been added to both plots indicating the approximate scale of supergranulation and giant-cells. The *blue dotted* line added to (**a**) approximately over-plots the numerical spectrum on that observed

approximately ten) in giant-cell scale motions in the simulations compared to those in observations.

3 Where do the Observed Horizontal Motions Originate?

One of the most important aspects of the solar convection zone is its steep stratification. The solar convection zone is about 210 Mm deep, and across that depth the density changes by a factor of about one million and the pressure by about 800 times that. The density scale height at the base of solar convection zone is nearly half its depth while in the surface layers it is about 150 km. By 10 Mm below the solar photosphere (5% the depth of the solar convection zone), the density and pressure have already changed by factors of more than 3.5×10^3 and 80×10^3 respectively.

This has profound influence on the convective flows. Convective motions must abide by this mean stratification, so by mass conservation only a tiny fraction of the upwelling fluid from below makes it into the photosphere above, the rest must overturn. Over each scale height, the density of rising fluid must decrease by a factor of $1/e$, implying that $1 - 1/e$ of the mass must overturn. If the flow geometry is approximated as vertical cylinders of radius r with height equal to the density scale height H_ρ, then mass continuity requires that $2\pi r H_\rho \rho u_h = \pi r^2 \rho u_z$, where u_h and u_z are the vertical and horizontal flow velocities respectively. Taking the flow amplitudes to be equal in the two directions yields an expression for the characteristic horizontal length scale of the flow at each height, $d = 2r \approx 4H_\rho$

[29]. It has been suggested that this length scale may also be take to be the integral, or driving scale, of the motions [44], and with this assumption, a simple mixing length type argument can reproduce the spectrum of the horizontal motions in the photosphere of highly nonlinear radiative hydrodynamic simulations [25].

The success of such a model in reproducing the photospheric power in radiative hydrodynamic simulations suggests that the scales of horizontal motions in the photosphere of the Sun also reflect the amplitude of the driving at depth, that the buoyancy driving in the deep layers of the Sun must be significantly lower than that in the simulations, and that it is this disparity that is responsible for the convective conundrum mismatch. Since the density scale height in the Sun is about 5 Mm ($4H_\rho = 20$ Mm) at a depth of 10 Mm, [25] tested this hypothesis by conducting a numerical experiment in which the convective velocities in the deep layers of a simulation were reduced without changing the mean density stratification, since, as per discussion above, it is fundamental to the surface spectrum. This was done by adding a depth dependent artificial flux term to the energy equation. It carried the full solar flux below 10 Mm and none of the flux at heights above that. In the resulting solution, the deep interior of the domain becomes more adiabatically stratified; the Schwarzschild convection criterion, $d \ln \rho / d \ln p - d \ln \rho / d \ln p|_{\rm ad}$, a direct measure of buoyancy instability assuming displacements occur in pressure equilibrium, changes by 20% right below the artificial flux divergence layer and about 1% in the deeper layers. The resulting reduction in buoyancy driving of the deep flow yields reduced flow amplitudes there, mirrored by reduced horizontal flow amplitudes in the photosphere (Fig. 2); the amplitude of the motions in the deep layers are reduced by a factor of 2.5 (Fig. 2a) and the photospheric horizontal velocity spectrum derived using Coherent Structure Tracking [41] of photospheric motions more nearly matches that of observations (Fig. 2b).

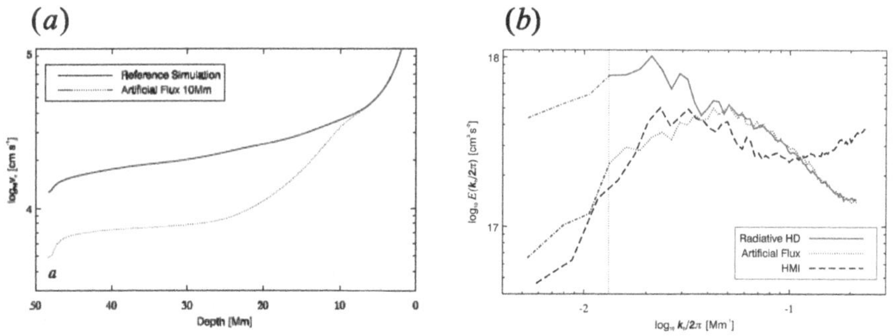

Fig. 2 In (**a**), the convective velocities (rms) are reduced by a factor of 2.5 in the deep layers of a simulation in which the convective driving is constrained to occur in the upper 10 Mm of the domain by the use of an artificial flux term (see text and [25]). Figure from [24]. In (**b**), photospheric velocity power computed using Coherent Structure Tracking techniques on the artificial flux simulation (*dotted blue*), HMI observations (*dashed black*), and an radiative hydrodynamic simulation without the artificial flux term (*solid red*). From [25]

4 Why 10 Mm?

The photospheric horizontal velocity spectrum reflects a hierarchy of driving scales with depth. Low wavenumber motions are driven deep in the convection zone and imprint as horizontal flows in the surface layers. In fully nonlinear radiative hydrodynamic simulations this results in monotonically increasing power above the supergranular scales, but in the Sun it does not. This suggests that in the Sun, low wavenumber flows deep in the convection zone are more weakly driven than they are in the models. Artificially reducing the driving in the models at depths below 10 Mm recovers the solar spectrum, but why 10 Mm in the Sun?

A number influences may contribute to the suppression of flows in the lower solar convection zone. The divergence of the radiative flux is nonzero there, and the consequent internal heating may lead to a slightly sub-adiabatic mean gradient. While many investigations of convection driven by uniform internal heating have been made in the context of terrestrial mantel convection, only limited work in the solar context, with a depth (or more correctly temperature and density) dependent internal heating profile, has been undertaken [21, 10]. Rotation may also be critical. If the amplitude of the large scale convective motions at depth can be sufficiently reduced so that they are subject to rotational constraint at solar rotation rates, these constraints will further limit their amplitudes relative to smaller scales [11]. Additionally, the Sun is essentially magnetized, its dynamics can likely not be accurately modeled without self consistent inclusion of magnetic fields, and both large and small scale fields in the lower convection zone may play a role in determining convective amplitudes.

More fundamentally we suggest that convection in the Sun is even more nonlocal than simulations suggest, that the convection zone is even more adiabatically stratified in the mean than simulations achieve, and that it is the fluid cooled by photospheric radiative losses that dominates the transport of low entropy fluid to the lower convection zone where it is radiatively heated. We suggest that granular downflows, merged downflows, or the vortical structures that result from their instabilities [33, 7, 1], transfer the radiatively cooled fluid from the photosphere to the radiatively heated lower convection zone without re-stratification of the adiabatic interior (e.g., [39, 43, 8, 5]). This is possible if the transit of the downflowing fluid occurs adiabatically, if the transit time of the cool material across the convection zone is shorter than the thermal diffusion time across the flow structures, if solar convection below the photospheric radiative boundary layer occurs in the high Peclet number regime. Estimates for the radiative diffusivity in the solar convection zone range from about 10^5 to 10^7 cm^2 s^{-1} [27]. The characteristic diffusion timescale of a plume with spatial scale 300 km (i.e., the approximate thickness of the photospheric radiative boundary layer) is thus between 3 and 285 years. Taking the transit time to be on the order of the turnover time of the largest convective cells ($\sim$1 month), cold material should traverse the convection zone without exchanging significant amounts of heat with the surrounding medium.

A consequence of this adiabatic transit of the cool material across the solar convection zone is that no re-stratification of the adiabatic background state occurs. The degree of superadiabaticity at any depth is determine only by the presence of the cool material via its geometric contribution to the mean. Entropy is conveniently additive, so for adiabatically stratified upflows, even with mixing but in the absence of diffusion, the mean gradient is determined by the downflow entropy fluctuations and their filling factor as induced by photospheric radiative cooling [8]. Since the ratio of the photospheric density to the density with depth decreases rapidly, the downflow contribution to the mean also decreases rapidly with depth. This implies that the interior stratification of the Sun should be extremely close to adiabatic throughout, with buoyancy driving occurring only within a relatively thin superadiabatic layer determined by the thermodynamic properties of the granular downflows in the uppermost layers. We suggest that it is $4H_\rho$ in these layers determines the scale of the horizontal motions observed.

The likely reason that the simulations do not achieve this highly nonlocal solution, but instead show a mixing length like forcing with depth, is the degree to which re-stratification of the layer, away from an adiabatic mean, occurs at the non-solar values of horizontal diffusion employed. Even when the resolution of the simulation is increased (with numerical or explicit diffusivities reduced), the problem persists because the horizontal scales of the flow structures are consequently also reduced, leaving the ratio of the advective to diffusive timescales little changed. It is important to note, that when the Prandtl number is explicitly increased (the thermal diffusivity is lowered while the viscous diffusion is held constant so that the flow structure sizes are comparable) either uniformly [32] or only in the horizontal direction [2], the convective amplitudes in the lower portions of the simulation domains are reduced, as expected for less re-stratification. While the Prandtl number based on molecular diffusivities in the Sun is very low, the values adopted in these experiments may capture some effects of small scale magnetic fields on turbulent diffusivities [19, 32]. Another possibility is that, because of the rapid change in the radiative opacity in the upper layers of the Sun, thermal structures (downflow plumes) generated in the photosphere are on scales much larger than the thermal diffusion scales just below, allowing the Sun to access the high Peclet number flow regime there. Thus, local area radiative magnetohydrodynamic simulations (and recent ambitious global models [20] which include an upper radiative boundary) may be able to reproduce granular flow structures in the photosphere even while they cannot achieve appropriately low thermal diffusion of those structures in the deeper layers below, as realized in the Sun.

5 Some Preliminary Checks

It is possible to check some aspects of this proposal: that the mean superadiabaticity of the solar convection zone is determined by the presence of granular downflows penetrating an otherwise adiabatically stratified domain, and that the depth to which

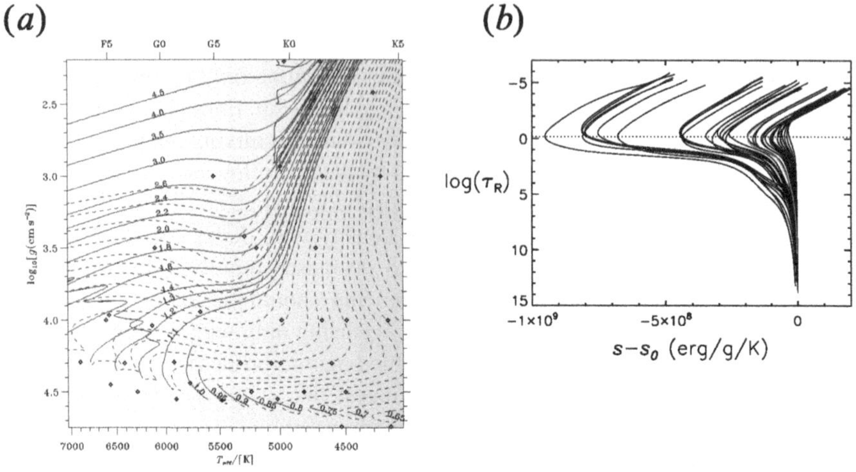

Fig. 3 Thirty-seven three-dimensional stellar surface convection models covering a range of atmospheric parameters from [45]. In (**a**), their positions on an effective temperature T_{eff}, surface gravity $\log(g)$ plot. In (**b**), their mean entropy deficit profiles as a function of optical depth τ. That for the Sun plotted in *green*

that layer extends determines the largest buoyantly driven scale and thus the peak in the horizontal spectrum (supergranulation) observed in the photosphere.

The first hypothesis, that the mean state is determined by the superposition of an isentropic background and adiabatically descending downflows (see [45] and references therein for previous formulations of this idea), can be tested in part using a series of three-dimensional radiative hydrodynamic models of the uppermost convective layers of stars. We [37] employ the 37 surface convection models of [45], which capture those layers for a range of main-sequence and more evolved stars, all with solar elemental abundance, which have been relaxed to statistically steady thermodynamic states. Across these layers of limited depth, the transit time of downflow plumes is shorter than the diffusion time across them. The position of the models on an effective-temperature surface-gravity, $T_{\text{eff}} - \log(g)$, plane (a Kiel diagram) are shown in Fig. 3a.

The mean entropy profiles in these models, relative to that of the upflows at the bottom of their domains, s_0, are plotted in Fig. 3b as a function of optical depth τ. For adiabatic motions, pure advection by downflows of the entropy deficit induced by surface cooling into an isentropic interior should result in mean entropy at each depth given by $\overline{s} = s_d f_d + (1 - f_d)s_{\text{ad}}$, where s_d and f_d are the filling factor and entropy of the downflows and s_{ad} is the entropy of the background. If the background is isentropically stratified, the mean state is completely determined by the filling factor and entropy fluctuations of the downflows in the stellar photospheres. In this case, s_d is taken as the downflow entropy measured at $\tau = 1$, s_{ad} is taken to be the entropy of the upflows at the bottom of the domain s_0, and the downflow filling factor with depth is modeled as $f_d = f_{d_{\tau=1}} \left(\overline{\rho}_{\tau=1}/\overline{\rho}(z)\right)^m$, where $f_{d_{\tau=1}}$ and $\overline{\rho}_{\tau=1}$

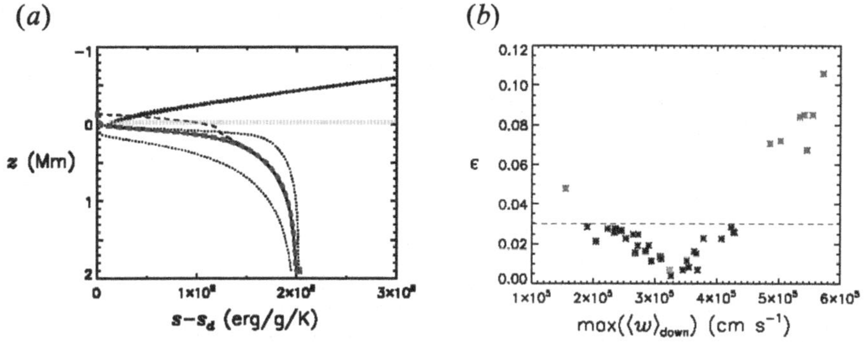

Fig. 4 In (**a**), solar entropy profile with depth derived from simple superposition of downflow plumes into an adiabatic background where the entropy and filling factor of the plumes are measured in the photosphere only (*blue dashed* curve). This is superimposed on mean entropy from radiative hydrodynamic simulation (*black solid*) curve. Horizontal line (*dotted cyan*) indicates position of photosphere. In (**b**), rms difference (as a fraction of s_0) between simplified plume models and simulation mean states. *Green* asterisk indicates solar model

are measured in the photosphere. Because only adiabatic mixing into an isentropic background occurs, $m = 1$; without diffusion, the mean entropy is the sum over isentropic upflows and downflows whether they have been mixed or not. The only correction that has to be made to this simple model, is for non-adiabaticity in the photosphere, where the transition from convective to radiative heat transport occurs. A simple correction can be made, $\overline{s} = \overline{s}\,(1 - F_{\rm rad}/F_*)$, where $F_{\rm rad}$ is the radiative flux at each depth and F_* is the stellar flux, equal to the maximum in $F_{\rm rad}$ in the model.

This simple model, based on the entropy and filling factor of the downflows measured in the simulation photosphere and the adiabatic descent of these through an otherwise isentropically stratified background, reproduces the mean state of the solar simulation well. Figure 4a shows the profile of the simple model (*blue dashed* curve) superimposed on the mean entropy profile in the simulation itself (*solid black*). The rms difference between the two curves below $\tau = 1$ (indicated by the horizontal *cyan* line) is only 0.68% of s_0. The model also works quite well (rms differences of $< 3\%$) for many of the other stellar surface layer models in the simulation suite. The differences show trends (Fig. 4b), trends however that are not fully understood. For low effective temperature stars (low downflow velocity red asterisk on Fig. 4b), the difference between the simple model and simulation mean state appears to be associated with both the radiative correction made (too large by the simple formulation) and the entropy fluctuations of the downflows (too warm in the photosphere). For low $\log(g)$ stars (high downflow velocity red asterisks on Fig. 4b) the downflow plumes in the simulations are either cooler or have higher filling factor in the photosphere than the simple model needs to reproduce the mean profile at depth. Understanding the physical origins underlying these trends is the focus of current research.

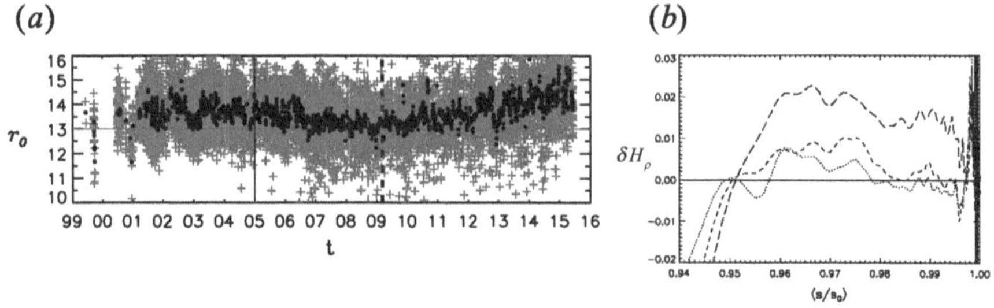

Fig. 5 In (**a**), change in supergranular radius with time, measured using Ca II K images from the Precision Solar Photometric Telescope (data and methods described in [13]). In (**b**), change in the density scale height (referenced to a fluctuation (local) dynamo only simulation) in the upper superadiabatic region of solar surface layer simulations differing in mean magnetic flux density: 30G (*dotted*), 60G (*dashed*), and 120G (*long dashed*)

Observational support for the second hypothesis, that the depth to which the upper superadiabatic layer extends determines the largest buoyantly driven scale and thus the peak in the photospheric horizontal spectrum (supergranulation), is found in the curious change in the supergranular scale with solar cycle. Unlike expectations based on magneto-convection simulations, supergranules are larger when the Sun is more active ([26], Fig. 5a). This temporal variation is consistent with changes in the depth of the upper superadiabatic layer seen in simulations with differing mean magnetic field. Radiative magnetohydrodynamic simulations of photospheric convection show the superadiabatic layer moving to greater depth with increased mean magnetic field strength [35]. As seen in Fig. 5b, solutions computed in $24^2 \times 7.5$ Mm domains, extending from 6 Mm below to 1.5 Mm above $\tau = 1$, and with varying degree of mean magnetic flux density (flux balanced local dynamo, 30, 60, and 120G mean field) show that equivalent values of mean entropy occur at higher values of $4H_\rho$ as the mean flux density increases. If a similar adjustment occurs in the Sun in response to increased magnetism, slightly larger scales would be preferred when it is more active. Understanding the physical origins of these changes in the mean entropy profile is the focus of current research.

6 Conclusion

This paper has explored our current understanding of solar convection, the convective conundrum and its possible solution. The steep stratification of the Sun requires that only a tiny fraction of fluid upwelling from the deep layers of the convection zone makes it to the photosphere. Given the enormous heat capacity of the material at the base of the convection zone, the enthalpy fluctuations required to convectively carry the solar luminosity there are very small and, integrated over the material that makes it into the photosphere, insufficient to support radiative losses. There are at least two possible ways to overcome this: (1) Continuous re-stratification of the domain by diffusive mixing of hot and cold plasma that leads to a mixing like hierarchy of scales at each depth so that the upflow enthalpy continuously changes with height and the upflows play a role in heat transport. This is the solution numerical simulations achieve. Or, (2) Adiabatic translation of the photospherically cooled downflowing material to the bottom of the convection zone with the heat flux maintained by this highly nonlocal transport. The mixing length like solution achieved by all numerical simulations leads to large scale horizontal flow amplitudes in the photosphere that exceed those observed. Strictly nonlocal transport by downflows is currently hinted at, but unachieved by numerical simulations, likely because of excess horizontal diffusion (non-adiabatic heating) of the plumes with depth.

It is very difficult for stellar convection models to achieve the effective Peclet number of flows in solar convection zone below the photosphere, very short advective to diffusive timescales. One possible way may be to employ a high horizontal Prandtl number in the surface layers where cooling is important, so that the downflow thermal structures are embedded in a larger scale flow structures. This approach may be difficult to calibrate for stellar modeling. Another possibility may be to use an explicitly two-fluid or Lagrangian tracer modeling approach [38] that separates downflows and upflows and minimizes diffusion between them. It may still be difficult with this approach to reach the low horizontal diffusion rates needed.

Acknowledgments This work was partially supported by National Science Foundation grant No. 1616538.

References

1. Anders, E. H., Lecoanet, D., & Brown, B. P. (2019). Entropy rain: Dilution and compression of thermals in stratified domains. *The Astrophysical Journal, 884*(1), 65. https://doi.org/10.3847/1538-4357/ab3644, arXiv:1906.02342.
2. Bekki, Y., Hotta, H., & Yokoyama, T. (2017). Convective velocity suppression via the enhancement of the subadiabatic layer: Role of the effective Prandtl number. *The Astrophysical Journal, 851*(2), 74. https://doi.org/10.3847/1538-4357/aa9b7f, arXiv:1711.05960.
3. Berrilli, F., Del Moro, D., Russo, S., Consolini, G., & Straus, T. (2005). Spatial clustering of photospheric structures. *The Astrophysical Journal, 632*(1), 677–683. https://doi.org/10.1086/432708
4. Berrilli, F., Scardigli, S., & Giordano, S. (2013). Multiscale magnetic underdense regions on the solar surface: Granular and mesogranular scales. *Solar Physics, 282*(2), 379–387. https://doi.org/10.1007/s11207-012-0179-2, arXiv:1208.2669.
5. Brandenburg, A. (2016). Stellar mixing length theory with entropy rain. *The Astrophysical Journal, 832*(1), 6. https://doi.org/10.3847/0004-637X/832/1/6, arXiv:1504.03189.
6. Cattaneo, F., Lenz, D., & Weiss, N. (2001). On the origin of the solar mesogranulation. *The Astrophysical Journal Letters, 563*(1), L91–L94. https://doi.org/10.1086/338355
7. Clyne, J., Mininni, P., Norton, A., & Rast, M. (2007). Interactive desktop analysis of high resolution simulations: Application to turbulent plume dynamics and current sheet formation. *New Journal of Physics, 9*(8), 301. https://doi.org/10.1088/1367-2630/9/8/301

8. Cossette, J. F., & Rast, M. P. (2016). Supergranulation as the largest Buoyantly driven convective scale of the Sun. *The Astrophysical Journal Letters, 829*(1), L17. https://doi.org/10.3847/2041-8205/829/1/L17, arXiv:1606.04041.
9. Duvall, J. T. L., & Birch, A. C. (2010). The vertical component of the supergranular motion. *The Astrophysical Journal Letters, 725*(1), L47–L51. https://doi.org/10.1088/2041-8205/725/1/L47
10. Elliott, J. R. (2000). Three-dimensional numerical simulations of compressible solar convection in Cartesian geometry. *The Astrophysical Journal, 539*(1), 469–479. https://doi.org/10.1086/309215
11. Featherstone, N. A., & Hindman, B. W. (2016). The emergence of solar supergranulation as a natural consequence of rotationally constrained interior convection. *The Astrophysical Journal Letters, 830*(1), L15. https://doi.org/10.3847/2041-8205/830/1/L15, arXiv:1609.05153.
12. Gastine, T., Yadav, R. K., Morin, J., Reiners, A., & Wicht, J. (2014). From solar-like to antisolar differential rotation in cool stars. *Monthly Notices of the Royal Astronomical Society, 438*(1), L76–L80. https://doi.org/10.1093/mnrasl/slt162, arXiv:1311.3047.
13. Goldbaum, N., Rast, M. P., Ermolli, I., Sands, J. S., & Berrilli, F. (2009). The intensity profile of the solar supergranulation. *The Astrophysical Journal, 707*(1), 67–73. https://doi.org/10.1088/0004-637X/707/1/67, arXiv:0909.3310.
14. Hart, A. B. (1954). Motions in the Sun at the photospheric level. IV. The equatorial rotation and possible velocity fields in the photosphere. *Monthly Notices of the Royal Astronomical Society, 114*, 17. https://doi.org/10.1093/mnras/114.1.17
15. Hathaway, D. H., Upton, L., & Colegrove, O. (2013). Giant convection cells found on the Sun. *Science, 342*(6163), 1217–1219. https://doi.org/10.1126/science.1244682
16. Hathaway, D. H., Teil, T., Norton, A. A., & Kitiashvili, I. (2015). The Sun's photospheric convection spectrum. *The Astrophysical Journal, 811*(2), 105. https://doi.org/10.1088/0004-637X/811/2/105, arXiv:1508.03022.
17. Herschel, W. (1801). Observations tending to investigate the nature of the Sun, in order to find the causes or symptoms of its variable emission of light and heat; with remarks on the use that may possibly be drawn from solar observations. *Philosophical Transactions of the Royal Society of London Series I, 91*, 265–318.
18. Hotta, H., Rempel, M., & Yokoyama, T. (2014). High-resolution calculations of the solar global convection with the reduced speed of sound technique. I. The structure of the convection and the magnetic field without the rotation. *The Astrophysical Journal, 786*(1), 24. https://doi.org/10.1088/0004-637X/786/1/24, arXiv:1402.5008.
19. Hotta, H., Rempel, M., & Yokoyama, T. (2015). Efficient small-scale dynamo in the solar convection zone. *The Astrophysical Journal, 803*(1), 42. https://doi.org/10.1088/0004-637X/803/1/42, arXiv:1502.03846.
20. Hotta, H., Iijima, H., & Kusano, K. (2019). Weak influence of near-surface layer on solar deep convection zone revealed by comprehensive simulation from base to surface. *Science Advances, 5*(1), 2307.
21. Hurlburt, N. E., & Weiss, N. O. (2000). The influence of internal heating on nonlinear compressible convection. In AAS/Solar Physics Division Meeting #31, AAS/Solar Physics Division Meeting (Vol. 31, p 05.06).
22. Leighton, R. B., Noyes, R. W., & Simon, G. W. (1962). Velocity fields in the solar atmosphere. I. Preliminary report. *The Astrophysical Journal, 135*, 474. https://doi.org/10.1086/147285
23. Leitzinger, M., Brandt, P. N., Hanslmeier, A., Pötzi, W., & Hirzberger, J. (2005). Dynamics of solar mesogranulation. *Astronomy & Astrophysics, 444*(1), 245–255. https://doi.org/10.1051/0004-6361:20053152
24. Lord, J. W. (2014). *Deep Convection, Magnetism and Solar Supergranulation*. PhD thesis, University of Colorado at Boulder.
25. Lord, J. W., Cameron, R. H., Rast, M. P., Rempel, M., & Roudier, T. (2014). The role of subsurface flows in solar surface convection: Modeling the spectrum of supergranular and larger scale flows. *The Astrophysical Journal, 793*(1), 24. https://doi.org/10.1088/0004-637X/793/1/24, arXiv:1407.2209.

26. McIntosh, S. W., Leamon, R. J., Hock, R. A., Rast, M. P., & Ulrich, R. K. (2011). Observing evolution in the supergranular network length scale during periods of low solar activity. *The Astrophysical Journal Letters, 730*(1), L3. https://doi.org/10.1088/2041-8205/730/1/L3, arXiv:1102.0303.
27. Miesch, M. S. (2005). Large-scale dynamics of the convection zone and tachocline. *Living Reviews in Solar Physics, 2*(1), 1. https://doi.org/10.12942/lrsp-2005-1
28. Nordlund, A. (1982). Numerical simulations of the solar granulation. I. Basic equations and methods. *Astronomy & Astrophysics, 107*, 1–10.
29. Nordlund, Å., Stein, R. F., & Asplund, M. (2009). Solar surface convection. *Living Reviews in Solar Physics, 6*(1), 2. https://doi.org/10.12942/lrsp-2009-2
30. November, L. J. (1989). The vertical component of the supergranular convection. *The Astrophysical Journal, 344*, 494. https://doi.org/10.1086/167818
31. November, L. J., Toomre, J., Gebbie, K. B., & Simon, G. W. (1981). The detection of mesogranulation on the Sun. *The Astrophysical Journal Letters, 245*, L123–L126. https://doi.org/10.1086/183539
32. O'Mara, B., Miesch, M. S., Featherstone, N. A., & Augustson, K. C. (2016). Velocity amplitudes in global convection simulations: The role of the Prandtl number and near-surface driving. *Advances in Space Research, 58*(8), 1475–1489. https://doi.org/10.1016/j.asr.2016.03.038, arXiv:1603.06107.
33. Rast, M. P. (1998). Compressible plume dynamics and stability. *Journal of Fluid Mechanics, 369*(1), 125–149. https://doi.org/10.1017/S0022112098001736
34. Rast, M. P. (2003). The scales of granulation, mesogranulation, and supergranulation. *The Astrophysical Journal, 597*(2), 1200–1210. https://doi.org/10.1086/381221
35. Rast, M. P., & Rempel, M. (2020). In preparation.
36. Rast, M. P., & Toomre, J. (1993). Compressible convection with ionization. I. Stability, flow asymmetries, and energy transport. *The Astrophysical Journal, 419*, 224. https://doi.org/10.1086/173477
37. Rast, M. P., & Trampedach, R. (2020). In preparation.
38. Rempel, M. (2020). Personal communication.
39. Rieutord, M., & Zahn, J. P. (1995). Turbulent plumes in stellar convective envelopes. *Astronomy & Astrophysics, 296*, 127.
40. Rieutord, M., Roudier, T., Malherbe, J. M., & Rincon, F. (2000). On mesogranulation, network formation and supergranulation. *Astronomy & Astrophysics, 357*, 1063–1072.
41. Roudier, T., Rieutord, M., Malherbe, J. M., Renon, N., Berger, T., Frank, Z., et al. (2012) Quasi full-disk maps of solar horizontal velocities using SDO/HMI data. *Astronomy & Astrophysics, 540*, A88. https://doi.org/10.1051/0004-6361/201118678, arXiv:1203.0514.
42. Simon, G. W., & Leighton, R. B. (1964). Velocity fields in the solar atmosphere. III. Large-scale motions, the chromospheric network, and magnetic fields. *The Astrophysical Journal, 140*, 1120. https://doi.org/10.1086/148010
43. Spruit, H. C. (1997). Convection in stellar envelopes: A changing paradigm. *Memorie della Societa Astronomica Italiana, 68*, 397–413. arXiv:astro-ph/9605020.
44. Stein, R. F., Lagerfjård, A., Nordlund, Å., Georgobiani, D., Benson, D., & Schaffenberger, W. (2009). Supergranulation scale convection simulations. In *Astronomical Society of the Pacific Conference Series* (Vol. 415, p. 63).
45. Trampedach, R., Asplund, M., Collet, R., Nordlund, Å., & Stein, R. F. (2013). A grid of three-dimensional stellar atmosphere models of solar metallicity. I. General properties, granulation, and atmospheric expansion. *The Astrophysical Journal, 769*(1), 18. https://doi.org/10.1088/0004-637X/769/1/18, arXiv:1303.1780

G Modes and the Solar Core

Yvonne Elsworth

Abstract It has always been a goal of Helioseismology to produce detailed maps of the deep interior of the Sun. Here I report on previous efforts to detect the g modes that would unlock the secrets of the solar core. Currently, there is limited success but efforts to detect them are on-going.

1 Solar Oscillations

In order to make observations of the interior of a star it is necessary to have a probe that is sensitive to the conditions in the interior. Global oscillations of the star, in this case the Sun, provide such a probe. These oscillations are sensitive to the sound speed of the wave as it passes through the Sun. What is actually measured is the frequency of the mode of oscillation which is due to an average transit time over the whole region of the Sun through which the wave has passed. Thus no one mode can provide localised information but the combination of many modes can. Michael Thompson was a master at extracting the localised information from a set of mode frequencies.

There are two classes of oscillations that one can expect to detect. They are pressure modes and gravity modes which respond to different physical phenomena. Pressure modes (p modes) have pressure gradients as the restoring force whereas gravity modes (g modes) have buoyancy as the restoring force. Pressure modes can travel through any region where the general conditions for oscillations are met. This is not to say that all p modes exist everywhere in the Sun. Depending on the parameters of the oscillation mode, there is localisation. The so-called high-degree modes are found in the outer regions of the Sun and only the very low-degree modes can penetrate through to the centre. The sensitivity of a p mode to conditions at any

Y. Elsworth (✉)
School of Physics & Astronomy, University of Birmingham, Birmingham, UK
e-mail: y.p.elsworth@bham.ac.uk

M. J. P. F. G. Monteiro et al. (eds.), *Dynamics of the Sun and Stars*, Astrophysics and Space Science Proceedings 57,
https://doi.org/10.1007/978-3-030-55336-4_24

point in the Sun is dependent on the time it takes the wave to pass through the region. Because the speed of sound is higher where the density is higher, p-mode waves pass relatively quickly through the central regions and spend rather longer in the outer regions. The fact that different p modes are sensitive to different regions in the Sun allows the interior of the Sun to be mapped using inversion techniques such as those pioneered by Michael Thompson. However, the short dwell time in the centre limits the precision available about the conditions there. Furthermore, there are many fewer modes that can penetrate to the core than are available for the outer layers. This also limits the precision available.

The solution to the challenge of getting high precision to the parameters of the solar core is to use modes that are most strongly localised in the core. This would be provided by g modes. But there is a problem. Because these modes have buoyancy as their restoring force, they cannot exist in a convective region. All solar-like oscillators are driven by convection in their outer regions and this convection zone effectively blocks the g-mode oscillations. The g modes that exist in the centre of the Sun are evanescent in the convection zone. Their amplitude is exponentially attenuated through the convection zone and hence the remaining amplitude at the surface is very small. There is a process whereby they can, in principle, be detected at the solar surface which is if the g mode couples to a p mode giving a so-called mixed mode. Nevertheless, the amplitudes expected for the mode are very small and the frequencies are low, much lower than those at which the strongest p modes are seen. It is the search for these modes that forms the subject of this paper.

2 Mixed Modes Detected in Other Stars

Mixed modes and g modes have been detected in stars that are different from the Sun. Here we note just two of these types of stars. In both cases the observations of the modes had huge impact on the subject.

2.1 g-Mode Oscillations in Compact Objects

White dwarf stars are the natural end state for low and medium mass stars. When found in binary systems they can be the source of supernova events. Both p-mode and g-mode oscillations can in principle be observed in white dwarf stars but the dominant observations are for the g modes. The p-mode periodicities are expected to be very short—time periods of around a few seconds. A very extensive review of this class of objects is to be found in Winget and Kepler [17]. Many peaks that can be associated with g modes have been found. A typical example is in the spectrum of GD 358 which is a DB star with a Helium atmosphere. In this case, the frequencies are around 1.5 mHz and strong changes in the mode amplitudes are seen on timescales of weeks and months. Not all the observed modes are real, independent modes as many hundreds of linear combination frequencies are seen in the spectrum. The wide range of data has solved many problems and opened up many more as increasing levels of sophistication in theory is applied.

2.2 *Mixed-Mode Oscillations in Red Giant Stars*

Because of the large radius of these stars and the consequently long sound travel times across the star, the typical p-mode periods are in the region of hours. This made ground-based observations very difficult. The advent of data from the CoRoT and *Kepler* satellites, made observations of oscillation in red-giant stars a practical option. The satellites also provide long-duration datasets lasting a few months for CoRoT and up to 4 years for *Kepler*. With the long duration comes frequency resolution and the signal-to-noise to detect faint, coherent oscillations. An extensive review of the seismology of red-giant stars is to be found in Hekker and Christensen-Dalsgaard [11]. In this case, the central density in the star is sufficiently high that there is strong coupling between the p modes and the g modes at frequencies where the p modes are strongly excited by the outer convection zone leading to mixed modes. Consequently, the mixed modes are visible in the acoustic spectrum and are located between the strong p modes. Observations of these modes allowed inferences to be made on the interior rotation rate and the evolutionary state of the star—has core-Helium burning commenced. These observations have revolutionised the field.

3 The Search for g Modes in the Sun

The observation of g modes or mixed modes in stars have greatly increased the knowledge obtained about the stellar interiors. It is not surprising therefore that there have been many attempts to make similar observations for the Sun. As indicated previously, the pure g modes are trapped in the radiative interior of the Sun and are evanescent in the convection zone and consequently have very small amplitudes at the surface. Mode coupling in the region of the strongest p modes (frequencies around 3 mHz) is not a consideration because the strongest g modes are at much lower frequencies. Thus any search for g modes has to concentrate on finding weak, coherent modes at low frequencies where the instrumental and solar noise contributions to the acoustic spectrum are significant. Just because the problem is difficult has been no reason not to try.

3.1 Early Observations

In the early days of helioseismology there were several detections of what might have been g-mode oscillations of the Sun. Two observations, using different methods, were published in the same issue of *Nature* in 1976. Severny et al. [15] observed the difference between the radial velocity of a central portion and a circular annulus of the solar disk, and Brookes et al. [4] looked for tiny variations in the Doppler velocity of the Sun. They both found a signal with a period of 160 min and an amplitude of a few metres per second. It was later shown by Brown et al. [5] that the two observations had consistent phase. These were soon followed up by further investigations by Scherrer et al. [13] and Delache and Scherrer [6].

However, it became difficult to understand these observations as g-mode oscillations and the current thinking is that the observed structure is an artefact of the ground-based measurement and is actually the 9th harmonic of the day [8]. It is still possible, that they were genuine features on the Sun that are not visible today. Astronomy is an observational and not experimental science. However, there was one really positive outcome to these publications which was that the Global Oscillation at Low Frequency (GOLF) instrument on the SoHo satellite was proposed and selected to find g modes by observations from the tranquility of space[7].

3.2 Upper Limits on g-Mode Detection

In the late 1990s, an international collaboration was assembled with the aim of using the full range of available oscillations data on the Sun, and guided by theoretical expectations, to either find the elusive g modes or, at the minimum, set an upper limit to their amplitude. A report from this *Phoebus* collaboration[2] indicated that they could not detect any g-mode signal and put an upper limit to the level of any such undetected signal. The statement was that at 200 μHz the signal could not be larger than 10 mm s^{-1} in velocity or equivalently it must be less than 0.5 ppm in intensity. These values were consistent with theoretical limits. After the GOLF team joined the collaboration, there were renewed efforts to detect individual g modes still with negative conclusions[3]. After the work of 2000 which had been aimed at detecting single g modes, the approach changed to one of looking for groups of modes with a view to improving the statistics. There have been several claimed detections.

3.3 Needles in the Haystack

Turck-Chièze and collaborators[16] used 1290 days of GOLF data in the frequency range above 150 μHz to look for multiplets rather than single modes. They used frequentist methods to set thresholds for detections and set a confidence limit of 90%

that what was seen was not due to noise. Under certain theoretical assumptions, it is possible to predict the separation of the modes in these multiplets. Their conclusion was that they had detections of 4 multiplets that were potentially of interest. They were cautious about claiming them as definitive detections but did suggest that if they were genuine then perhaps the rotation rate at the centre of the Sun was very different from that in the radiative zone, or perhaps there were significant magnetic fields. Inspection of the spectrum suggests that the authors were right to be cautious although there remains the possibility that the detections are genuine.

3.4 Search for Repeated Structure

A different approach was taken by García et al. in 2007[10]. It is known that the g modes should be equally spaced in period (not frequency). Hence, if the spectrum is expressed as a function of period rather than frequency and then Fourier transformed, the resulting second spectrum should show peaks in power for any regular structure in the original spectrum.

They used almost 10 years of velocity observations of GOLF data (11 April 1996 to 21 October 2005) to form such a second spectrum and found a significant feature at a period of 24 min. Various statistical tests appeared to confirm the robustness of the detection. However, if this peak is to be associated with asymptotic period spacing of the dipole modes, then it assumes that the inner core of the Sun is rotating 3 to 5 times faster than the regions a bit further out. The results also require that the lifetime of the g modes is not long as is theoretically expected but is considerably shorter. For these reasons, although the detection is very robust there is some hesitation in associating the peak with solar g modes.

3.5 Influence on Other Parameters

A recent approach to the problem has been to look at the impact of any g mode on other features within the acoustic spectrum. The feature targeted is the so-called large spacing. This is the separation in frequency between successive modes of the same degree but with an order different by unity. The large separation is a measure of the sound travel time across the Sun. Fossat and co-workers[9] stated the expectation that g modes would cause periodic changes in the deep solar structure and that this in turn would impose regular changes on the large frequency separation. No individual g mode would produce a measurable effect, but the expectation is that more than 100 g modes of degree 2 would produce a detectable effect. The GOLF data were split up into short segments of 8 h which were Fourier analysed to find the location in frequency of the peak due to the large spacing. In fact, due to the odd/even mode locations, the peak is found at half the large spacing. They then look for regular structure in the time series formed by the location of this peak. The conclusions that they draw from the analysis is that they have detected the presence of g modes and that the data imply a core that rotates nearly 4 times faster than the mean in the radiative zone. They also claim to be able to identify individual g modes in the original spectrum.

This, if true is an exciting result. However, several groups [14, 1] have tried to reproduce the results and have failed. Furthermore, not only did Scherrer and Gough [12] fail to reproduce the results, they also showed that, theoretically, Fossat et al. should not have been able to detect the dipole modes that they claim.

4 What Next

Although an uncontroversial detection of solar g modes remains elusive, the work published since the first detections has not been in vain. There is no doubt that there have been improvements in both the way of analysing the data and also in the instrumentation to collect the data. These changes have led to an increasing number of low-degree, low-order p modes being detected. These are very useful to constrain the models of the internal structure of the Sun. They have also raised questions about the nature of the modes. Questions like, are they really very lightly damped, what are the right predictions for the frequencies of the modes are very important. To try to answer these and other questions, the Phoebus group has reformed.

It is also useful to remember that until a new observational window has been opened, the feeling can be that it is all too difficult. Nevertheless, it is rather sad that Michael Thompson did not live to apply his beautiful inversion techniques to robust detections of solar g modes.

References

1. Appourchaux, T., & Corbard, T. (2019). *Astronomy & Astrophysics, 624*, 106A.
2. Appourchaux, T., Fröhlich, C., Andersen, B., Berthomieu, G., Chaplin, W. J., Elsworth, Y., et al. (2000). *The Astrophysical Journal, 538*, 401A.
3. Appourchaux, T., Belkacem, K., Broomhall, A.-M., Chaplin, W. J., Gough, D. O., Houdek, G., et al. (2010). *The Astronomy and Astrophysics Review, 18*, 197.
4. Brookes, J. R., Isaak, G. R., & van der Raay, H. B. (1976). *Nature, 259*, 92.
5. Brown, T. M., Stebbins, R. T., & Hill, H. A. (1978). *The Astrophysical Journal, 223*, 324.
6. Delache, P., & Scherrer, P. H. (1983). *Nature, 306*, 651.
7. Domingo, V., Fleck, B., & Poland, A. I. (1995). *Space Science Reviews, 72*, 81.
8. Elsworth, Y. P., Jefferies, S. M., McLeod, C. P., New, R., Palle, P. L., van der Raay, H. B., et al. (1989). *The Astrophysical Journal, 338*, 557.
9. Fossat, E., Boumier, P., Corbard, T., Provost, J., Salabert, D., Schmider, F. X., et al. (2017). *Astronomy & Astrophysics, 604*, A40.
10. García, R. A., Turck-Chièze, S., Jiménez-Reyes, S. J., Ballot, J., Pallé, P.L., Eff-Darwich, A., et al. (2007). Science, 316, 1591.
11. Hekker, S., & Christensen-Dalsgaard, J. (2017). *The Astronomy and Astrophysics Review, 25*, 1.

12. Scherrer, P. H., & Gough, D. O. (2019). *The Astrophysical Journal, 877*, 42.
13. Scherrer, P. H., Wilcox, J. M., Kotov V. A., Severny, A. B., & Tsap, T. T. (1979). *Nature, 277*, 635.
14. Schunker, H., Schou, J., Gaulme, P., & Gizon, L. (2018). *Solar Physics, 293*, 95S.
15. Severny, A. B., Kotov, V. A., & Tsap, T. T. (1976). *Nature, 259*, 87.
16. Turck-Chièze, S., García, R. A., Couvidat, S., Ulrich, R.K., Bertello, L., Varadi, F., et al. (2004). *The Astrophysical Journal, 604*, 455.
17. Winget, D. E., & Kepler, S. O. (2008). *Annual Review of Astronomy and Astrophysics, 46*, 157.

Inverse Analysis of Asteroseismic Data: A Review

Earl P. Bellinger, Sarbani Basu, and Saskia Hekker

Abstract Asteroseismology has emerged as the best way to characterize the global and internal properties of nearby stars. Often, this characterization is achieved by fitting stellar evolution models to asteroseismic observations. The star under investigation is then assumed to have the properties of the best-fitting model, such as its age. However, the models do not fit the observations perfectly. This is due to incorrect or missing physics in stellar evolution calculations, resulting in predicted stellar structures that are discrepant with reality. Through an inverse analysis of the asteroseismic data, it is possible to go further than fitting stellar models, and instead infer details about the actual internal structure of the star at some locations in its interior. Comparing theoretical and observed stellar structures then enables the determination of the locations where the stellar models have discrepant structure, and illuminates a path for improvements to our understanding of stellar evolution. In this invited review, we describe the methods of asteroseismic inversions, and outline the progress that is being made towards measuring the interiors of stars.

E. P. Bellinger (✉)
Stellar Astrophysics Centre, Department of Physics and Astronomy, Aarhus University, Aarhus, Denmark

School of Physics, UNSW Sydney, Sydney, NSW, Australia
e-mail: bellinger@phys.au.dk

S. Basu
Department of Astronomy, Yale University, New Haven, CT, USA
e-mail: sarbani.basu@yale.edu

S. Hekker
Max Planck Institute for Solar System Research, Göttingen, Germany

Stellar Astrophysics Centre, Department of Physics and Astronomy, Aarhus University, Aarhus, Denmark
e-mail: hekker@mps.mpg.de

M. J. P. F. G. Monteiro et al. (eds.), *Dynamics of the Sun and Stars*, Astrophysics and Space Science Proceedings 57,
https://doi.org/10.1007/978-3-030-55336-4_25

1 Introduction

In stars that are like the Sun, turbulent motions near to the stellar surface randomly excite acoustic modes that oscillate throughout the entire star. The oscillation frequencies of these modes depend on the star's internal chemical composition and structure, and particularly on the speeds at which sound travels through the star. The sound speed at any given location in the stellar interior is dictated by the local properties of the stellar plasma, and specifically its density, pressure, and compressibility. Detection of a sufficient number of these acoustic oscillation modes thus permits inferences of the sound speed—and hence the structure—at various locations within the star. As the structure of stars is the principal prediction made by the theory of stellar evolution, asteroseismic measurements of stellar interiors can therefore be used to test the theory. This analysis is the topic of the present review.

This is the first review on asteroseismic structure inversions in the post-*Kepler* era (for earlier reviews, see [12, 43]). Here we focus on techniques that have been developed and applied to actual stellar data. This review is divided into several, mostly independent sections. Section 2 gives a brief mathematical introduction to inverse problems and defines the scope of the review. Section 3 describes how to obtain a reference model through an evolutionary inversion, which is generally done before performing a structure inversion. Section 4 states the structure inversion problem and describes the difficulties and limitations imposed by asteroseismic data. Finally, Sect. 5 summarizes current results and gives an outlook on future measurements of stellar interiors.

For a general introduction to asteroseismology, we refer the reader to the two textbooks and numerous review articles on the topic (e.g., [2, 14, 41, 44, 56, 40, 21, 58, 57, 50, 1]).

2 A Primer on Inverse Problems

Finish the sequence: 1, 2, 3, 4, _.

This task cannot be done with certainty. Although the answer would naturally seem to be 5, other solutions are also possible, such as:

- 6 (a sequence known as Ulam's numbers [80])
- 7 (numbers n such that $(68 \cdot 10n + 7)/3$ is prime)
- 10 (counting in base 5)
- 17 (n such that $n \cdot \varphi(n)$ is a palindrome, where φ is Euler's totient function)
- -42π (a sequence we have just invented).

Indeed, an infinity of solutions exist, each with infinitely many possible justifications.

This is often the situation with inverse problems. Broadly speaking, inverse problems can be posed for every data-generation scheme, such as those listed above.

The *forward problem* is to generate data using the scheme (for example, listing the natural numbers). The *inverse problem* is to determine the generating scheme itself from the data, or to estimate the parameters of an assumed scheme.

Inverse problems are characteristically, though not always, *ill-posed*. A well-posed problem satisfies three requirements [55]:

1. **Existence**: a solution exists,
2. **Uniqueness**: there is exactly one solution, and
3. **Stability**: the solution changes continuously with changes to the input.[1]

Ill-posed problems—problems that violate any of these conditions—tend to be difficult or impossible to solve. In order to obtain a solution, assumptions must be made, or the problem must be altered. An example of a well-posed inverse problem is to find the line corresponding to a few points belonging to that line. An ill-posed inverse problem would be to find a line after any of the points have been perturbed by noise. As no line will pass through all of the points, an assumption must be made, such as assuming the noise is Gaussian-distributed and accepting the least-squares solution.

For more information, the reader is referred to textbooks and review articles on inverse problems (e.g., [61, 65, 77]).

2.1 *Inverse Problems in Asteroseismology*

There are numerous inverse problems in asteroseismology. These include inversions to infer a star's:

1. **Evolutionary history**: age, mass, initial composition, and evolutionary parameters such as mixing efficiencies, assuming the theory of stellar evolution is approximately correct (e.g.,[69, 26, 27, 29, 31, 6, 7, 59, 35, 54, 64, 76, 4, 73, 25, 60, 3, 23, 24]).
2. **Internal structure**: specifically, the isothermal sound speed at various locations around the stellar core [28, 30].

---

[1]Consider the system of linear equations

$$x = a + b \tag{1}$$

$$y = a + b \cdot (1 + \epsilon) \tag{2}$$

where ϵ is an arbitrarily small positive quantity and the rest are real numbers. One can imagine that x, y are values observed in nature (i.e., they are the data) and a, b are the hidden parameters that we wish to infer (i.e., they form the data-generation scheme). If we observe that $(x, y) = (2, 2)$ then clearly $(a, b) = (2, 0)$. Yet if $(x, y) = (2, 2 + \epsilon)$ then $(a, b) = (1, 1)$. An arbitrarily small change to the observed value of y has resulted in a completely different solution. The system is unstable.

3. **Radial differential rotation**: rates of rotation of the stellar core and surface (e.g.,[45, 46, 47, 32, 49, 5]).
4. **Latitudinal differential rotation**: surface rotation rates at different latitudes [33, 19, 20].
5. **Integrated quantities**: any quantity that takes its value over some or all of the stellar interior [67], such as the star's
   a. *Mean density*: $(3/4\pi)M/R^3$, where M is the mass of the star and R is its radius [72].
   b. *Acoustic radius*: $\int c^{-1}\,\mathrm{d}r$, where $c = (\Gamma_1 P/\rho)^{1/2}$ is the speed of sound, Γ_1 is the adiabatic compressibility, P is pressure, ρ is density, and r is the distance to the stellar center [37].
   c. *Sound speed gradient*: $\int r^{-1}(\mathrm{d}c/\mathrm{d}r)\,\mathrm{d}r$, which is a proxy for stellar age on the main sequence because the sound speed gradient depends on the mean molecular weight, which changes over time due to hydrogen fusion [38, 39].

These problems are ill-posed and therefore difficult to solve—some due to existence, some due to uniqueness, some due to stability, and some due to a combination of these. This review focuses on the first two of these problems, and especially the second. That being said, essentially all of these problems can be solved using the techniques discussed in this review.

3 Evolutionary Inversions

The first step for most inversions is to obtain a reference model. This is in fact itself an inverse problem. Given observations of a star, we seek to determine its age, internal structure, and other parameters using the theory of stellar evolution.

Stellar evolution theory can be regarded as the function $f_{\text{evol}}(\tau, \mathbf{x}) = \mathbf{y}$, where τ is the age of the star and $\mathbf{x} =$ [mass, chemical composition, etc.] are the initial conditions. Other parameters may include anything relevant to the evolution, such as the mixing length and overshooting parameters for convection, diffusion coefficients, nuclear reaction rates, opacities, stellar companions, and so on. The forward problem is to compute a stellar model[2] that has observables $\mathbf{y} =$ [effective temperature, metallicity, oscillation frequencies, etc.]. Other observables may include the radius, luminosity, and so on.

We seek to invert this function: given observations $\mathbf{y}$, we wish to infer $f_{\text{evol}}^{-1}(\mathbf{y}) = [\tau, \mathbf{x}]$. However, this inverse function is not guaranteed to exist because f_{evol} is not

[2]There are numerous software implementations for computing f_{evol}; one open-source and freely available one-dimensional stellar evolution code is MESA [66]. The global oscillation modes of the resulting stellar model is generally computed using a separate code. Two freely available codes for computing linear adiabatic stellar oscillation frequencies are ADIPLS [42] and GYRE [79].

necessarily injective. Broadly speaking, there are three approaches for solving the evolution inversion problem:

1. **Machine learning**: an approximation to the inverse function f_{evol}^{-1} is learned from a large number of examples of τ, $\mathbf{x}$, and $\mathbf{y}$ [69, 26, 27, 29, 31, 6, 7, 59].
2. **Optimization**: sometimes referred to as repeated forward modeling, a likelihood function that quantifies how well a given model fits the data is maximized, often along with priors on the parameters of the model (e.g., [35, 54, 64, 76, 4, 73, 25, 60, 3]).
3. **Scaling relations**: by assuming homology with the Sun (e.g., [57]), the age and mass of a star can be deduced using calibrated scaling relations [23, 24]. The chemical composition can be approximated using a galactic chemical evolution law (e.g., [4]). These parameters can then be used with f_{evol} to obtain a reference model.

In past decades, when observations consisted of only the effective temperature, metallicity and sometimes luminosity, this problem was ill-posed because the solution lacked uniqueness, i.e., many stellar models satisfied the same constraints. With the advent of asteroseismology, this problem is ill-posed because the solution lacks existence: there are no stellar models which fit all the asteroseismic data [28, 30]. This is the case even for the Sun and models of solar evolution (e.g., [13, 81]).

This motivates the structure inversion problem. If no stellar model has the correct structure, then what is the correct structure?

4 Sound Speed Inversions

The problem at hand is to deduce the internal structure of a star, insofar as it is possible, from asteroseismic measurements, without assuming the theory of stellar evolution. This problem is generally cast as one of deducing the differences in the internal sound speed profile between the star and a suitably chosen reference model, which can be achieved through an analysis of the differences in their frequencies of oscillation. This approach leads directly to an assessment of the quality of stellar evolution models, and particularly whether the models have the correct structure at the locations within the star where the inverse analysis is successful. This problem is not fundamentally different from the structure inversion problem in helioseismology (e.g., [53, 13]) and geoseismology [8, 9]. However, inverting asteroseismic data comes with unique difficulties and limitations over helioseismic inversions [12, 14].

The fact that the solar disk can be resolved permits the detection of solar oscillation modes with spherical degrees of $\ell \lessapprox 250$ [13]. In contrast, cancellation effects limit asteroseismic detections to modes with $\ell \leq 3$ (e.g., [2, 48, 76, 63]). Although all acoustic oscillation modes have upper turning points near to the stellar surface, only the high degree modes have lower turning points in the outer parts of the star (see Figure 4.4 of [28]). Thus, unlike the Sun (e.g., [18]), only the near-core region ($\sim 0.05 - 0.35R$ away from the center) can be probed with asteroseismic inversions.

The structure inversion problem is complicated by the fact that frequencies of stellar oscillations do not depend exclusively on the sound speed; the effects of a second variable, such as the density or adiabatic compressibility, must be considered as well (e.g., [2, Section 3.3.3.2]). Unfortunately, asteroseismic mode sets are too limited to resolve (or suppress) the effects of a second variable [14, Chapter 10]. A path forward was realized very early [52]. By assuming an equation of state, it is possible to explain oscillation frequencies in terms of the abundance of helium and any function of pressure and density, such as the squared isothermal sound speed $u = P/\rho$. The reason this is helpful is because the helium abundance only substantially affects the oscillation frequencies in ionization zones, which are located very near to the stellar surface where asteroseismic inversions are not sensitive anyway. Hence, asteroseismic data probe the u profile of the stellar core.

A further difficulty concerns the fundamental parameters of the star. The solar fundamental parameters, and particularly the solar mass and radius, are precise, accurate, and known independently of helioseismology. For stars, these parameters must be estimated, generally also from the same asteroseismic data with which one wishes to perform the inverse analysis. The relative uncertainties of asteroseismic mass and radius estimates are about 4% and 2%, respectively (e.g., [29]). Thus, the model we choose as reference may not have the same mass and radius as the star. This has two consequences. The first is in the mode frequencies themselves: the frequencies scale with the stellar root mean density, and so a difference there results in a constant relative difference between the frequencies of the star and of the stellar model. Secondly, a difference in mass or radius affects the u profile itself. From a dimensional analysis of the equations of stellar structure, one finds $P \propto M^2/R^4$ and $\rho \propto M/R^3$, and hence $u \propto M/R$. Thus, asteroseismic structure inversions are actually sensitive to the dimensionless structure $u' = uR/M$ [12, Figures 9 and 10]. Hereinafter, u' will be referred to simply as the sound speed.

A final complication is that theoretical and observed mode frequencies are well-known to systematically differ due to deficiencies in modeling the near-surface layers of stars (e.g., [10]). Generally, these differences are merely fit and subtracted. Though clearly not the best solution, it does not otherwise present a major hurdle here.

Once a suitable reference model is selected, one can perturb the equations of adiabatic stellar oscillations and linearize them around the reference model (e.g., [12, 2, 53, 62, 13, 14, 36, 22, 71]). From this, one obtains the central equations of the asteroseismic structure inversion—a set of equations, one for each observed oscillation mode, that express how the mode frequencies depend on the stellar structure and composition:

$$\underbrace{\frac{\delta\nu_i}{\nu_i}}_{\text{difference in oscillation frequency}} = \int \overbrace{K_i^{(u',Y)} \frac{\delta u'}{u'} + K_i^{(Y,u')} \delta Y}^{\text{differences in structure \& composition}} \,\mathrm{d}r + \underbrace{F_{\text{surf}}(\nu_i)}_{\text{surface term}} + \overbrace{\epsilon_i}^{\text{measurement errors}}, \qquad i = 1 \ldots N. \tag{3}$$

Here the index i refers to a mode whose frequency ν_i has been measured. The difference in any quantity between the star and the reference model is denoted by δ. The kernels K_i for a particular mode are functions derived from the perturbation analysis that relate u' and the fractional helium abundance Y to that mode's frequency (for examples, see [22, Figures 1.21 and 1.22]). The surface term F_{surf} is some function of frequency (e.g., [11]). The unknown measurement error, given by ϵ_i, is assumed to be drawn from normal distribution with zero mean and standard deviation σ_i.

The principal assumption of a structure inversion is that the mode kernels of the reference model serve as sufficient approximations to the actual mode kernels of the star. As mode kernels do not differ much between even rather different models [15], the inversion results can be considered independent of evolutionary considerations, despite usually using a model calculated using stellar evolution theory as reference.

The problem now at hand is to determine the $\delta u'/u'$ (or, equivalently, u') profile from the data using these N equations. This problem is ill-posed for two reasons. The first is uniqueness: infinitely many sound-speed profiles fit the data equally well (e.g., [53]). These include nonphysical solutions, such as those with negative sound speeds. The second is stability: as the modes trace similar paths through the star, the kernels are nearly linearly dependent, and so a small perturbation to the mode frequencies can lead to a disproportionate change in the result (recall Eqs. 1 and 2). Thus the problem must be changed. A way forward was discovered by geologists Backus and Gilbert [8, 9]: the method of *optimally localized averages* (OLA).

4.1 The Method of Optimally Localized Averages

OLA is a method that can be used to measure the sound speed at various locations within a star. The method works by combining the mode kernels belonging to the observed modes into a new kernel, called the *averaging kernel*, which by design is only sensitive to one region of the star, called the *target radius*. If the creation of an averaging kernel is successful, it can be used as an instrument to measure the sound speed at the target radius. However, the creation of an averaging kernel must be balanced against the amplification of the data errors and the contamination of the effects of a differing helium abundance. It is important to note that the OLA method does not involve fitting the data; in fact, the construction of the averaging kernel does not make use of the oscillation frequencies at all; it uses only their uncertainties.

The averaging kernel $\mathscr{K}$ is formed by taking a linear combination $\mathbf{c}$ of the mode kernels: $\mathscr{K} = \sum_i^N c_i K_i^{(u',Y)}$. If this linear combination is chosen such that $\mathscr{K}$ integrates to unity and only has amplitude around the target radius, then

$\int \mathscr{K}\,\delta u'/u'\,\mathrm{d}r$ provides a localized average, denoted $\langle \delta u'/u' \rangle$, of the sound speed at the target radius. Now consider a sum over all the equations in Eq. 3:

$$\underbrace{\left\langle \frac{\delta u'}{u'} \right\rangle}_{\text{localized average}} \simeq \sum_i^N c_i \overbrace{\frac{\delta \nu_i}{\nu_i}}^{\text{observed data}} = \underbrace{\int \mathscr{K}\,\frac{\delta u'}{u'}\,\mathrm{d}r}_{\text{averaging term}} + \int \overbrace{\mathscr{C}\,\delta Y}^{\text{cross term}}\,\mathrm{d}r + \sum_i^N \underbrace{c_i\,F_{\text{surf}}(\nu_i)}_{\text{surface effects}} + \sum_i^N \overbrace{c_i\,\epsilon_i}^{\text{errors}}. \tag{4}$$

The same linear combination that forms the averaging kernel is applied to all terms, such as the cross-term kernel $\mathscr{C} = \sum_i c_i K_i^{(Y,u')}$. Thus, $\mathbf{c}$ must be chosen in such a way that suppresses these other effects.

A common way of solving the OLA problem is known as Subtractive OLA (SOLA, [68]). This method defines a *target kernel*, which is typically a Gaussian function peaked at the target radius. Free parameters control the width of the target kernel, the suppression of the cross-term kernel, and the amplification of uncertainty (e.g., [70]). SOLA then finds the $\mathbf{c}$ that makes the averaging kernel resemble the target kernel, insofar as it is possible (e.g., [22]). Examples of averaging kernels constructed using SOLA with *Kepler* data [63] are given in Fig. 1. The final averaging kernel is reasonably well-localized and suffices to estimate $\delta u'/u'$ at the target radius.

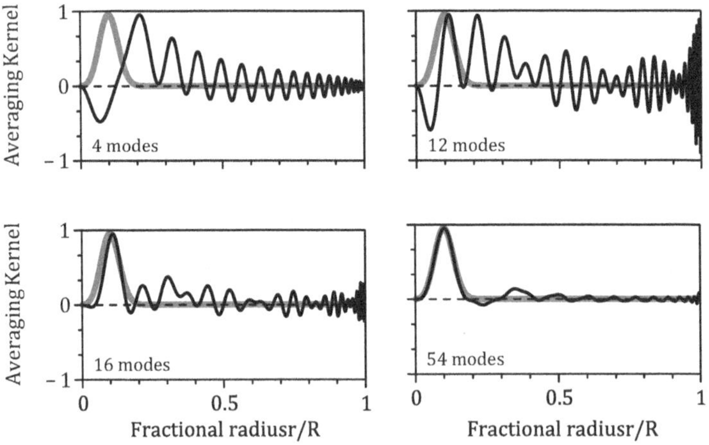

Fig. 1 Averaging kernels at a target radius of $0.1\,R$ constructed using four differently sized mode sets from the solar-like oscillator 16 Cyg A [28]. The target kernel is indicated by the gray line

5 Summary of Current Results and Future Outlook

Prior to the widespread detection of oscillations in solar-like stars by space missions such as *Kepler*, a number of early works investigated the prospect of asteroseismic inversions using faux data generated from stellar models [52, 51, 16, 34, 75, 17, 78, 74]. These works are summarized in the first review on asteroseismic inversions [12]. Generally, these early attempts were optimistic about the number of mode frequencies that would be observed and the uncertainties on those observations. Most critically, the assumed uncertainties on the mass and radius of the target star were generally too small.

To date, asteroseismic structure inversions have been applied to three actual stars: the solar analogs 16 Cyg A and B [28] and a star with a small convective core, KIC 6225718 [30]. In the former case, the dimensional u profile was sought by inverting many reference models spanning the uncertainties in mass and radius. A procedure called inversions-for-agreement was devised, which had the benefit of using the data to select the parameters of the inversion.

In each of the three cases, there was tension between the predicted and observed internal stellar structures. The model of the close solar analog 16 Cyg B performed the best, while the star with a convective core—the least solar-like star—showed the greatest differences. For this star, alternate models employing differing physics in the evolution of the star were calculated. The input physics of convective core overshoot and atomic diffusion were tested against the asteroseismic observations. However, no set of input physics was found that could explain the observed structure.

The field of asteroseismic inversions is still in its infancy with many interesting results yet to come. Several more stars from legacy *Kepler* data are still amenable to inverse analysis; this is work in preparation from the present authors. By analyzing a large number of stars, it will be possible to rigorously test theories of stellar evolution across a variety of non-solar regimes and pinpoint the areas of stellar modeling that are currently deficient. Future space missions will furthermore yield orders-of-magnitude more detections of solar-like oscillations, at which point inverse analyses will likely need to be applied at a large scale rather than star-by-star.

As a final note, the stars discussed here are all firmly on the main sequence. Structure inversions of evolved solar-like stars, whose non-radial acoustic modes couple with gravity modes in the core and give insights into the very deep interiors of stars, are sure to be a promising source of discovery.

Acknowledgments These proceedings are dedicated to the memory of Michael J. Thompson. Funding for the Stellar Astrophysics Centre is provided by The Danish National Research Foundation (Grant agreement no.: DNRF106). S.H. acknowledges funding from the European Research Council under the European Community's Seventh Framework Programme (FP7/2007-2013)/ERC grant agreement no 338251 (StellarAges). S.B. acknowledges partial support from NSF grant AST-1514676 and NASA grant NNX13AE70G.

References

1. Aerts, C. (2019). Probing the interior physics of stars through asteroseismology. e-prints. arXiv:1912.12300.
2. Aerts, C., Christensen-Dalsgaard, J., & Kurtz, D. W. (2010). *Asteroseismology*. New York: Springer.
3. Aerts, C., Molenberghs, G., Michielsen, M., Pedersen, M. G., Björklund, R., Johnston, C., et al. (2018). *The Astrophysical Journal Supplement Series, 237*(1), 15. https://doi.org/10.3847/1538-4365/aaccfb
4. Aguirre, V. S., Lund, M. N., Antia, H. M., Ball, W. H., Basu, S., Christensen-Dalsgaard, J., et al. (2017). *The Astrophysical Journal, 835*(2), 173. https://doi.org/10.3847/1538-4357/835/2/173
5. Ahlborn, F., Bellinger, E. P., Hekker, S., Basu, S., & Angelou, G. C. (2020). *Astronomy & Astrophysics, 639*, A98. https://doi.org/10.1051/0004-6361/201936947
6. Angelou, G. C., Bellinger, E. P., Hekker, S., & Basu, S. (2017). *The Astrophysical Journal, 839*(2), 116. https://doi.org/10.3847/1538-4357/aa6a54
7. Angelou, G. C., Bellinger, E. P., Hekker, S., Mints, A., Elsworth, Y., Basu, S., et al. (2020). *Monthly Notices of the Royal Astronomical Society, 493*(4), 4987. https://doi.org/10.1093/mnras/staa390
8. Backus, G. E., & Gilbert, J. F. (1967). *Geophysical Journal, 13*(1–3), 247. https://doi.org/10.1111/j.1365-246X.1967.tb02159.x
9. Backus, G., & Gilbert, J. (1968). *Geophysical Journal, 16*, 169. https://doi.org/10.1111/j.1365-246X.1968.tb00216.x
10. Ball, W. H. (2017). *European Physical Journal Web of Conferences* (vol. 160, p. 02001). https://doi.org/10.1051/epjconf/201716002001
11. Ball, W. H., & Gizon, L. (2014). *Astronomy & Astrophysics, 568*, A123. https://doi.org/10.1051/0004-6361/201424325
12. Basu, S. (2003). *Astrophysics and Space Science, 284*(1), 153 .
13. Basu, S. (2016). *Living Reviews in Solar Physics, 13*(1), 2. https://doi.org/10.1007/s41116-016-0003-4
14. Basu, S., & Chaplin, W. J. (2017). *Asteroseismic data analysis: foundations and techniques*. Princeton: Princeton University Press.
15. Basu, S., Pinsonneault, M. H., & Bahcall, J. N. (2000). *The Astrophysical Journal, 529*(2), 1084. https://doi.org/10.1086/308302
16. Basu, S., Christensen-Dalsgaard, J., Monteiro, M. J. P. F. G., & Thompson, M. J. (2001). In A. Wilson & P. L. Pallé (Eds.), *SOHO 10/GONG 2000 Workshop* (vol. 464, pp. 407–410).
17. Basu, S., Christensen-Dalsgaard, J., & Thompson, M. J. (2002). In B. Battrick, F. Favata, I. W. Roxburgh, & D. Galadi (Eds.), *Stellar structure and habitable planet finding* (vol. 485, pp. 249–252).
18. Basu, S., Chaplin, W. J., Elsworth, Y., New, R., & Serenelli, A. M. (2009). *The Astrophysical Journal, 699*(2), 1403. https://doi.org/10.1088/0004-637X/699/2/1403
19. Bazot, M., Nielsen, M. B., Mary, D., Christensen-Dalsgaard, J., Benomar, O., Petit, P., et al. *Astronomy & Astrophysics, 619*, L9 (2018). https://doi.org/10.1051/0004-6361/201834251
20. Bazot, M., Benomar, O., Christensen-Dalsgaard, J., Gizon, L., Hanasoge, S., Nielsen, M., et al. *Astronomy & Astrophysics, 623*, A125 (2019). https://doi.org/10.1051/0004-6361/201834594
21. Bedding, T. R. (2014). *Solar-like oscillations: An observational perspective* (p. 60). Cambridge: Cambridge University Press.
22. Bellinger, E. P. (2018). *Inverse Problems in Asteroseismology*. Ph.D. thesis, Max Planck Institute for Solar System Research. http://hdl.handle.net/11858/00-1735-0000-002E-E5BB-A
23. Bellinger, E. P. (2019). *Monthly Notices of the Royal Astronomical Society, 486*(4), 4612. https://doi.org/10.1093/mnras/stz714

24. Bellinger, E. P. (2020). *Monthly Notices of the Royal Astronomical Society, 492*(1), L50. https://doi.org/10.1093/mnrasl/slz178
25. Bellinger, E. P., & Christensen-Dalsgaard, J. (2019). *The Astrophysical Journal Letters, 887*(1), L1. https://doi.org/10.3847/2041-8213/ab43e7
26. Bellinger, E. P., Angelou, G. C., Hekker, S., Basu, S., Ball, W. H., Guggenberger, E. (2016). *The Astrophysical Journal, 830*, 31. https://doi.org/10.3847/0004-637X/830/1/31
27. Bellinger, E. P., Angelou, G. C., Hekker, S., Basu, S., Ball, W. H., Guggenberger, E. (2017). *European Physical Journal Web of Conferences, 160*, 05003. https://doi.org/10.1051/epjconf/201716005003
28. Bellinger, E. P., Basu, S., Hekker, S., & Ball, W. H. (2017). *The Astrophysical Journal, 851*(2), 80. https://doi.org/10.3847/1538-4357/aa9848
29. Bellinger, E. P., Hekker, S., Angelou, G. C., Stokholm, A., & Basu, S. (2019). *Astronomy & Astrophysics, 622*, A130. https://doi.org/10.1051/0004-6361/201834461
30. Bellinger, E. P., Basu, S., Hekker, S., & Christensen-Dalsgaard, J. (2019). *The Astrophysical Journal, 885*(2), 143. https://doi.org/10.3847/1538-4357/ab4a0d
31. Bellinger, E. P., Kanbur, S. M., Bhardwaj, A., & Marconi, M. (2020). *Monthly Notices of the Royal Astronomical Society, 491*(4), 4752. https://doi.org/10.1093/mnras/stz3292
32. Benomar, O., Takata, M., Shibahashi, H., Ceillier, T., & García, R. A. (2015). *Monthly Notices of the Royal Astronomical Society, 452*(3), 2654. https://doi.org/10.1093/mnras/stv1493
33. Benomar, O., Bazot, M., Nielsen, M. B., Gizon, L., Sekii, T., Takata, M., et al. (2018). *Science, 361*(6408), 1231. https://doi.org/10.1126/science.aao6571
34. Berthomieu, G., Toutain, T., Gonczi, G., Corbard, T., Provost, J., & Morel, P. (2001). In A. Wilson & P. L. Pallé (Eds.), *SOHO 10/GONG 2000 Workshop* (vol. 464, pp. 411–414).
35. Brown, T. M., Christensen-Dalsgaard, J., Weibel-Mihalas, B., & Gilliland, R. L. (1994). *The Astrophysical Journal, 427*, 1013. https://doi.org/10.1086/174208
36. Buldgen, G. (2017). *Development of Inversion Techniques in Asteroseismology*. Ph.D. thesis, Université de Liège. http://hdl.handle.net/2268/213523
37. Buldgen, G., Reese, D. R., Dupret, M. A., & Samadi, R. (2015). *Astronomy & Astrophysics, 574*, A42. https://doi.org/10.1051/0004-6361/201424613
38. Buldgen, G., Reese, D. R., & Dupret, M. A. (2015). *Astronomy & Astrophysics, 583*, A62. https://doi.org/10.1051/0004-6361/201526390
39. Buldgen, G., Salmon, S. J. A. J., Reese, D. R., & Dupret, M. A. (2016). *Astronomy & Astrophysics, 596*, A73. https://doi.org/10.1051/0004-6361/201628773
40. Chaplin, W. J., & Miglio, A. (2013). *Annual Review of Astronomy and Astrophysics, 51*(1), 353. https://doi.org/10.1146/annurev-astro-082812-140938
41. Christensen-Dalsgaard, J. (2004). *Solar Physics, 220*(2), 137. https://doi.org/10.1023/B:SOLA.0000031392.43227.7d
42. Christensen-Dalsgaard, J. (2008). *Astrophysics and Space Science, 316*(1–4), 113. https://doi.org/10.1007/s10509-007-9689-z
43. Christensen-Dalsgaard, J. (2012). *Astronomische Nachrichten, 333*(10), 914. https://doi.org/10.1002/asna.201211836
44. Cunha, M. S., Aerts, C., Christensen-Dalsgaard, J., Baglin, A., Bigot, L., Brown, T. M., et al. (2007). Astronomy and Astrophysics Review **14**(3–4), 217. https://doi.org/10.1007/s00159-007-0007-0
45. Deheuvels, S., García, R. A., Chaplin, W. J., Basu, S., Antia, H. M., Appourchaux, T., et al. (2012). *The Astrophysical Journal, 756*(1), 19. https://doi.org/10.1088/0004-637X/756/1/19
46. Deheuvels, S., Doğan, G., Goupil, M. J., Appourchaux, T., Benomar, O., Bruntt, H., et al. (2014). *Astronomy & Astrophysics, 564*, A27. https://doi.org/10.1051/0004-6361/201322779
47. Deheuvels, S., Ballot, J., Beck, P. G., Mosser, B., Østensen, R., García, R. A., et al. (2015). *Astronomy & Astrophysics, 580*, A96. https://doi.org/10.1051/0004-6361/201526449
48. Dziembowski, W. (1977). *Acta Astronomica, 27*, 203.
49. Eggenberger, P., den Hartogh, J. W., Buldgen, G., Meynet, G., Salmon, S. J. A. J., Deheuvels, S. (2019). *Astronomy & Astrophysics, 631*, L6. https://doi.org/10.1051/0004-6361/201936348

50. García, R. A., & Ballot, J. (2019). *Living Reviews in Solar Physics, 16*(1), 4. https://doi.org/10.1007/s41116-019-0020-1
51. Gough, D. O. (1998). In H. Kjeldsen & T. R. Bedding (Eds.), *The First MONS Workshop: Science with a Small Space Telescope* (p. 33)
52. Gough, D. O., & Kosovichev, A. G. (1993). *Initial asteroseismic inversions. ASP conference series, ASPCS* (vol. 40, p. 541).
53. Gough, D. O., & Thompson, M. J. (1991). *The inversion problem* (pp. 519–561). Tucson: University of Arizona Press.
54. Guenther, D. B., & Brown, K. I. T. (2004). *The Astrophysical Journal, 600*(1), 419. https://doi.org/10.1086/379644
55. Hadamard, J. (1902). *Princeton University Bulletin, 13*, 49.
56. Hekker, S. (2013). *Advances in Space Research, 52*(9), 1581. https://doi.org/10.1016/j.asr.2013.08.005
57. Hekker, S. (2019). Scaling relations for solar-like oscillations: a review. e-prints. arXiv:1907.10457.
58. Hekker, S., &Christensen-Dalsgaard, J. (2017). *Astronomy and Astrophysics Review, 25*(1), 1. https://doi.org/10.1007/s00159-017-0101-x
59. Hendriks, L., & Aerts, C. (2019). *Publications of the Astronomical Society of the Pacific, 131*(1004), 108001. https://doi.org/10.1088/1538-3873/aaeeec
60. Jørgensen, A. C. S., & Angelou, G. C. (2019). *Monthly Notices of the Royal Astronomical Society, 490*(2), 2890. https://doi.org/10.1093/mnras/stz2825
61. Kirsch, A. (2011). *An introduction to the mathematical theory of inverse problems*. New York: Springer.
62. Kosovichev, A. G. (2011). *Advances in global and local helioseismology: An introductory review* (vol. 832, p. 3). Berlin: Springer. https://doi.org/10.1007/978-3-642-19928-8_1
63. Lund, M. N., Aguirre, V. S., Davies, G. R., Chaplin, W. J., Christensen-Dalsgaard, J., Houdek, G., et al. (2017). *The Astrophysical Journal, 835*(2), 172. https://doi.org/10.3847/1538-4357/835/2/172
64. Metcalfe, T. S., Creevey, O. L., & Christensen-Dalsgaard, J. (2009). *The Astrophysical Journal, 699*(1), 373. https://doi.org/10.1088/0004-637X/699/1/373
65. Neto, F. D. M., & Neto, A. J. d. S. (2012). *An introduction to inverse problems with applications*. New York: Springer.
66. Paxton, B., Bildsten, L., Dotter, A., Herwig, F., Lesaffre, P., & Timmes, F. (2011). *The Astrophysical Journal Supplement Series, 192*(1), 3. https://doi.org/10.1088/0067-0049/192/1/3
67. Pijpers, F. P. (1998). *Monthly Notices of the Royal Astronomical Society, 297*(3), L76. https://doi.org/10.1046/j.1365-8711.1998.01801.x
68. Pijpers, F. P., & Thompson, M. J. (1994). *Astronomy & Astrophysics, 281*, 231.
69. Pulone, L., & Scaramella, R. (1997). *Neural Nets WIRN VIETRI-96* (pp. 231–236). New York: Springer.
70. Rabello-Soares, M. C., Basu, S., & Christensen-Dalsgaard, J. (1999). *Monthly Notices of the Royal Astronomical Society, 309*(1), 35. https://doi.org/10.1046/j.1365-8711.1999.02785.x
71. Reese, D. R. (2018). In T. L. Campante, N. C. Santos, & M. J. P. F. G. Monteiro (Eds.), *Asteroseismology and exoplanets* (vol. 49, p. 75). https://doi.org/10.1007/978-3-319-59315-9_4
72. Reese, D. R., Marques, J. P., Goupil, M. J., Thompson, M. J., & Deheuvels, S. (2012). *Astronomy & Astrophysics, 539*, A63. https://doi.org/10.1051/0004-6361/201118156
73. Rendle, B. M., Buldgen, G., Miglio, A., Reese, D., Noels, A., Davies, G. R., et al. (2019). *Monthly Notices of the Royal Astronomical Society, 484*(1), 771. https://doi.org/10.1093/mnras/stz031
74. Roxburgh, I. W. (2002). In B. Battrick, F. Favata, I. W. Roxburgh, & D. Galadi (Eds.), *Stellar structure and habitable planet finding* (vol. 485, pp. 75–85).
75. Roxburgh, I. W. , & Vorontsov, S. V. (2002). In B. Battrick, F. Favata, I. W. Roxburgh, & D. Galadi (Eds.), *Stellar structure and habitable planet finding* (vol. 485, pp. 337–339).

76. Silva Aguirre, V., Davies, G. R., Basu, S., Christensen-Dalsgaard, J., Creevey, O., Metcalfe, T. S., et al. (2015). *Monthly Notices of the Royal Astronomical Society, 452*(2), 2127. https://doi.org/10.1093/mnras/stv1388
77. Tenorio, L. (2001). *SIAM Review, 43*(2), 347.
78. Thompson, M. J., & Christensen-Dalsgaard, J. (2002). In B. Battrick, F. Favata, I. W. Roxburgh, & D. Galadi (Eds.), *Stellar structure and habitable planet finding* (vol. 485, pp. 95–101).
79. Townsend, R. H. D., & Teitler, S. A. (2013). *Monthly Notices of the Royal Astronomical Society, 435*(4), 3406. https://doi.org/10.1093/mnras/stt1533
80. Ulam, S. (1962). *Proceedings of symposia in applied mathematics* (vol. 14, pp. 215–224). Providence: American Mathematical Society.
81. Zhang, Q. S., Li, Y., & Christensen-Dalsgaard, J. (2019). *The Astrophysical Journal, 881*(2), 103. https://doi.org/10.3847/1538-4357/ab2f77

Diagnostics from Solar and Stellar Glitches

Margarida S. Cunha

Abstract Sudden changes in the internal structure of stars, placed at the interface between convective and radiative regions, regions of partial ionisation, or between layers that have acquired different chemical composition as a result of nuclear burning, often produce specific signatures in the stars' oscillation spectra. Through the study of these signatures one may gain information on the physical processes that shape the regions that produce them, including diffusion and chemical mixing beyond the convectively unstable regions, as well as information about the helium content of stars. In this talk, I will review important theoretical and observational efforts conducted over the years towards this goal. I will emphasise the potential offered by the study of acoustic, gravity, and mixed modes observed in stars of different mass and evolutionary stages, at a time when space-based data is allowing us to build on the knowledge gained from the study of the Sun and white dwarfs, where these efforts have long been undertaken, extending the methods developed to stars across the HR diagramme.

1 Glitch, Glitch Signature, and Mode Trapping

In Helio- and Asteroseismology, the term glitch usually refers to *a structural variation that is perceived as abrupt by the waves travelling across it*. Under this interpretation, a significant local variation in the structure of a star is a glitch only if it occurs within the wave's propagation cavity and the scale in which that variation takes place is comparable or smaller than the local wavelength of the mode under consideration. Figure 1 illustrates this concept based on two stellar models. On the left panel, the buoyancy frequency, N, in a 6 $M_\odot$ main-sequence (MS) model is seen

M. S. Cunha (✉)
Instituto de Astrofísica e Ciências do Espaço, Universidade do Porto, CAUP, Porto, Portugal

School of Physics and Astronomy, University of Birmingham, Birmingham, UK
e-mail: mcunha@astro.up.pt

M. J. P. F. G. Monteiro et al. (eds.), *Dynamics of the Sun and Stars*, Astrophysics and Space Science Proceedings 57,
https://doi.org/10.1007/978-3-030-55336-4_26

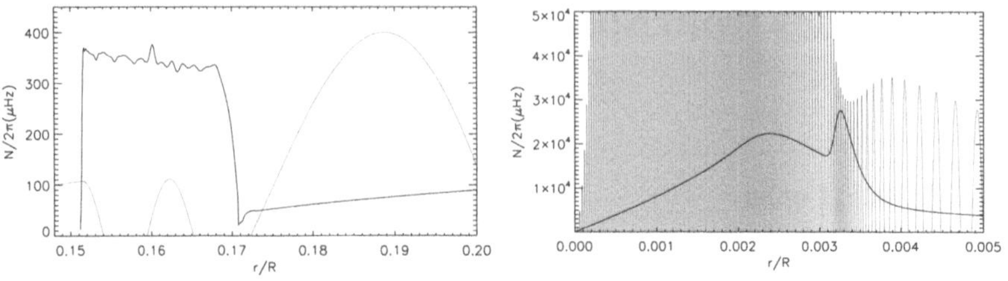

Fig. 1 Glitch (left) and no glitch (right). Detail of the buoyancy frequency in a M=6M$_\odot$ main-sequence model (left, thick line) and a M=1M$_\odot$ RGB model (right, thick line). For comparison the eigenfunction $r^3\delta p$ multiplied by an arbitrary factor is show in each case (thin lines; where δp is the Lagrangian pressure perturbation) for frequencies that may be observed in stars represented by these models, respectively, $\nu = 5.4\,\mu$Hz and $\nu = 51\,\mu$Hz (see [11], for further model details)

to vary on a short scale at a fractional radius of $r/R = 0.17$. This variation, which results from the chemical gradient built at the edge of the retracting convective core (see [11] for details) is perceived as a glitch by oscillations with frequencies that may be observed in such a star. On the other hand, on the right panel it is shown how N varies in the inner layers of a red giant branch (RGB) model. A significant local variation of N is also seen, at $r/R \approx 0.0033$, associated to the hydrogen-shell burning region, but because the modes that may be observed in this star have a rather short wavelength at that location, that variation does not constitute a glitch.

The phase of a wave is perturbed at a glitch and, as a result, the mode frequency is also perturbed with respect to the frequency it would have in the absence of the glitch. That perturbation shows a cyclic behaviour in frequency for pressure (p-) modes and in period for gravity (g-) modes. Perturbations to the frequencies resulting from the presence of a glitch are usually named *glitch signatures*. Examples of expected glitch signatures on the frequencies of g- and p-modes can be identified by computing the first period differences (or period spacings) and second frequency differences [14], respectively, as illustrated in Fig. 2. The scale of the cyclic behaviour of the glitch signature provides information on the location of the glitch within the star, with glitches located closer to the center of the mode propagation cavity inducing signatures that vary on shorter scales than glitches located closer to the edges of the cavity (e.g., [17, 10]). Moreover, the way the signature's amplitude varies with frequency depends on the structural quantity exhibiting the glitch, as well as on the shape and width of the glitch itself (e.g., [23, 11]). Finally, glitches can significantly change the energy distribution of the modes throughout the star, resulting in a concentration of mode energy in a particular region, a phenomena that is usually named *mode trapping*.

Signatures of glitches are seen in stellar pulsators across the Hertzsprung–Russell (HR) diagram. Their diagnostic potential is unique, allowing us to probe directly localised features within the stars, whose origin and shape often depend on physical processes that are not fully understood or on the unknown abundance of a particular chemical element. Some early works attempting to make use of this diagnostic

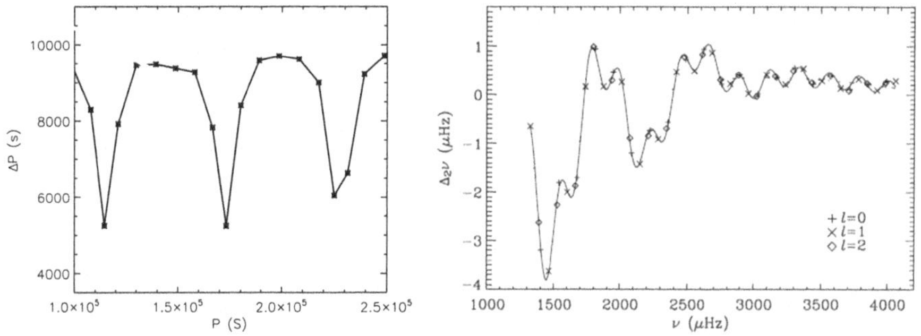

Fig. 2 Left: Dipolar period spacings, ΔP, for the MS model shown in Fig. 1. In the absence of a buoyancy glitch ΔP would be approximately constant [29]. Right: second frequency differences ($\Delta_2\nu$) [14] for a model of the Sun. The variations in $\Delta_2\nu$ result from the combined effect of the glitches at the base of the convective zone (variations on the shortest scale) and at the Helium I and II ionisation zones (variations on larger scales). Adapted from Houdek and Gough [17], by permission of Oxford University Press on behalf of the Royal Astronomical Society

potential in the Sun date back to the late 1980s, when Michael Thompson was still a PhD student [15, 30, 34]. In these works, the authors discussed the signature that would be left on the solar frequencies by a toroidal magnetic field confined at the base of the solar convective zone. While no signature of such a confined magnetic field was found in the data at that time, or in later data [2], the idea of looking for glitch signatures in stellar pulsations remained, with signatures of glitches from different physical origins being found and studied both in the Sun and other stars over the following decades. In what follows I will summarise a selection of results from these studies.

2 Acoustic Glitches

Glitches in the sound speed or sound-speed derivatives are usually termed *acoustic glitches*. The most common examples in solar-like stars are the glitch located at the base of the convective envelope, resulting from the rapid transition in the temperature gradient, and the glitch located at the helium second ionisation region, resulting from the drop in the first adiabatic exponent. Together they offer the opportunity to establish the depth of stellar convective envelopes, to improve our understanding of overshooting into the radiative regions, and to infer the abundance of helium, a key element for stellar evolution. This important diagnostic potential has been recognised by Michael Thompson and collaborators, as well as by other authors, in the 1990s, leading to a series of studies related to the inference of information from the acoustic glitch signatures seen in the solar data and to theoretical discussions of the potential impact of observing these signatures in other solar-like stars (e.g. [14, 17, 23, 2, 3, 28, 24, 4, 22]).

The signature from the base of a convective envelop depends critically on how overshoot modifies the temperature gradient at the transition. Earlier studies compared the glitch signature observed in the solar data with that derived from models with no overshoot, where the first derivative of the temperature gradient (hence, the second derivative of the sound speed) is discontinuous and that from models incorporating a simple overshoot prescription where the adiabatically stratified region is extended into the radiative region, leading to a discontinuity in the temperature gradient (hence, in the first derivative of the sound speed) [23, 3, 28]. In all cases the glitch signature in the solar data was found to be comparable or smaller than that from models with no overshooting (hence, smaller than from models including an overshoot region), favouring the models without overshooting and setting a maximum limit to the overshoot region of less than one tenth of the pressure scale height. However, the authors acknowledged that the adopted overshoot prescription was likely to be unrealistic and that a more meaningful picture than no overshooting would be one in which overshooting takes place, but where the transition in the temperature gradient at the base of the convective envelope is less abrupt. This possibility gained support from a variety of overshooting models, including those derived from 3D numerical simulations, non-local convection and semi-analytic plume models, which seemed to allow for a significantly smoother transition (e.g., [7, 27, 12]). With this possibility in mind [9] compared the glitch signature from models with a range of possible temperature gradient profiles at the transition between the convective and radiative regions. An example of one such profile is shown by the continuous line on the left panel of Fig. 3. On the right panel, the signature found in the solar data is compared with signatures derived from the different model sequences considered in the same work.

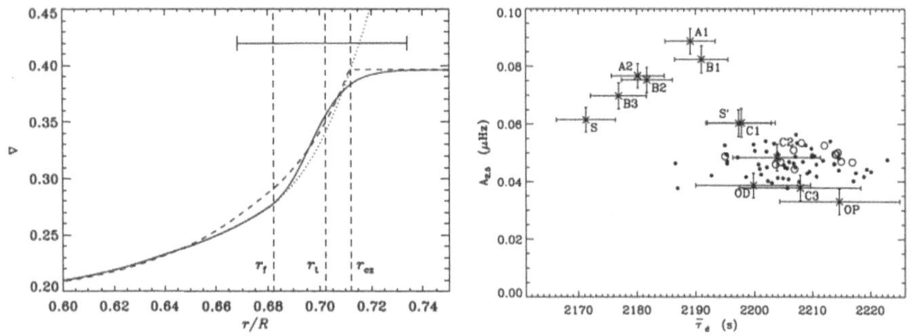

Fig. 3 Left: Temperature gradients (d ln T/d ln P) for model S [8] without overshooting (dashed curve) and for a second model (C2- continuous line) considered in [9], whose glitch signature is found to be consistent with that observed in the solar data. The dotted line is the radiative gradient of model C2. Right: Comparison of two properties, $A_{2.5}$ and $\bar{\tau}$, inferred from the glitch signatures of the series of models considered by Christensen-Dalsgaard et al.[9] (stars) and from the solar data (open and filled circles, derived from 1-year and 72-d data sets, respectively). See [9] for details. Reproduced from [9] by permission of Oxford University Press/on behalf of the Royal Astronomical Society

The comparison is based on $A_{2.5}$ which is related with the amplitude of the glitch signature at fixed frequency, and $\bar{\tau}$, related to the location of the glitch (see [9] for precise definitions). The results show that the series of models with a smoother transition (series C) are more consistent with the solar data. The standard model S [8], with no overshooting, as well as other models with sharp transitions (series A and B), all produce signatures with amplitudes that are significantly larger than the one found in the solar data.

With the advent of space-based data, the detection of glitch signatures from the base of convective envelopes in other solar-like stars became possible [19, 20, 31]. From studying these signatures it is possible to establish how the extent of stellar convective envelopes depends on stellar properties such as mass, age, and metallicity, further constraining stellar evolution and stellar dynamo theory. Space-based data offer also the opportunity to constrain the helium abundance in solar-like stars, where it cannot be directly determined from the analysis of spectroscopic data. This element has a significant influence on how a star evolves. The fact that its abundance is only weakly constrained (usually based of galactic evolution models) thus leads to significant uncertainties in the evolution path of a star and, due to the mass-helium degeneracy in main-sequence stellar evolution, to large uncertainties in the determination of the stellar mass, given a large uncertainty in the helium abundance. The opportunity to constrain the helium abundance directly through the study of the glitch signature associated with the depression in the adiabatic exponent at the location of the helium second ionisation is, therefore, a very interesting one. The underlying idea is that the amplitude of the helium glitch signature is predominantly a function of the helium abundance in the helium ionisation zone and, thus, can be calibrated against models where the helium abundance in known. An example of this was provided by Verma et al. [32], where the authors used the amplitude of the helium glitch signatures derived from the oscillations of the binary system 16Cyg A and B to infer the current helium abundance in their envelopes. The results are illustrated in Fig. 4 for the case of 16Cyg A, reproduced from their work, the upper panel showing the fit to the glitch signature derived from the observations and the lower panel showing the star-models comparison of the amplitudes. The spread in the model results at fixed helium is mostly associated with differences in other model properties, such as mass, effective temperature, and luminosity (same colour) or in the chemical mixture considered (different colours). Despite this spread, it is clear that the amplitude of the glitch signature is very strongly dependent of the helium abundance, allowing the authors to derive the envelope helium abundance in both stars. The work was later extended to the study of the *Kepler* legacy sample [33]. Interestingly, in [31] the authors have shown that incorporating the information inferred on the glitch amplitude in the forward modelling of stars can help removing biases in the inferred stellar parameters resulting from the helium-mass degeneracy. This is illustrated in Fig. 5 reproduced from [31] which shows the Chi-squared maps of the stellar parameters mass and radius resulting from two different forward modelling attempts of the data for the Sun-as-a-star prepared by Lund et al. [18] (where the noise level was made similar to that of the data for the *Kepler* legacy sample stars). The left panel shows the results

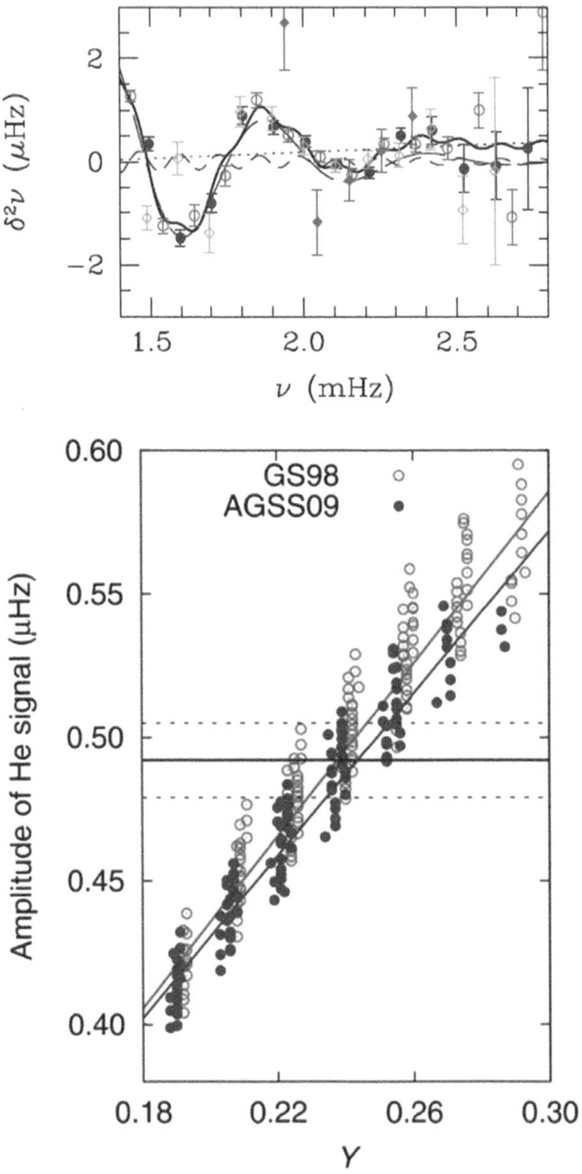

Fig. 4 Upper panel: Second frequency differences derived from data collected by the NASA *Kepler* satellite for the solar-like star 16 Cyg A (blue, red, green and magenta symbols for mode degrees $l = 0, 1, 2$ and 3, respectively). The full line shows the fit to the data using the expected functional form of the signature seen in the data, while the long-dashed and dashed lines show the contributions to the signature from the helium and the base of the convective envelope glitches, respectively. The dotted line is a smooth, non-glitch like component included in the fitted expression. Bottom panel: comparison of the amplitude of the glitch signature inferred from the data (thick horizontal line with 1-σ uncertainties given by the dashed lines) with the amplitudes derived from two model series with chemical mixtures from [16] (red) and [1] (blue). Figures from Verma et al. [32], ©AAS. Reproduced with permission

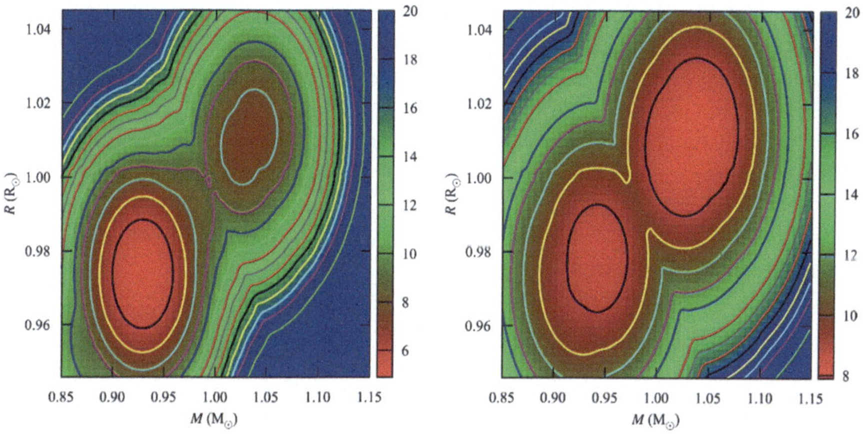

Fig. 5 Chi-square maps for the parameters mass and radius derived from the forward modelling of data for the Sun-as-a-star with noise levels similar to those found in stars from the *Kepler* legacy sample, where a lower chi-square (red) indicates a better fit to the data. The difference in the results shown in the two panels stems from the fact that the amplitude of the glitch signature was incorporated in the fitting procedure used on the right panel, but not in that used on the left panel. Figure from [31], ©AAS. Reproduced with permission

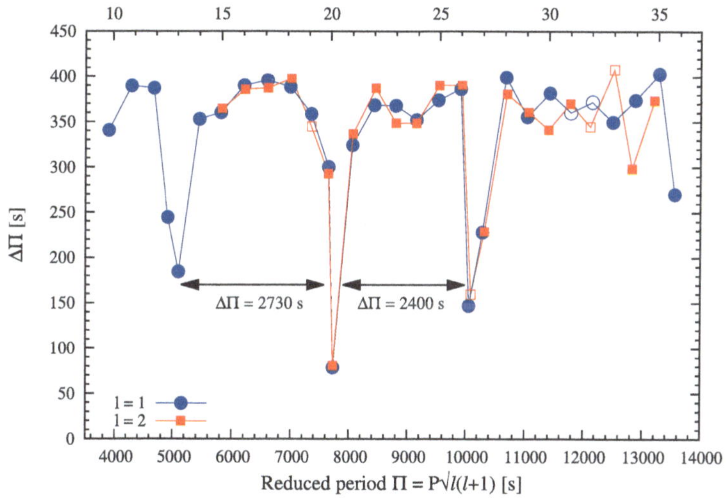

Fig. 6 Reduced period spacings computed from modes of degree $l = 1$ (blue) and degree $l = 2$ (red) for the subdwarf B star KIC10553698A where the reduced period is defined by $\Pi = P\sqrt{l(l+1)}$. The deviation from a constant value is induced by a glitch in N. Figure from [26], reproduced with permission, ©ESO

from the modelling conducted without considering the information on the amplitude of the signature of the helium glitch, while the right panel shows the results when that information is incorporated explicitly. In the absence of the glitch information a lower radius and mass are preferred. However, the glitch amplitude adds information on the helium abundance content, leading to results that are in better agreement with the solar values.

3 Buoyancy Glitches

Buoyancy glitches refer to glitches in the buoyancy frequency, N. They affect the pulsation properties of g modes and mixed modes and are, thus, important in the context of g-mode pulsators, such as main-sequence intermediate-mass, Subdwarf B, and white dwarf stars, and mixed-mode pulsators, such as red giants, both during the RGB ascent and during the helium-core-burning (HeCB) phase. Through the analysis of buoyancy glitches it is possible to address key questions in the physics of g-mode and mixed-mode pulsators, related to the chemical stratification in compact pulsators [6] and to the mixing conditions at the edge of convective cores in core-convective hydrogen-burning stars and in HeCB stars [5].

Amongst the most promising recent results in this context is the detection of glitch signatures in HeCB stars, both in red giants [25, 35] and in subdwarf B stars [26, 13], which are thought to hold information on the chemical stratification in the innermost layers of these stars. That stratification depends on the details of the mixing beyond the helium-burning core, which is largely unknown and represents one of the key open questions in present-day stellar astrophysics [5]. Figures 6 and 7 show examples of glitch signatures found in two HeCB stars, namely, a subdwarf B star and a red-giant star, respectively. Ideally, one would hope to infer the information contained in stellar glitch signatures in a model independent way. This is even more important in the case of the buoyancy glitches discussed here, as the models of stars in the HeCB phase are particularly complex and, at the same time, less robust and less tested than those of low-mass main-sequence stars. An important step towards this goal is the incorporation of the glitch effect in the asymptotic analysis of the frequencies (or periods) of g and mixed modes. The glitch-induced perturbation to the periods of g modes has been theoretically addressed in the context of white dwarfs and main-sequence intermediate-mass stars in the past [6, 21]. Recently, analytical expressions describing directly the behaviour of the period spacings in the presence of buoyancy glitches of different amplitudes and functional forms in pure g-mode pulsators were presented by Cunha et al. [11]. The authors also extended the work by Cunha et al. [10], where the first asymptotically-based analytical expression for period spacings of mixed modes in the presence of buoyancy glitches had been presented, by considering different possible functional forms for the glitches in red-giant pulsators. An example of the fit of the analytical expression derived by Cunha et al. [11] to the data on a HeCB red-giant star is shown by the red curve in Fig. 7. Through the fit, the authors were able to infer the location and amplitude of the glitch in the red-giant star [35].

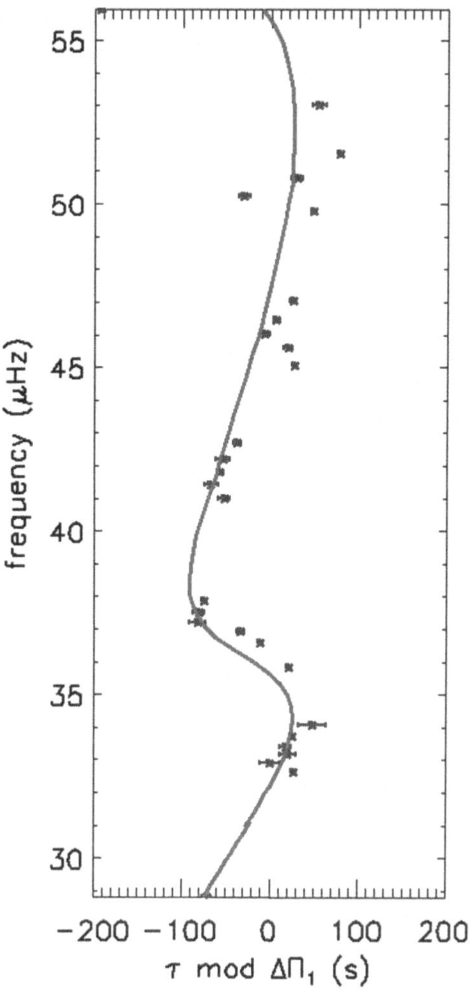

Fig. 7 Stretched period echelle diagram for the HeCB red-giant star KIC 009332840. The stretched periods correspond to the g-mode periods that would be observed if no coupling existed between the gravity and acoustic waves (see [25] for an exact definition). They would be constantly spaced if no glitch was present (in a spherically symmetric star). The deviation of the blue symbols (representing the data) from a vertical alignment is induced by a glitch in N. The red line shows the best fit of the asymptotic expression derived taking into account the glitch effect [11] to the data. Figure from [35]

4 Concluding Remarks

The study of glitches allows us to probe specific layers within stars whose structural details are often shaped by physical processes that are most challenging to model. Moreover, it allows us to infer information on the abundance of helium, a key element for stellar evolution that cannot otherwise be constrained in cool stars. As a consequence, the study of glitches offers a unique potential to further advance our understanding of stellar structure, dynamics, and evolution. Its foundation was established in the early days of helioseismology. Michael Thompson has undoubtedly understood the potential of this technique and played a major role in its development, contributing, along with a few others in the field, to set the first stones that paved the way for it to be more generally applied. Following on those earlier developments, and also on developments of similar techniques in the context of the study of g modes in white dwarfs, the study of glitches has recently gained a new breath, thrusted by the exquisit space-born long asteroseismic data sets made available by the NASA satellite *Kepler*, and shall continue to develop into the future, in particular with the expected launch of the PLATO mission, by ESA.

Acknowledgments The author would like to express her profound gratitude to the Scientific Organizing Committee for the opportunity to speak at the conference honouring the life and work of Michael Thompson. MSC is supported in the form of a work contract funded through Fundação para a Ciência e Tecnologia (FCT). This work was supported by FCT through the research grants UID/FIS/04434/2019, UIDB/04434/2020 and UIDP/04434/2020 and by FCT through national funds (PTDC/FIS-AST/30389/2017) and by FEDER—Fundo Europeu de Desenvolvimento Regional through COMPETE2020—Programa Operacional Competitividade e Internacionalização (POCI-01-0145-FEDER-030389).

References

1. Asplund, M., Grevesse, N., Sauval, A. J., & Scott, P. (2009). The chemical composition of the sun. *Annual Review of Astronomy and Astrophysics, 47*(1), 481–522.
2. Basu, S. (1997, July). Seismology of the base of the solar convection zone. *Monthly Notices of the Royal Astronomical Society, 288*(3), 572–584.
3. Basu, S., Antia, H. M., Narasimha, D. (1994). Helioseismic measurement of the extent of overshoot below the solar convection zone. *Monthly Notices of the Royal Astronomical Society, 267*, 209.
4. Basu, S., Mazumdar, A., Antia, H. M., & Demarque, P. (2004). Asteroseismic determination of helium abundance in stellar envelopes. *Monthly Notices of the Royal Astronomical Society, 350*(1), 277–286.
5. Bossini, D., Miglio, A., Salaris, M., Pietrinferni, A., Montalban, J., Bressan, A., et al. (2015). Uncertainties on near-core mixing in red-clump stars: effects on the period spacing and on the luminosity of the AGB bump. *Monthly Notices of the Royal Astronomical Society, 453*, 2290–2301.
6. Brassard, P., Fontaine, G., Wesemael, F., & Hansen, C. J. (1992). Adiabatic properties of pulsating DA white dwarfs. II – Mode trapping in compositionally stratified models. *The Astrophysical Journal Supplement Series, 80*, 369–401.
7. Brummell, N. H., Clune, T. L., & Toomre, J. (2002). Penetration and overshooting in turbulent compressible convection. *The Astrophysical Journal, 570*(2), 825–854.
8. Christensen-Dalsgaard, J., Dappen, W., Ajukov, S. V., Anderson, E. R., Antia, H. M., Basu, S., et al. (1996). The current state of solar modeling. *Science, 272*, 1286–1292.
9. Christensen-Dalsgaard, J., Monteiro, M. J. P. F. G., Rempel, M., & Thompson, M. J. (2011). A more realistic representation of overshoot at the base of the solar convective envelope as seen by helioseismology. *Monthly Notices of the Royal Astronomical Society, 414*(2), 1158–1174.
10. Cunha, M. S., Stello, D., Avelino, P. P., Christensen-Dalsgaard, J., & Townsend., R. H. D. (2015). Structural glitches near the cores of red giants revealed by oscillations in g-mode period spacings from stellar models. *The Astrophysical Journal, 805*, 127.
11. Cunha, M. S., Avelino, P. P., Christensen-Dalsgaard, J., Stello, D., Vrard, M., Jiang, C., et al. (2019). Analytical modelling of period spacings across the HR diagram. *Monthly Notices of the Royal Astronomical Society, 490*(1), 909–926.

12. Deng, L., & Xiong, D. R. (2008). On the boundaries of a convective zone and the extent of overshooting. In L. Deng, & K. L. Chan (Eds.), *The art of modeling stars in the 21st century. IAU symposium* (vol. 252, pp. 83–88).
13. Ghasemi, H., Moravveji, E., Aerts, C., Safari, H., & Vučković, M. (2017). The effects of near-core convective shells on the gravity modes of the subdwarf B pulsator KIC 10553698A. *Monthly Notices of the Royal Astronomical Society, 465*(2), 1518–1531.
14. Gough, D. O. (1990). Comments on helioseismic inference. In Y. Osaki & H. Shibahashi (Eds.), *Progress of seismology of the sun and stars. Lecture notes in physics* (vol. 367, p. 283). Berlin: Springer.
15. Gough, D. O., & Thompson, M. J. (1988). Magnetic perturbations to stellar oscillation eigenfrequencies. In J. Christensen-Dalsgaard & S. Frandsen (Eds.), *Advances in helio- and asteroseismology. IAU Symposium* (vol. 123, p. 155)
16. Grevesse, N., & Sauval, A. J. (1998). Standard solar composition. *Space Science Reviews, 85*, 161–174.
17. Houdek, G., & Gough, D. O. (2007). An asteroseismic signature of helium ionization. *Monthly Notices of the Royal Astronomical Society, 375*, 861–880.
18. Lund, M. N., Aguirre, V. S., Davies, G. R., Chaplin, W. J., Christensen-Dalsgaard, J., Houdek, G., et al. (2017). Standing on the shoulders of dwarfs: the Kepler asteroseismic LEGACY sample. I. Oscillation mode parameters. *The Astrophysical Journal, 835*(2), 172.
19. Mazumdar, A., Michel, E., Antia, H. M., & Deheuvels, S. (2012). Seismic detection of acoustic sharp features in the CoRoT target <ASTROBJ>HD 49933</ASTROBJ>. *Astronomy & Astrophysics, 540*, A31.
20. Mazumdar, A., Monteiro, M. J. P. F. G., Ballot, J., Antia, H. M., Basu, S., Houdek, G., et al. (2014). Measurement of acoustic glitches in solar-type stars from oscillation frequencies observed by Kepler. *The Astrophysical Journal, 782*(1), 18.
21. Miglio, A., Montalbán, J., Noels, A., & Eggenberger, P. (2008). Probing the properties of convective cores through g modes: high-order g modes in SPB and γ Doradus stars. *Monthly Notices of the Royal Astronomical Society, 386*, 1487–1502.
22. Monteiro, M. J. P. F. G., & Thompson, M. J. (2005). Seismic analysis of the second ionization region of helium in the Sun – I. Sensitivity study and methodology. *Monthly Notices of the Royal Astronomical Society, 361*, 1187–1196.
23. Monteiro, M. J. P. F. G., Christensen-Dalsgaard, J., & Thompson, M. J. (1994). Seismic study of overshoot at the base of the solar convective envelope. *Astronomy and Astrophysics, 283*, 247–262.
24. Monteiro, M. J. P. F. G., Christensen-Dalsgaard, J., & Thompson, M. J. (2000). Seismic study of stellar convective regions: the base of the convective envelope in low-mass stars. *Monthly Notices of the Royal Astronomical Society, 316*, 165–172.
25. Mosser, B., Vrard, M., Belkacem, K., Deheuvels, S., & Goupil, M. J. (2015). Period spacings in red giants. I. Disentangling rotation and revealing core structure discontinuities. *Astronomy & Astrophysics, 584*, A50.
26. Østensen, R. H., Telting, J. H., Reed, M. D., Baran, A. S., Nemeth, P., & Kiaeerad, F. (2014). Asteroseismology revealing trapped modes in KIC 10553698A. *Astronomy & Astrophysics, 569*, A15.
27. Rempel, M. (2004). Overshoot at the base of the solar convection zone: a semianalytical approach. *The Astrophysical Journal, 607*(2), 1046–1064.
28. Roxburgh, I. W., & Vorontsov, S. V. (1994). Seismology of the solar envelope – the base of the convective zone as seen in the phase shift of acoustic waves. *Monthly Notices of the Royal Astronomical Society, 268*, 880.
29. Tassoul, M. (1980). Asymptotic approximations for stellar nonradial pulsations. *The Astrophysical Journal Supplement Series, 43*, 469–490.
30. Thompson, M. J. (1988, December). Evidence for a thin perturbative layer near the base of the solar convection zone. In E. J. Rolfe (Ed.), *Seismology of the sun and sun-like stars. ESA special publication* (vol. 286, pp. 321–324)

31. Verma, K., Raodeo, K., Antia, H. M., Mazumdar, A., Basu, S., Lund, M. N., et al. Seismic measurement of the locations of the base of convection zone and helium ionization zone for stars in the Kepler seismic LEGACY sample. *The Astrophysical Journal, 837*(1), 47.
32. Verma, K., Faria, J. P., Antia, H. M., Basu, S., Mazumdar, A., Monteiro, M. J. P. F. G., et al. (2014). Asteroseismic estimate of helium abundance of a solar analog binary system. *The Astrophysical Journal, 790*(2), 138.
33. Verma, K., Raodeo, K., Basu, S., Silva Aguirre, V., Mazumdar, A., Mosumgaard, J. R., et al. (2019). Helium abundance in a sample of cool stars: measurements from asteroseismology. *Monthly Notices of the Royal Astronomical Society, 483*(4), 4678–4694.
34. Vorontsov, S. V. (1988, January) A search of the effects of magnetic field in the solar 5-minute oscillations. In J. Christensen-Dalsgaard & S. Frandsen (Eds.), *Advances in helio- and asteroseismology. IAU symposium* (vol. 123, p. 151)
35. Vrard, M., & Cunha, M. S. (2019). Influence of structural discontinuities present in the core of red-giant stars on the observed mixed-mode pattern and characterization of their properties. *arXiv e-prints*. arXiv:1911.05175.

Dynamo States with Strikingly Different Symmetries Coexisting in Global Solar Simulations

Loren I. Matilsky and Juri Toomre

Abstract We present new global 3D MHD simulations of the solar convection zone that exhibit two distinct states of the dynamo coexisting simultaneously. We discuss our results in terms of the solar active longitudes and hemispheric asymmetry.

Despite the intense turbulence in the solar convection zone (CZ), the Sun's magnetism displays strikingly ordered time-dependent behavior. In particular, sunspots show equatorward migration in their sites of emergence and reverse their polarity sense every 11 years, forming the well-known magnetic butterfly diagram. Sunspots also sometimes appear at preferred azimuthal locations on the Sun, a phenomenon known as *active longitudes*. Finally, sunspots preferentially cluster in either the northern or southern hemispheres, a feature called hemispheric asymmetry [2].

The methods and results of the work presented here are described with considerably more detail in another paper [3]. We use the open-source software Rayleigh [1] to solve the equations of MHD in a 3D, rotating spherical shell encompassing the lower three density scale heights of the solar CZ. Our models rotate at three times the solar rate and have a solar luminosity injected. The diffusivities vary with radius like $\overline{\rho}(r)^{-1/2}$, where $\overline{\rho}(r)$ is the background density. The thermal Prandtl number (Pr) is unity. We explore dynamo cases with different magnetic Pr—the one in the model presented here being unity.

The simulated dynamo's magnetic field is dominated by the azimuthal component, which consists initially of a fourfold set of "wreaths", two in each hemisphere, of alternating polarity (Fig. 1, *left*). Later in the simulation the field occupies one hemisphere preferentially and is made up of two opposite-polarity "partial wreaths"

L. I. Matilsky (✉) · J. Toomre
JILA and Department of Astrophysical and Planetary Sciences, University of Colorado Boulder, Boulder, CO, USA
e-mail: loma3853@colorado.edu

M. J. P. F. G. Monteiro et al. (eds.), *Dynamics of the Sun and Stars*, Astrophysics and Space Science Proceedings 57,
https://doi.org/10.1007/978-3-030-55336-4_27

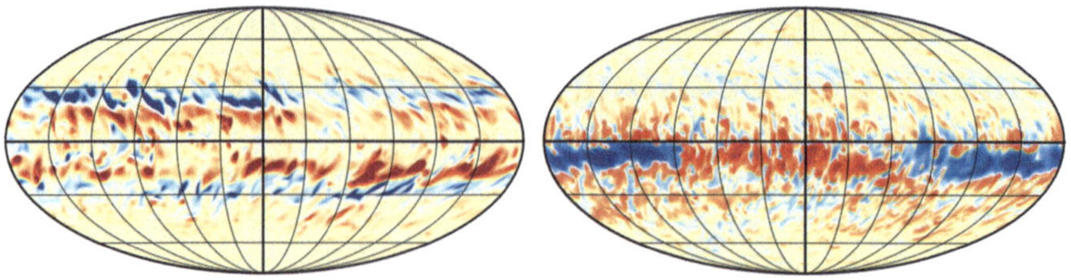

Fig. 1 Snapshots of the azimuthal magnetic field B_ϕ at two different depth/time pairs, shown in global Mollweide view. (*Left*) The field deep in the shell at an early time (the fourfold wreath). (*Right*) The field in the middle of the shell at a later time (the partial-wreath pair). Red tones indicate positive B_ϕ and blue tones indicate negative B_ϕ

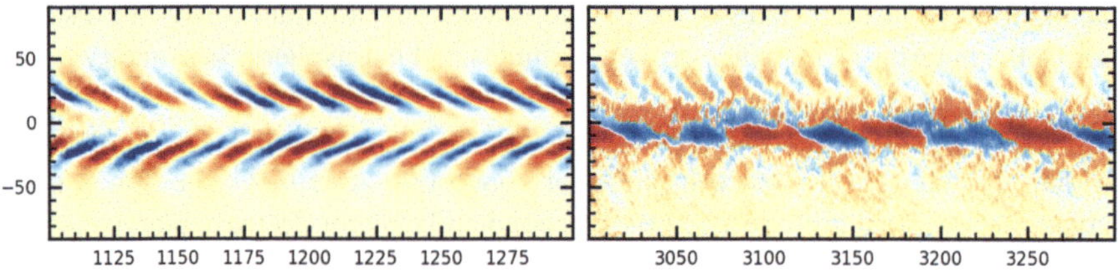

Fig. 2 Time-latitude diagrams of the longitude-averaged azimuthal magnetic field $\langle B_\phi \rangle$ near the bottom of the CZ (*left*) and mid-CZ (*right*). Time is in rotation periods and latitude in degrees

that each wrap roughly 180° around the sphere (Fig. 1, *right*). Over the course of the simulation, the system alternates chaotically between these two states.

Both cycles seem to constitute an $\alpha\Omega$-type dynamo [3]. In particular, the equatorward propagation in the fourfold-wreath cycle is due to an α-effect arising from the characteristic sense of twist in the interior wreaths. This result agrees with a recent study of Joy's Law, which suggests that the tilt of bipolar regions is established in the deep interior, and not by the Coriolis force acting on rising flux ropes [4].

The fourfold-wreath structure migrates equatorward and flips polarity regularly every ~25 rotations, reminiscent of the solar sunspot cycle (Fig. 2, *left*). The partial-wreath structure migrates very slightly poleward and changes its dominant polarity every ~75 rotations (Fig. 2, *right*). Even when interrupted by the partial-wreath cycle, the fourfold-wreath cycle persists with the same period, but at a lower amplitude.

The fourfold-wreath cycle is similar to the solar butterfly diagram. With regard to the partial-wreath cycle, some studies suggest active longitudes appear on opposite sides of the Sun. The partial-wreath structure in our simulation correspondingly has sites of the strongest field separated by ~180°. Our simulation's "active hemisphere" also switches between north and south over a long timescale, reminiscent of the observations of sunspot number asymmetry [2].

Our work suggests that spherical-shell convection can sustain intricate dynamos that exhibit strikingly different global structures for the magnetism simultaneously, and that such structures may be reflected in observations of the solar magnetic field. We acknowledge that the simulations presented here are just beginning to capture the complex dynamics of the highly nonlinear and turbulent Sun. However, it is clear that our dynamo simulations yield some reasonable large-scale behavior, which may provide insight into the physics of the solar interior.

Acknowledgments Supported by NASA grants 80NSSC18K1127 and 80NSSC19K1428.

References

1. Featherstone, N. A., & Hindman, B. W. (2016). *The Astrophysical Journal, 818*, 32.
2. Hathaway, D. H. (2015). *Living Reviews in Solar Physics, 12*, 4.
3. Matilsky, L. I., & Toomre, J. (2020). *The Astrophysical Journal, 892*, 106.
4. Stenflo, J. O., & Kosovichev, A. G. (2012). *The Astrophysical Journal, 745*, 129.

Influence of Turbulence on an Essentially Nonlinear Dynamo Mechanism

Jacob B. Noone Wade and Nicholas Brummell

Abstract Through direct numerical simulations, we study the influence of turbulence on a novel essentially nonlinear dynamo (END) mechanism. This dynamo operates through the interaction of a velocity shear with magnetic buoyancy mitigated by secondary Kelvin-Helmholtz instabilities. As opposed to the traditional mean-field or kinematic theory treatment, this END paradigm only exhibits dynamo action if the initial magnetic configuration is finite and above a critical threshold. We first study the effects of the smallest scales of turbulence by injecting noise into the magnetic fields. We then seek to study the effect of a wider range of scales by analyzing the interactions of actual resolved convection. To achieve a setup that is perhaps more realistic, we ultimately embed the END mechanism in a two-layer penetrative simulation. These preliminary results indicate that turbulence in many forms does little to hinder the END mechanism and therefore this example of an END at least appears to be quite robust.

An essentially nonlinear dynamo (END) is a subcritical dynamo triggered when the initial magnetic field strength is sufficiently large. An example of an END was presented in [2]. Through direct numerical simulations of the fully nonlinear compressible MHD equations in a non-rotating, convectively stable Cartesian box with a forced velocity shear, that paper demonstrates an END driven by interactions between shear and magnetic buoyancy instabilities. This system is embedded in a quiescent non-turbulent setting while for many realistic applications one may

J. B. Noone Wade (✉) and N. Brummell
University of California Santa Cruz, Santa Cruz, CA, USA
e-mail: jnoonewa@ucsc.edu; brummell@soe.ucsc.edu

M. J. P. F. G. Monteiro et al. (eds.), *Dynamics of the Sun and Stars*, Astrophysics and Space Science Proceedings 57,
https://doi.org/10.1007/978-3-030-55336-4_28

expect a turbulent environment. Thus the goal of this study is to assess the effects of turbulence on the mechanisms involved in the generation of this END. We will refer to case A2 from [2] as the canonical run, which will serve as a basis for comparison. Here, dynamo action occurs through complex interactions between the shear and a primary magnetic buoyancy instability mitigated by secondary Kelvin-Helmholtz instabilities. The shear repeatedly generates strong toroidal (B_x) structures which rise and then feedback on the shear through secondary Kelvin-Helmholtz instabilities.

As a first step to understand this ENDs robustness to turbulence, we have injected uniformly distributed random noise of varying amplitudes into the magnetic field. The implementation of such noise is simple and computationally inexpensive when compared with simulating resolved turbulence. By confining the noise to a k-range of [30,50], we can isolate small scale turbulence that is reasonably well-separated from the laminar END mechanism. For the range of noise amplitudes we have tested, the system still generates strong toroidal structures that behave similarly to the canonical case, although some variations in polarity reversals are evident (Fig. 1). Here we were careful to keep the amplitudes low enough so dynamics are not driven by noise.

An obvious next step may be to embed the END into a convective environment. However, it is well known that for the right parameters, convection can drive a kinematic dynamo. Therefore, it would be difficult to understand if dynamo action would be dominated by the END mechanism or by convection. In an effort to separate between the two mechanisms, we construct a two-layer penetrative system where we have a convectively unstable upper layer sitting atop a quiescent environment in the bottom layer where we drive the END as in the canonical run. We analyze this system through a series of simulations. First, a simulation with only convection in the top layer reveals that convective eddies overshooting into the stable layer below extend to the location of peak shear forcing, which is where strong toroidal structures are formed. Second, we have run a case with only the END in the lower layer, finding that with no convection in the top layer, the magnetic structures formed from the END mechanism rise and diffuse away similarly to the canonical case. Finally, we combine the two simulations with convection in the top layer and the canonical set up in the bottom layer. Observing the time and horizontally averaged magnetic energy, we find evidence for the operation of the END in the lower layer despite the interaction with convection (Fig. 2). We also notice an additional concentration of magnetic energy in the top layer that was not present in the simulation with convection turned off (Fig. 2). This may be an example of a "stoked kinematic dynamo" in the upper layer fed by the END in the lower layer [1], which warrants further investigation.

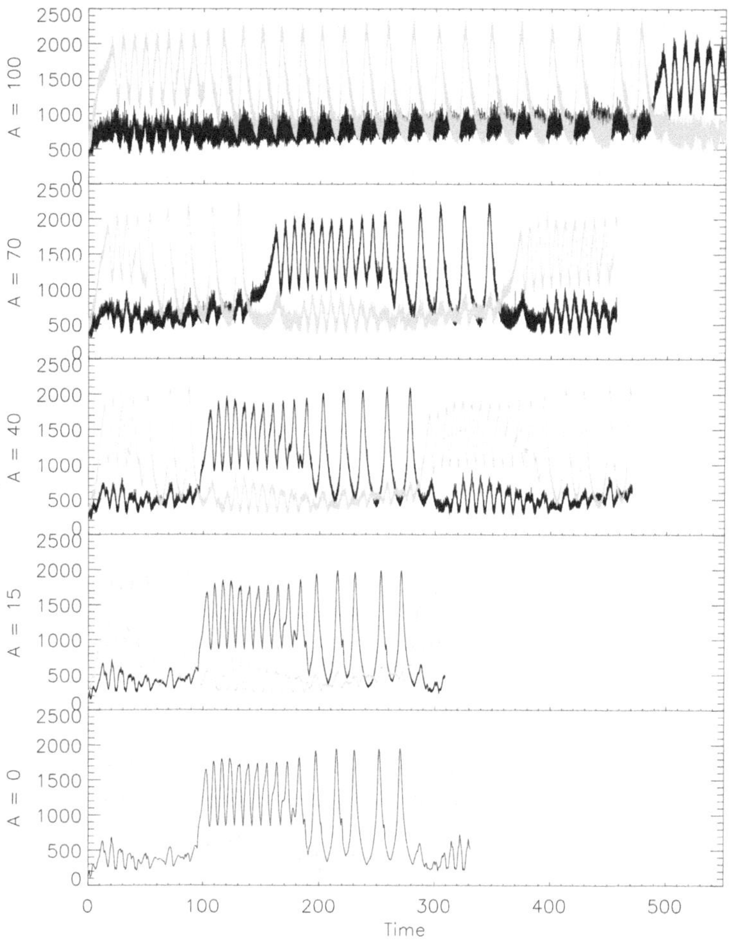

Fig. 1 B_x vs t for varying noise amplitudes, A. Dark shades are $\max B_x$, light shades are $\min|B_x|$. $A = 0$ is the canonical run

These investigations have led to the conclusion that this END mechanism is robust to many forms of turbulence. While the fundamental characteristics of the END may be altered, neither small scale noise or larger scale fully resolved convection interacting with the END mechanism hinder the generation of long term sustained magnetic energy. However, our models need to be extended to higher degrees of turbulence with a wider spectrum of scales.

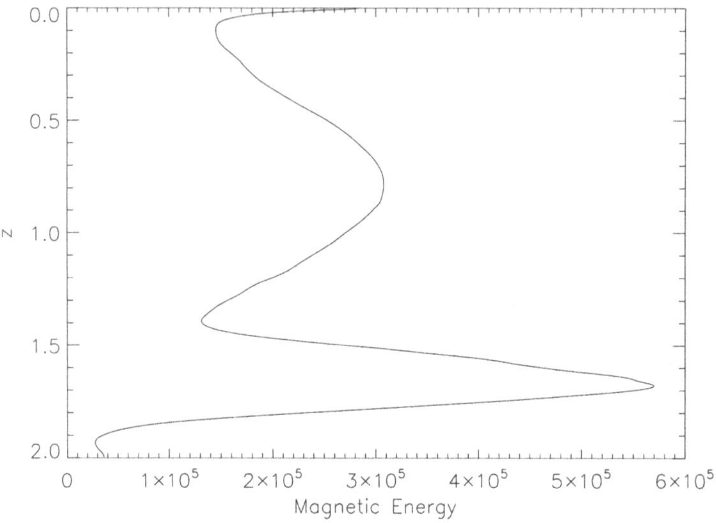

Fig. 2 Time and horizontally averaged magnetic energy vs depth. Bottom peak in energy is END, top peak in energy is "stoked dynamo"

References

1. Byington, B. M., Brummell, N. H., Stone, J. M., & Gough, D. O. (2014). *New Journal of Physics, 16*, 083002.
2. Cline, K. S., Brummell, N. H., & Cattaneo, F. (2003). *The Astrophysical Journal, 599*(2), 1449.

Part IV
Stellar Dynamics

The Tachocline Revisited

Pascale Garaud

Abstract The solar tachocline is a shear layer located at the base of the solar convection zone. The horizontal shear in the tachocline is likely turbulent, and it is often assumed that this turbulence would be strongly anisotropic as a result of the local stratification. What role this turbulence plays in the tachocline dynamics, however, remains to be determined. In particular, it is not clear whether it would result in a turbulent eddy diffusivity, or anti-diffusivity, or something else entirely. In this paper, we present the first direct numerical simulations of turbulence in horizontal shear flows at low Prandtl number, in an idealized model that ignores rotation and magnetic fields. We find that several regimes exist, depending on the relative importance of the stratification, viscosity and thermal diffusivity. Our results suggest that the tachocline is in the *stratified turbulence* regime, which has very specific properties controlled by a balance between buoyancy, inertia, and thermal diffusion.

1 The Turbulent Tachocline

Characterizing and understanding the solar tachocline was one of Michael Thompson's many fundamental contributions to the subject of solar physics [9]. Discovered in the late 1980s [1], the tachocline is now thought to play a fundamental role in the solar dynamo because it combines substantial radial and horizontal shear while being at the interface between the convection zone and the radiation zone. The first model of the solar tachocline [8] was, however, purely hydrodynamical. Noting that the horizontal shear in the tachocline is likely unstable (see [11, 3]), Spiegel and Zahn argued that the latter should be turbulent. In addition, the strong stratification

P. Garaud (✉)
Department of Applied Mathematics, Baskin School of Engineering, UC Santa Cruz, Santa Cruz, CA, USA
e-mail: pgaraud@ucsc.edu

M. J. P. F. G. Monteiro et al. (eds.), *Dynamics of the Sun and Stars*, Astrophysics and Space Science Proceedings 57,
https://doi.org/10.1007/978-3-030-55336-4_29

would cause the turbulence to be highly anisotropic, so the transport of angular momentum in the horizontal direction should be much larger than in the vertical direction. With these assumptions, they were able to propose a simple steady-state model in which the tachocline effectively operates as a turbulent internal boundary layer, across which the latitudinal shear in the convection zone is rapidly reduced down to the negligible levels observed in the radiation zone below.

To model this idea mathematically, Spiegel and Zahn argued that the transport of angular momentum in the tachocline should be governed by the following equation,

$$r^2 \sin^2\theta \frac{\partial \tilde{\Omega}}{\partial t} + 2\Omega_0 r \cos\theta \sin\theta u_\theta \simeq \frac{\nu_h}{\sin\theta} \frac{\partial}{\partial \theta}\left(\sin^3\theta \frac{\partial \tilde{\Omega}}{\partial \theta}\right), \tag{1}$$

where vertical advection terms as well as vertical turbulent transport terms have been neglected. In this equation, r and θ are the radius and co-latitude, respectively, Ω_0 is the mean rotation rate of the Sun while $\tilde{\Omega}$ is the deviation away from that mean, u_θ is the latitudinal velocity, and ν_h is the turbulent horizontal momentum diffusivity. It is important to note that the term on the r.h.s. is a model for the effect of the turbulence—whether the latter actually behaves in this manner is a question that one should attempt to answer. This equation must be completed with a model for the dynamics of the meridional flow, which can be obtained by considering geostrophic balance and thermal equilibrium:

$$\frac{1}{\rho}\frac{\partial \tilde{p}}{\partial r} = g\frac{\tilde{T}}{T}, \quad 2\Omega_0 r \cos\theta \tilde{\Omega} = \frac{1}{\rho r \sin\theta}\frac{\partial \tilde{p}}{\partial \theta}, \tag{2}$$

$$\text{and } \frac{N^2 T}{g} u_r = \frac{\kappa_T}{r^2}\frac{\partial}{\partial r}\left(r^2 \frac{\partial \tilde{T}}{\partial r}\right), \tag{3}$$

where ρ and T are the mean background density and temperature in the tachocline, $\tilde{p}$ and $\tilde{T}$ are perturbations away from the background means, respectively, N is the buoyancy frequency, g is gravity, and κ_T is the local thermal diffusivity. When combined together with incompressibility, these equations ultimately lead to a fourth order differential equation for $\tilde{\Omega}$, which after some further simplifications (notably, a boundary layer approximation) reads

$$\frac{\partial^4 \tilde{\Omega}}{\partial r^4} = -\frac{\nu_h}{\kappa_T}\left(\frac{N}{2\Omega_0}\right)^2 \left(\frac{\mu}{r_t}\right)^4 \tilde{\Omega}, \tag{4}$$

where μ is the eigenvalue of an associated latitudinal eigenvalue problem (and is approximately equal to 5), and r_t is the radius of the base of the convection zone. Spiegel and Zahn showed that the solution of this equation that satisfies appropriate boundary conditions in the radial direction has a characteristic thickness

$$h = \frac{3\pi}{2\mu}\left(\frac{\Omega_0}{N}\right)^{1/2}\left(\frac{\kappa_T}{\nu_h}\right)^{1/4} r_t \simeq 1.7\left(\frac{\nu_h}{\nu}\right)^{-1/4} r_t, \tag{5}$$

using values commonly associated with the tachocline, namely $\Omega_0 \simeq 3 \times 10^{-6}\,\mathrm{s}^{-1}$, $N \simeq 10^{-3}\,\mathrm{s}^{-1}$, $\nu \simeq 20\,\mathrm{cm}^2/\mathrm{s}$ and $\kappa_T \simeq 2 \times 10^7\,\mathrm{cm}^2/\mathrm{s}$. We therefore see that in order to match observations, where h is estimated to be a few percent of r_t, the turbulent viscosity ν_h in this model needs to be about 8 orders of magnitude larger than the microscopic viscosity ν. A posteriori verification of the conditions under which the model applies also reveals that the ratio of the horizontal to vertical turbulent viscosity should be much larger than $(r_t/h)^2 \sim 10^4$ for the model to be valid, therefore requiring $\nu_r \ll 10^{-4}\nu_h \sim 10^4\nu \sim 10^5\mathrm{cm}^2/\mathrm{s}$.

The Spiegel and Zahn model was later criticized by Gough and McIntyre [4], who argued by analogy with observations of strongly stratified turbulence in the Earth's atmosphere, that

> [...] horizontal turbulence controls the distribution of angular momentum in such a way as to drive the system away from, not towards, uniform rotation. Meteorologists once called this negative viscosity.

In particular, they argued that the use of a turbulent viscosity prescription adopted by Spiegel and Zahn (leading to Eq. 1) is inappropriate, and proposed an alternative model of the tachocline that involves the presence of a large-scale primordial magnetic field embedded in the radiation zone.

Today, the question of whether turbulent angular momentum transport in the tachocline is diffusive or anti-diffusive remains open. For *purely* two-dimensional flows, Tobias, Diamond and Hughes [10] confirmed that turbulence in the presence of rotation drives large-scale shear rather than quenches it. What form it takes in a more realistic three-dimensional system, and whether this conclusion holds or not, remains to be determined. In this paper, I will present recent work on the subject of horizontal shear instabilities in strongly stratified low Prandtl number flows, that will go part-way towards answering these questions.

2 Horizontal Shear Instabilities in Stars

Horizontal shear instabilities in stellar interiors were recently studied by Cope et al. [2] in a series of numerical experiments. This section summarizes their results, and presents additional experiments that provide important clues on the nature and relevance of horizontal shear instabilities in stars in general and in the tachocline in particular.

2.1 Model Description

In Cope et al. [2], we consider what is possibly the simplest model for stratified horizontal shear instabilities. We ignore the effect of curvature, rotation, magnetic fields, and compositional stratification, to consider a uniformly thermally stratified, and otherwise triply-periodic Cartesian domain with z pointing in the upward direction. A body force $\mathbf{F}$ is applied to drive a flow in the streamwise direction x, that varies in the spanwise direction y, so $\mathbf{F} = F_0 \sin(ky)\mathbf{e}_x$. The system of equations governing the flow, in the Spiegel-Veronis-Boussinesq approximation [7] (which is valid in the tachocline) are:

$$\frac{\partial \mathbf{u}}{\partial t} + \mathbf{u} \cdot \nabla \mathbf{u} = -\rho^{-1}\nabla \tilde{p} + \alpha \tilde{T}\mathbf{g} + \nu \nabla^2 \mathbf{u} + \rho^{-1}\mathbf{F}, \tag{6}$$

$$\nabla \cdot \mathbf{u} = 0, \tag{7}$$

$$\frac{\partial \tilde{T}}{\partial t} + \mathbf{u} \cdot \nabla \tilde{T} + u_z \left(\frac{\partial T}{\partial z} - \frac{\partial T_{ad}}{\partial z} \right) = \kappa_T \nabla^2 \tilde{T}, \tag{8}$$

where $\mathbf{u} = (u_x, u_y, u_z)$, $\mathbf{g} = -g\mathbf{e}_z$ is gravity, α is the coefficient of thermal expansion, roughly equal to $1/T$, $\partial T/\partial z$ is the background mean temperature gradient, and $\partial T_{ad}/\partial z = -g/c_p$ is the background adiabatic temperature gradient, where c_p is the specific heat at constant pressure. The quantities $\tilde{T}$ and $\tilde{p}$ are perturbations away from the background state.

The equations are non-dimensionalized using the shear lengthscale k^{-1} and the anticipated horizontal velocity of the turbulent flow U, obtained by assuming a balance between the forcing and the inertial terms on a scale k^{-1}, so

$$kU^2 = \rho^{-1} F_0 \to U = \left(\frac{F_0}{k\rho} \right)^{1/2}. \tag{9}$$

The temperature scale is taken to be $k^{-1}\left(\frac{\partial T}{\partial z} - \frac{\partial T_{ad}}{\partial z}\right)$. The nondimensional equations are then

$$\frac{\partial \hat{\mathbf{u}}}{\partial t} + \hat{\mathbf{u}} \cdot \nabla \hat{\mathbf{u}} = -\nabla \hat{p} + B\hat{T}\mathbf{e}_z + Re^{-1}\nabla^2 \hat{\mathbf{u}} + \sin y \mathbf{e}_x, \tag{10}$$

$$\nabla \cdot \hat{\mathbf{u}} = 0, \tag{11}$$

$$\frac{\partial \hat{T}}{\partial t} + \hat{\mathbf{u}} \cdot \nabla \hat{T} + \hat{u}_z = Pe^{-1}\nabla^2 \hat{T}, \tag{12}$$

where

$$Re = \frac{U}{k\nu}, \quad Pe = \frac{U}{k\kappa_T} = Pr Re, \text{ and } B = \frac{N^2}{k^2 U^2}, \tag{13}$$

are the Reynolds number, Péclet number and buoyancy parameters, respectively. Note that in stars, Pr is always much smaller than one (it is for instance of order 10^{-6} in the solar tachocline). In addition, Re is always very large. As a result, we only consider cases where $Pr \ll 1$ and $Re \gg 1$. The remaining parameters, B and Pe, can be either large or small. In the limit of low Péclet number, an interesting asymptotic reduction of the equations can be made [6, 5], whereby the dominant balance in the temperature equation is given by

$$\hat{u}_z = Pe^{-1}\nabla^2\hat{T}. \tag{14}$$

Substituting this into the momentum equation leads to

$$\frac{\partial\hat{\mathbf{u}}}{\partial t} + \hat{\mathbf{u}}\cdot\nabla\hat{\mathbf{u}} = -\nabla\hat{p} + BPe\nabla^{-2}\hat{u}_z\mathbf{e}_z + Re^{-1}\nabla^2\hat{\mathbf{u}} + \sin y\mathbf{e}_x, \tag{15}$$

which is called the low Péclet number approximation [5]. We see that in that limit, the parameters B and Pe combine to form a single parameter, BPe. As such, we expect the flow dynamics to be controlled by the product BPe rather than by B and Pe individually whenever thermal diffusion is important.

2.2 *Low Péclet Number Results*

The study of Cope et al. [2] specifically focusses on the case of small Pe, using both the standard governing equations at low Pe (10)–(12) (here, low Pe is loosely defined as $Pe \leq 1$), as well as the low Péclet number approximation (15). The two are found to be in excellent agreement for $Pe \leq 0.1$.

2.2.1 Qualitative Results

In [2], we present a number of Direct Numerical Simulations (DNSs), evolving the governing equations from some initial conditions until a statistically stationary state is reached (the final state achieved is, to our knowledge, always independent of the initial conditions in this system). In all cases, the domain size is $4\pi \times 2\pi \times 2\pi$. When started from quiescent initial conditions ($\hat{\mathbf{u}} = 0$), the imposed force first drives a vertically-invariant flow that is sinusoidal in the y direction. Once it has reached a sufficient amplitude that flow becomes unstable to shear instabilities. The fastest-growing mode is always two-dimensional (2D, i.e. vertically invariant), and is therefore unaffected by the stratification. This 2D mode causes a horizontal meandering of the original shear flow. Depending on the parameter regime, other modes of instability are then also excited, which now depend on z. These modes cause a vertical modulation of the phase of the horizontal meanders, and thereby generate vertical shear (see Fig. 1). This vertical shear drives further dynamics that eventually saturate the primary instability.

We found that the system dynamics, once a statistically stationary state is achieved, only depend on Re and the product BPe, as expected from the argument above. We also found that there are (at least) four distinct regimes (in addition to the laminar solution at low Reynolds number) depending on the respective values

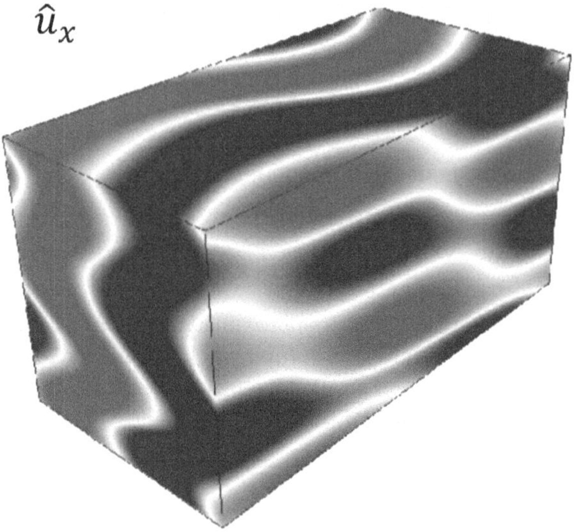

Fig. 1 Snapshot of the streamwise flow $\hat{u}_x$ in a simulation with $Re = 300$, $Pe = 0.1$ and $B = 30{,}000$, during the initial phase of exponential growth of the instability, showing the presence of horizontal meanders of the basic flow, that are vertically modulated. This creates shear in the vertical direction. Figure adapted from Cope et al. [2]

of Re and BPe, see Fig. 2. The boundaries between the various regimes can be determined from dominant balance arguments when possible, and empirically otherwise (see [2] for more detail). Snapshots of two simulations taken in the stratified turbulence regime and stratified intermittent regime are shown in Fig. 3. In the stratified turbulent regime, vertical shear instabilities ubiquitously develop between the meanders of the streamwise flow. The vertical size of the eddies is controlled by a balance between buoyancy, inertia, and thermal diffusion, see Sect. 2.2.2. Interestingly, we find that the eddies are relatively isotropic, with a horizontal scale that is commensurate with the vertical scale.

As stratification continues to increase, the system enters a regime where turbulence is only found intermittently (both in time and space). Turbulence within these localized patches is similar that found in the stratified turbulence regime. On the other hand, the effect of viscosity becomes important outside of these patches, which are increasingly sparse as BPe increases. For sufficiently large BPe, the system enters a viscously dominated regime.

2.2.2 Scaling Laws

In order to gain a more quantitative understanding of the dynamics of the various regimes identified in the low Péclet number case, we extracted various diagnostics from the DNSs. In particular, we measured the vertical eddy scale $\hat{l}_z$ (using the

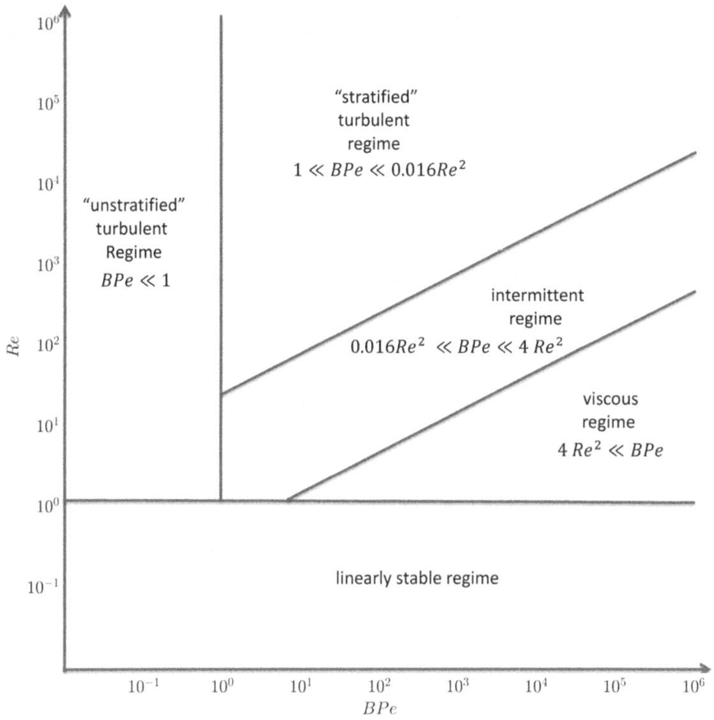

Fig. 2 Partitioning of parameter space in the low Péclet number limit ($Pe \lesssim 0.1$). Each regime (except the intermittent regime) corresponds to a well-defined dominant balance in the momentum equation, between the forcing, the inertial terms, the buoyancy term, and the viscous terms. The intermittent regime contains regions that are viscously-dominated as well as regions that are in the stratified turbulent regime. Figure from Cope et al. [2]

autocorrelation function of the vertical velocity field), the r.m.s. vertical velocity $\hat{u}_z^{\rm rms} = \langle \hat{u}_z^2 \rangle^{1/2}$, the r.m.s. temperature $\hat{T}^{\rm rms}$ and the mixing efficiency η, which is defined as

$$\eta = \frac{-B\langle \hat{u}_z \hat{T} \rangle}{-B\langle \hat{u}_z \hat{T} \rangle + Re^{-1}\langle |\nabla \hat{\mathbf{u}}|^2 \rangle}, \tag{16}$$

where $\langle \cdot \rangle$ denotes an average over the entire computational domain. These quantities were then time-averaged during the statistically stationary phase, and the results are shown in Fig. 4. The significance of η can be understood by noting that in a statistically stationary state, the kinetic energy equation (which is obtained by dotting the momentum equation by $\hat{\mathbf{u}}$ and integrating over the domain, using periodicity to eliminate boundary terms) reduces to

$$\langle \hat{\mathbf{u}} \cdot \hat{\mathbf{F}} \rangle = -B\langle \hat{u}_z \hat{T} \rangle + Re^{-1}\langle |\nabla \hat{\mathbf{u}}|^2 \rangle. \tag{17}$$

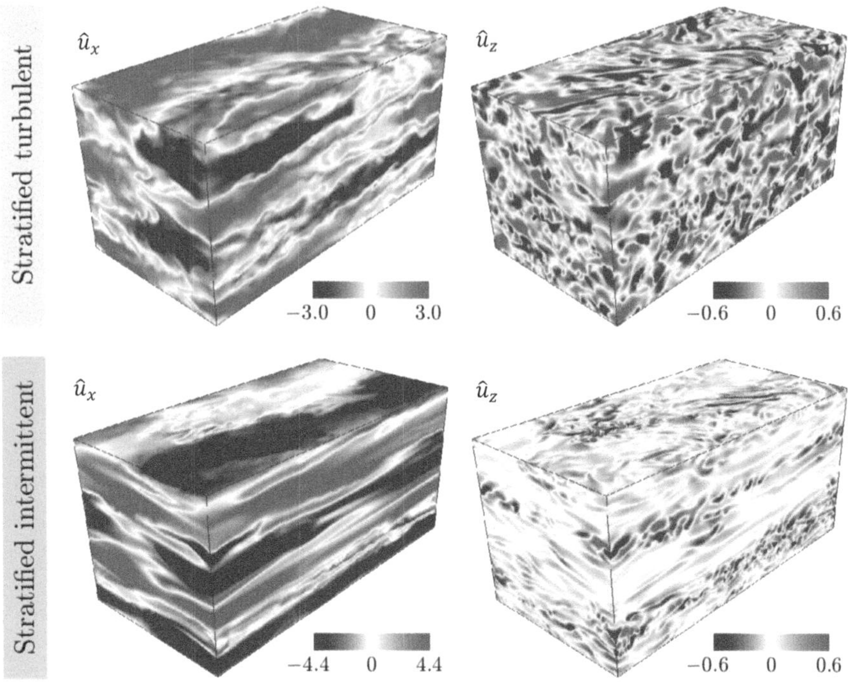

Fig. 3 Snapshots of $\hat{u}_x$ and $\hat{u}_z$ in the stratified turbulent regime (top, $B = 100$) and intermittent regime (bottom, $B = 10{,}000$) at $Re = 300$, $Pe = 0.1$. Note how the meanders of the streamwise flow are still visible even in the fully turbulent flow, and note the small vertical scale of the turbulent eddies in both cases. Figure adapted from Cope et al. [2]

We therefore see that the energy input into the system by the force $\hat{\mathbf{F}}$ (on the l.h.s.) is either converted into potential energy (first term on the r.h.s.) or viscously dissipated (second term on the r.h.s.). As such, the ratio η measures how efficiently the total energy input into the system is used to mix the background stratification. This quantity turns out to be an excellent diagnostic of the properties of the flow, as shown by Cope et al. [2].

In the stratified turbulent regime, which is of potential relevance to stellar interiors, we found that the vertical lengthscale of the turbulent eddies scales as $\hat{l}_z \sim (BPe)^{-1/3}$. This scaling can be understood from a simple dominant balance in the vertical component of the momentum equation between the nonlinear terms and buoyancy term, $\hat{\mathbf{u}} \cdot \nabla \hat{u}_z \simeq B\hat{T}$. Meanwhile from the low Péclet number thermal energy equation, we have $\hat{u}_z \simeq Pe^{-1}\nabla^2\hat{T}$. From a scaling perspective, these two equations result in

$$\frac{\hat{u}_h^{\mathrm{rms}}\hat{u}_z^{\mathrm{rms}}}{\hat{l}_z} \sim B\hat{T}^{\mathrm{rms}} \text{ and } \hat{u}_z^{\mathrm{rms}} \sim Pe^{-1}\frac{\hat{T}^{\mathrm{rms}}}{\hat{l}_z^2}, \tag{18}$$

where $\hat{u}_h^{\rm rms} \sim O(1)$ is the expected horizontal velocity in the non-dimensionalization chosen. Combined, this results in $\hat{l}_z \sim (BPe)^{-1/3}$, as observed.

In the same regime, we also observe that η is constant and roughly equal to 0.4, which shows that about 40% of the total energy input into the system is spent mixing the background stratification, while about 60% is dissipated viscously. By construction, the non-dimensional energy input rate is of order one, so we find that $B\langle \hat{u}_z \hat{T}\rangle = O(1)$ as well, which implies that $B\hat{u}_z^{\rm rms}\hat{T}^{\rm rms} = O(1)$. Combining this with (18) above, we then predict that

$$\hat{u}_z^{\rm rms} \sim (BPe)^{-1/6} \text{ and } \hat{T}^{\rm rms} \sim Pe(BPe)^{-5/6}. \quad (19)$$

Both scalings are indeed observed in high Reynolds number and low Péclet number simulations (see Fig. 4). The prefactor in both cases is found to be of order unity.

As discussed by Cope et al. [2], the stratified turbulent regime is valid for $Pe \lesssim 0.1$ and $1 \lesssim BPe \lesssim 0.0016Re^2$. When BPe increases beyond that threshold, the fraction of the domain that is turbulent decreases and is gradually replaced by laminar regions that are viscously dominated.

2.2.3 Mixing by Horizontal Shear Flows at Low Péclet Number

If a star is known to exhibit a horizontal shear flow with amplitude U and wavenumber k, then one can easily construct the parameters Re, Pe and B from (13). If the Péclet number is then found to be small, the scalings obtained above can be expressed dimensionally to yield

$$l_z \sim (BPe)^{-1/3}k^{-1} \sim \left(\frac{N^2}{U\kappa_T}\right)^{-1/3}, \quad (20)$$

$$u_z^{\rm rms} \sim (BPe)^{-1/6}U \sim \left(\frac{N^2}{U^7k^3\kappa_T}\right)^{-1/6}, \quad (21)$$

with proportionality constants of order unity. A vertical mixing coefficient can then be formed as

$$D \sim l_z u_z^{\rm rms} \sim \left(\frac{N^2}{Uk^3\kappa_T}\right)^{-1/2} Uk^{-1}. \quad (22)$$

2.3 *Low Prandtl Number/High Péclet Number Results*

Unfortunately, the results described in the previous section are not a priori applicable to the solar tachocline because the latter is not a low Péclet number shear layer, at

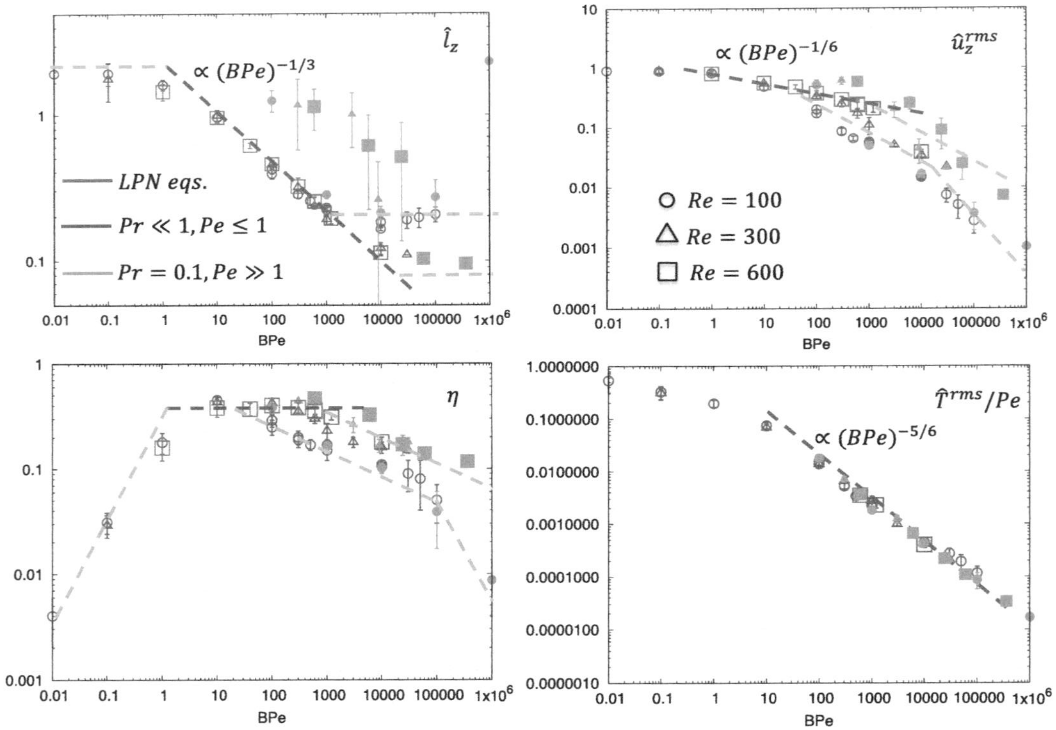

Fig. 4 Diagnostic quantities extracted from all available simulations. The open blue and red symbols correspond to simulations at very low Prandtl number and low Pe ($Pe \leq 1$), see figure for legend. The green filled symbols correspond to simulations with $Pr = 0.1$, high Pe. Errorbars are added in all cases, but are sometimes smaller than the symbol. The dashed lines correspond to the model scaling laws. Emphasis is put on the stratified turbulence regime (dark dashed lines), see [2] for information on the other regimes (pale dashed lines)

least when the Péclet number is computed using properties of the mean flow. Indeed, using a typical lengthscale $k^{-1} \simeq r_t/4$ and typical velocity $U = r_t \Delta\Omega$, where $r_t \simeq 5 \times 10^{10}$cm and $\Delta\Omega \simeq 3 \times 10^{-7}$s^{-1} is the difference between the angular velocity of the equator and the pole, we find that

$$Re \sim O(10^{13}), \quad Pe \sim O(10^7), \text{ and } B \sim O(10^6), \tag{23}$$

so clearly $Pe \gg 1$. In that limit, the low Péclet number approximation that is central to the derivation of the scaling laws presented in the previous section does not apply a priori. To see how having a large Péclet number modifies the results, we ran a number of simulations at $Pr = 0.1$ and large Reynolds number, so that $Pe = 0.1 Re$ remains large. This is computationally challenging, since a high Péclet number with a low Prandtl number requires an even larger Reynolds number, which demands very high resolution. The preliminary results presented here are therefore limited to $Re = 100(Pe = 10)$, $Re = 300(Pe = 30)$ and $Re = 600(Pe = 60)$.

Surprisingly, we found that many of the trends observed in low Péclet number flows continue to hold in this case, at least qualitatively speaking. In particular, we find that the instability of the base flow continues to give rise to vertically-modulated meandering of the streamwise velocity field, that then creates strong vertical shear. For weak to moderate stratification, the vertical shear becomes in turn unstable, causing vertical mixing. The same regimes can be identified (unstratified turbulence, stratified turbulence, intermittent and viscous). In the stratified turbulent regime, snapshots of the flow look qualitatively very similar to those obtained in the low Péclet number regime (see Fig. 5).

In each case, we extracted again quantities such as $\hat{l}_z$, $\hat{T}^{\rm rms}$, $\hat{u}_z^{\rm rms}$ and η. Results for moderate to large B are presented in Fig. 4 (green symbols) and reveal a number of very interesting findings. Crucially, we find for instance that the data for $\hat{T}^{\rm rms}$

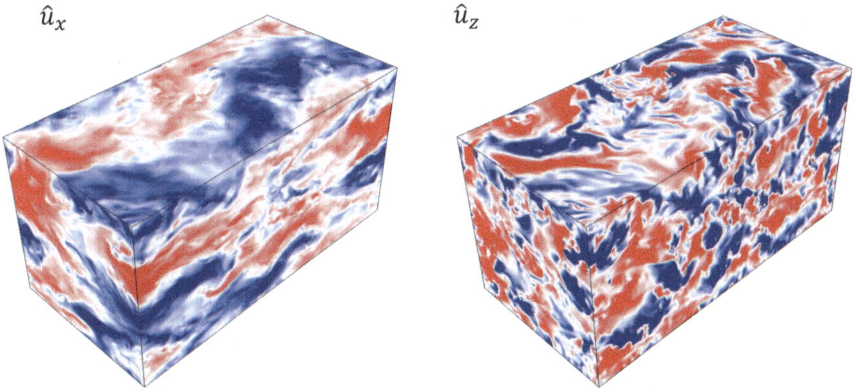

Fig. 5 Snapshots of the horizontal velocity $\hat{u}_x$ and vertical velocity $\hat{u}_z$ taken in the statistically stationary state of a simulation with $Re = 600$, $Pe = 60$ and $B = 10$, which is in the stratified turbulence regime

collapses onto the same approximate scaling of $\hat{T}^{\rm rms} \sim Pe(BPe)^{-5/6}$ found in the low Péclet limit, suggesting that the flow dynamics continue to be thermally diffusive even though Pe is large. We also find that the new high Péclet number data collapse with the low Péclet number data in the viscous regime (this is particularly apparent for the $Re = 100$ runs (circles), for which the viscous regime starts around $BPe \simeq 1000$). Finally, and most importantly, our results tentatively suggest that the *same* scaling laws apply in the stratified turbulent regime, albeit with a different pre-factor that depends on the Prandtl number.

These results are surprising at first, since the scaling laws derived in Sect. 2.2.2 require the low Péclet number approximation to the thermal energy equation (14) to hold, which should not be the case at high Pe. To understand why these low Péclet number scalings might still apply to the low Prandtl number/high Péclet number cases, it is important to remember that the derivation of the low Péclet approximation [5] relies on the assumption that the *turbulent* Péclet number $Pe_l = Ul/\kappa_T$ be small, where Pe_l is computed using the r.m.s. velocity of the fluid (for which U is a good approximation, at least in the horizontal direction) and the *actual* eddy scale l. By contrast, the input parameter Pe defined in (13) is based on the largest possible physical scale of the system, which is that of the imposed shear, k^{-1}. We have seen that the emergent scale of the eddies in the stratified turbulent regime is much smaller than k^{-1} in all directions, so it is quite likely that our simulations are in a regime where $Pe_l \lesssim 1$. This would explain why the emergent dynamics remain thermally diffusive even though $Pe \gg 1$.

Based on the very limited simulations available at high Péclet number/low Prandtl number, we therefore argue that the scalings in the stratified turbulent regime likely remain the same as those described and derived in Sect. 2.2.2, with the exception of constant prefactors that depend weakly on Pr (perhaps logarithmically so), and tend to the ones obtained in the low Péclet limit when $Pr \to 0$. The regime boundaries for the stratified turbulent regime remain to be determined, but it is quite likely that these will also only depend logarithmically on Pr, when $Pr \ll 1$. To test these predictions will require simulations at lower Prandtl number and ideally higher Péclet number, which will be very challenging computationally, but not impossible. If they are confirmed, this will have important consequences for stellar interiors.

3 Discussion

The findings presented in this paper suggest that horizontal shear instabilities in stars (i.e. at low Prandtl number) generate vertically modulated meanders of the basic flow, on vertical scales that are sufficiently thin to be thermally diffusive. As a result, diffusive vertical shear instabilities can develop, and give rise to small-scale turbulence and vertical mixing. The turbulence is relatively isotropic on the small scales, but becomes more anisotropic on the larger horizontal scales associated with the meanders, as illustrated in Figs. 3 and 5.

This is quite different from the effect of horizontal shear flows in geophysical systems, where the Prandtl number is large. Indeed, in that case thermal diffusion is always negligible if the flow is turbulent (since $\kappa_T \lesssim \nu$). This implies that secondary vertical shear instabilities cannot be excited if the stratification is strong, but are instead limited to localized regions where the stratification is weakened. By and large, the turbulence therefore remains almost two-dimensional, and in a rotating system, this two-dimensional turbulence would indeed have an anti-diffusive behavior, as seen in the Earth's atmosphere and invoked by Gough and McIntyre to argue against the Spiegel and Zahn model of the tachocline. Our simulations at low Prandtl number however demonstrate that the turbulence is clearly three-dimensional and almost isotropic on the small scales, and only become anisotropic on the larger scales. Whether this ultimately behaves in a diffusive or anti-diffusive manner in the presence of rotation therefore remains to be determined.

Tentatively, we propose new scaling laws for the vertical eddy scale, vertical velocity, and vertical mixing coefficient in horizontal shear instabilities in the stratified turbulent regime, given in Eqs. (21) and (22). The prefactors are of order unity when the Péclet number is small (see [2] for more detail), but might depend logarithmically on Pr if $Pr \ll 1$ but $Pe \gg 1$. If this is confirmed then simple order-of-magnitude estimates for l_z, $u_z^{\rm rms}$, $T^{\rm rms}$ and D in the tachocline are

$$l_z \sim O(10\text{km}),\, u_z^{\rm rms} \sim O(10\text{cm/s}),\, T^{\rm rms} \sim O(100\text{K}) \text{ and } D \sim O(10^7\text{cm}^2/\text{s}), \tag{24}$$

since the tachocline is likely in the stratified turbulent regime (with $BPe \sim 10^{13} \ll 0.002Re^2$). With this estimate, we see that one of the fundamental assumptions of the Spiegel and Zahn model may not be satisfied: indeed, if the vertical turbulent momentum diffusivity is of order $\nu_r \sim D \sim 10^7$ cm^2/s, then it is too large to neglect in Eq. (1), and would therefore invalidate the model. A possible solution to the problem (other than Gough and McIntyre's magnetic idea) is that the model holds with $\nu_h \gg 10^8\nu$, and that the actual tachocline is in fact much thinner than $\sim 0.01r_t$—this cannot be ruled out by observations.

Of course, much remains to be done to characterize mixing by shear instabilities in the tachocline (and in stars more generally). Several effects have indeed been neglected, that will have to be included before definite conclusions can be made. First, it will be important to include the effect of rotation. Indeed, while the latter is not expected to influence the turbulence on the smaller scales (where the Rossby number is large), it could alter or even suppress the development of the primary instability, which occurs on the larger scales. Second, magnetic fields will also need to be included, since they are expected to be present and significant in the tachocline. Finally, the tachocline has both large-scale vertical and horizontal shear, and the former may influence the nonlinear saturation of the latter.

Acknowledgments This work was initiated as a project at the Woods Hole Geophysical Fluid Dynamics summer program in 2018; we thank the GFD program for their support. P.G. is funded by the National Science Foundation (NSF AST 1814327), and wishes to thank L. Cope, C.P. Caulfield and D. Gough for stimulating discussions.

References

1. Christensen-Dalsgaard, J., & Schou, J. (1988). In E. J. Rolfe (Ed.), *Seismology of the sun and sun-like stars. ESA special publication* (vol. 286, pp. 149–153).
2. Cope, L., Garaud, P., & Caulfield, C. P. (2019). e-prints. arXiv:1911.09674.
3. Garaud, P. (2001). *Monthly Notices of the Royal Astronomical Society, 324*(1), 68.
4. Gough, D. O., & McIntyre, M. E. (1998). *Nature, 394*, 755.
5. Lignières, F. (1999). *Astronomy and Astrophysics, 348*, 933.
6. Spiegel, E. A. (1962). *Journal of Geophysical Research, 67*(3), 3063.
7. Spiegel, E. A., & Veronis, G. (1960). *The Astrophysical Journal, 131*, 442.
8. Spiegel, E. A., & Zahn, J. P. (1992). *Astronomy and Astrophysics, 265*, 106.
9. Thompson, M. J., Christensen-Dalsgaard, J., Miesch, M. S., & Toomre, J. (2003). *Annual Review of Astronomy and Astrophysics, 41*, 599.
10. Tobias, S. M., Diamond, P. H., & Hughes, D. W. (2007). *The Astrophysical Journal and Letters, 667*(1), L113.
11. Watson, M. (1980). *Geophysical & Astrophysical Fluid Dynamics, 16*(1), 285.

Seismic Signatures of Solar and Stellar Magnetic Activity

Ângela R. G. Santos

Abstract The 11-year solar activity cycle manifests itself through different phenomena, with the most obvious being the varying number of sunspots on the solar surface. As magnetic activity affects the propagation of acoustic waves, their properties are observed to vary in a cyclic way. In particular, activity-related frequency shifts in the Sun have been known for more than three decades. Such seismic signatures of magnetic activity are expected to be common among solar-type stars. Thanks to the advent of planet-hunting missions, we are now able to measure temporal variations in the mode properties of dozens of solar-type stars. As follows, helio- and asteroseismology provide an unique way to study solar and stellar magnetism.

1 Introduction

The properties of the acoustic modes are sensitive to the conditions of the medium where the waves propagate. As the magnetic activity changes those conditions, the acoustic-mode properties vary in close relationship with the proxies for solar activity. Such variations enclose information on the activity-related changes taking place in stellar interiors. Therefore, we can use helioseismology and asteroseismology to better understand the magnetic properties of the Sun and other stars.

In the Sun, activity-related frequency shifts have been known since the mid-80s. In 1985, [86] reported the first detection of seismic signatures of solar magnetic activity. Using solar observations taken at solar maximum and at near solar minimum of cycle 21, the authors found a frequency shift of about 0.42μHz between the two epochs. Since then, subsequent analyses have been enhancing our knowledge on the solar activity cycle and its properties (see Sect. 2).

Â. R. G. Santos (✉)
Space Science Institute, Boulder, CO, USA
e-mail: asantos@spacescience.org

M. J. P. F. G. Monteiro et al. (eds.), *Dynamics of the Sun and Stars*, Astrophysics and Space Science Proceedings 57,
https://doi.org/10.1007/978-3-030-55336-4_30

Magnetic activity cycles are also observed in solar-type stars (for example [84, 2]), such as low-mass main-sequence stars with convective outer layers. Stellar magnetic activity is known to depend on stellar age, with stars becoming less active as they evolve [83, 85] and, simultaneously, spin-down [77, 82, 55]. Solar-type stars are expected to exhibit acoustic oscillations [26] possibly sensitive to stellar magnetic activity similarly to the solar case. Thus, activity-related variations in the mode properties are expected to be typical of solar-type stars. In spite of the different challenges (see below), we are now able to evaluate temporal variations in the mode properties of dozens of stars. While currently its is not possible to individually establish the source for the variations for all stars, for three stars in particular the observed variations have been shown to be indeed connected to magnetic activity (see Sect. 3.2).

In these proceedings, we summarize the main findings on seismic signatures of solar and stellar magnetic activity. For a review on helioseismology of the solar cycle see [9]. For a recent review on asteroseismology of solar-type stars see [24].

2 Seismic Signatures of Solar Magnetic Activity

2.1 Activity-Related Frequency Shifts

As a result of the solar cycle, the properties of the acoustic modes are observed to vary in a cyclic way. In particular, the mode frequencies vary in phase with the activity level (for example [86, 19, 49, 10, 31]), i.e. frequencies increase with increasing activity. This is shown in Fig. 1 for solar cycle 23, where frequency shift refers to the variation in frequency relative to the average value.

Modes of different frequency and angular degree are affected differently, with activity-related frequency shifts increasing with frequency and angular degree (for example [49, 10, 31, 12, 20, 14]). The mode degree dependency is mainly related to the mode inertia E_{nl}, which decreases with increasing degree l (for example [12, 29]). One can correct for this dependency by multiplying the frequency shifts by the inertia ratio Q_{nl} defined as [17]

$$Q_{nl} = \frac{E_{nl}}{E_0(\nu_{nl})}, \tag{1}$$

where n is the radial order of the mode and E_0 corresponds to the mode inertia that a radial mode would have at the frequency ν_{nl}. E_{nl} is defined as [17]

$$E_{nl} = \frac{4\pi}{M_\odot} \int_0^{R_\odot} |\xi|^2 \rho r^2 dr = \frac{M_{nl}}{M_\odot} \tag{2}$$

where $R_\odot$ and $M_\odot$ are the solar radius and mass, ξ is the normalized displacement, ρ is the density, and M_{nl} is the modal mass.

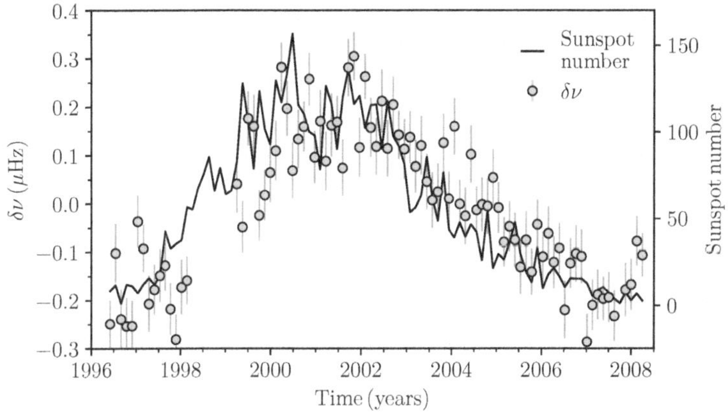

Fig. 1 Average frequency shifts $\delta\nu$ (grey symbols) for solar cycle 23 obtained by Santos et al. [73] using 90-day segments of VIRGO/SPM data (Variability of solar IRradiance and Gravity Oscillations on board SOlar and Heliospheric Observatory—SOHO—, where SPM stands for sunphotometers; [22, 23, 38]). For reference, the International Sunspot Number is shown in black, being averaged over 90 days. Sunspot data from NCEI/NOAA (National Centers for Environmental Information, part of National Oceanographic Data Center). Note that the so-called SOHO vacations took place between 1998 and 1999, hence the absence of frequency shifts for that time period

Between low-degree modes ($l \leq 3$), the mode inertia does not differ significantly. Nevertheless, activity-related frequency shifts still depend on angular degree for modes of $l \leq 3$, generally increasing with l (for example [13, 8, 66]). For low-degree modes, the l-dependency is, in fact, a consequence of the latitudinal distribution of magnetic activity and the sensitivity of the low-degree modes of different l to those latitudes. This opens the possibility of using the degree-dependence of the frequency shifts to infer active latitudes for other stars (see Sect. 3.2).

In the Sun, the activity-related frequency shifts increase with frequency. High-frequency modes are more sensitive to the outermost layers than low-frequency modes (for example [49, 38, 3]). Thus, the increasing frequency shifts with frequency suggest that their origin is related to near-surface changes in the solar magnetism, in particular changes taking place above the upper turning points of the affected modes. In fact, [3] determined that the perturbation leading to the activity-related frequency shifts is located in the outer 99.6% of the solar radius for solar cycle 23.

The different contributions to the activity-related frequency shifts and their relative impact are still not well understood. [18] investigated the impact of the solar global magnetic field and concluded that the indirect effects related to structural and thermal changes are dominant over the direct effect from the Lorentz force. However, cyclic changes in the global field are also accompanied by changes in the area coverage by active regions (regions of strong magnetic field), which can also contribute with both direct and indirect effects. The sunspot indirect effects on the oscillations were explored by Santos et al. [70], who concluded that these are

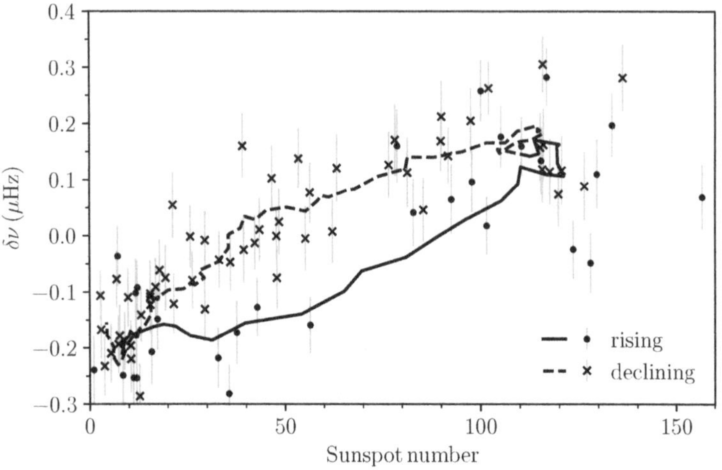

Fig. 2 Frequency shifts obtained by Santos et al. [73] from VIRGO/SPM data (same as in Fig. 1) as a function of the International Sunspot Number (NCEI/NOAA). The rising phase and declining phases of the solar cycle 23 are shown by the filled circles and crosses, respectively. For illustration purpose the solid and dashed lines show the smoothed frequency shifts. Note that the hysteresis pattern differs slightly for different activity proxies (for example [8, 35, 68])

small in comparison with the observed frequency shifts. Observationally, frequency shifts have been found to be better correlated with activity proxies more sensitive to the weak component of the magnetic field (for example [14, 79, 35]), yet both weak and strong components are important (for example [79, 35, 27, 30, 6, 71, 72]). Santos et al. [71] concluded that the contribution from sunspots should be around 30%. Also, [32] speculated that the high response of the high-frequency modes to activity during solar cycle 23 in comparison with the previous cycle could arise from an increase of weak-field ephemeral regions. In contrast, using seismic and magnetogram data, [16] studied the contribution from active regions, where spots are formed, and weak-field ephemeral regions. The authors concluded that the frequencies of the solar acoustic modes are more than three times more sensitive to active regions than to ephemeral regions.

In spite of the strong correlation between frequency shifts and other activity proxies, they are not related in the same way during the rising and declining phases of the cycle ([66, 35, 6, 37], Fig. 2). This arises from magnetic hysteresis resulting from the combination of the sensitivity of the modes to different latitudes and the equatorward migration of active regions over the solar cycle. Furthermore, there is a temporal offset between frequency shifts and other activity proxies, with frequency shifts being ahead (for example [66, 35]).

2.2 Quasi-Biennial Variations

Additionally to the 11-year cycle variation of the mode frequencies, they also vary on a quasi-biennial timescale (for example [8, 21, 75]). The quasi-biennial variations occur in all phases of the solar cycle having, however, higher amplitude during the activity maximum. This mid-term variation is found to be strongly correlated to other solar activity proxies, such as sunspot number and area coverage, and 10.7-cm flux [8, 6]. The quasi-biennial frequency variation shows a weaker dependence on frequency than the cycle variation, which may suggest that their source is located at different depths with that of the mid-term signal taking place deeper [8, 21, 76]. However, we note that the origin of the quasi-biennial variations are still in debate.

2.3 Variations in the Excitation and Damping of the Acoustic Modes

While the mode frequencies vary in phase with the solar activity, the amplitude of the acoustic modes varies in anti-phase with activity (for example [31, 38, 11, 64, 46]). The magnetic structures in the solar photosphere are known to be strong absorbers of magnetic activity (for example [34]). However, the mechanisms behind the changes in the mode excitation and/or damping are not well understood. On one hand, the reduced mode amplitudes due to activity have impact on the detection of acoustic oscillations in solar-type stars (see Sect. 3.2). On the other hand, the anti-correlated behaviour between mode frequencies and amplitudes helps to identify seismic signatures of magnetic cycles in stars other than the Sun (Sect. 3). Note, however, that mode amplitudes are generally harder to constraint than mode frequencies.

The linewidths of the solar acoustic modes increase with increasing activity (for example [31, 38, 11, 64, 46]). This behaviour is connected to an increase on the damping rates leading to a decrease on the mode lifetimes, which may result from changes in the convective motions [28].

2.4 Cycle-to-Cycle Variation

Activity-related frequency shifts have been measured for almost three complete solar cycles. This allows to investigate the changes in the solar magnetism happening between cycles. During solar cycle 22, the observed frequency shifts closely followed the activity proxies [3, 6, 32], while for solar cycle 23 there is a departure during the declining phase, which is later recovered during the rising phase of solar cycle 24 [66, 32]. In addition to the changes in the relation between frequency shifts and other activity proxies between different cycles, it also changes with frequency. The relative smaller shifts for low-frequency modes in solar cycle 23 in comparison with the previous cycle suggest that the observed activity-related frequency shifts originated at deeper layers for solar cycle 22 than for solar cycle 23 [3, 32]. The solar minimum preceding solar cycle 24 was unusually long and deep, which resulted in mode frequencies significantly lower than in the previous minimum (for example [7, 80, 36]). Solar cycle 24 is also weaker than previous recent cycles and the resulting frequency shifts for high-frequency modes are about 30% smaller in comparison with solar cycle 23 [66, 32, 5]. For low-frequency modes the frequency shifts do not change significantly between cycles 23 and 24, suggesting that the strength of magnetic field at layers below 1400 Km remained similar [66]. For solar cycle 24, the agreement between frequency shifts and activity proxies decreases significantly with decreasing frequency [66, 32], which may indicate that the solar dynamo nature is changing. Nevertheless, by investigating the frequency shift dependence on frequency, [5] concluded that the frequency shifts for different phases of the solar cycle, including the extended weak minimum of solar cycle 23, are caused by the same near-surface magnetic perturbation.

3 Seismic Signatures of Stellar Magnetic Activity

Activity-related variations in the acoustic-mode properties are likely common among solar-type stars. However the small amplitudes of the solar-type oscillations and their suppression by magnetic activity hamper the detection of solar-type oscillations and consequently of seismic signatures of magnetic activity (see Sects. 2.1 and 3.2).

The advent of the planet-hunting missions, such as CoRoT (Convection, Rotation, and planetary Transits [1]) and *Kepler* [4], allowed for a new opportunity to search and detect activity-related variations in the acoustic-mode properties.

3.1 CoRoT

Activity-related frequency shifts in the Sun were known for 25 years, when in 2010 using data collected by CoRoT, [25] reported the first detection of activity-related frequency shifts in a star other than the Sun. The authors measured temporal variations in the mode frequencies of the F-type star HD 49933, while a starspot proxy suggests an on-going magnetic activity cycle. Furthermore, the mode amplitudes of HD 49933 also endure a temporal variation which is in anti-phase relative to the frequency shifts. This anti-correlated behaviour between frequencies and amplitudes is consistent with the seismic signatures of magnetic activity in the Sun (see Sect. 2.3). Salabert et al. [65] determined that the frequency shifts increase with mode frequency, similarly to the solar frequency shifts, indicating a near-surface magnetic perturbation.

3.2 Kepler

In 2016, when *Kepler* main-mission was already concluded, [67] uncovered activity-related frequency shifts in KIC 10644253, which is an active solar analog. The authors found the observed frequency shifts varying in phase with the photometric activity proxy S_{ph}. S_{ph} is measured through the spot modulation seen in the *Kepler* long-cadence data as defined by Mathur et al. [50]. The spectroscopic data obtained with HERMES [58] are consistent with KIC 10644253 being about 18% chromospherically more active than the Sun. Similarly to the Sun and HD 49933 (Sect. 3.1), the high-frequency modes of KIC 10644253 are more sensitive to activity than the low-frequency modes.

So far, temporal frequency variations have been measured for a total of 93 *Kepler* solar-type stars [73, 62, 45, 69]. In particular, Santos et al. [73] found that about 60% of their target sample shows significant frequency variations, often accompanied by variations in the S_{ph}, acoustic mode heights, and/or characteristic timescale of the granulation. One of those targets is KIC 8006161,[1] a solar analog with near solar parameters except for its high metallicity. *Kepler* data revealed large temporal frequency shifts, which vary in anti-phase with the mode heights and approximately in phase with the S_{ph} (see Fig. 3). Karoff et al. [43] compiled nearly 20 years of spectroscopic observations of KIC 8006161 from the Mount Wilson HK project [2], the Keck telescope [33], and the Nordic Optical Telescope (NOT [40, 41]). The chromospheric activity measured through the S index shows a cyclic variation with a cycle period of about 7.4 years (see Fig. 3) and an amplitude more than twice that of the solar chromospheric emission. The strong activity cycle of the solar analog KIC 8006161 is interpreted as a consequence of the high metallicity which contributes to a deep convection zone [43]. By combining spectroscopic, photometric, and asteroseismic analysis, Karoff et al. [43] also showed that *Kepler* observed part of the rising phase of the cycle and that the observed variations in the mode properties of KIC 8006161 are related to its activity. Similarly to the Sun, the frequency shifts depend on mode frequency, with high-frequency modes experiencing larger variations than low-frequency modes [45, 69], and on angular degree [73, 43]. The degree-dependency of frequency shifts for low-degree modes reflects the latitudinal distribution of magnetic activity (see Sect. 2.1). Using the frequency shifts measured for individual angular degrees, [78] determined that the activity bands for KIC 8006161 are wider and located at higher latitudes than the solar activity bands.

[1] See [73] for the results on the remainder of the individual targets, such as KIC 5184732, KIC 7970740, KIC 8379927, and KIC 10644253.

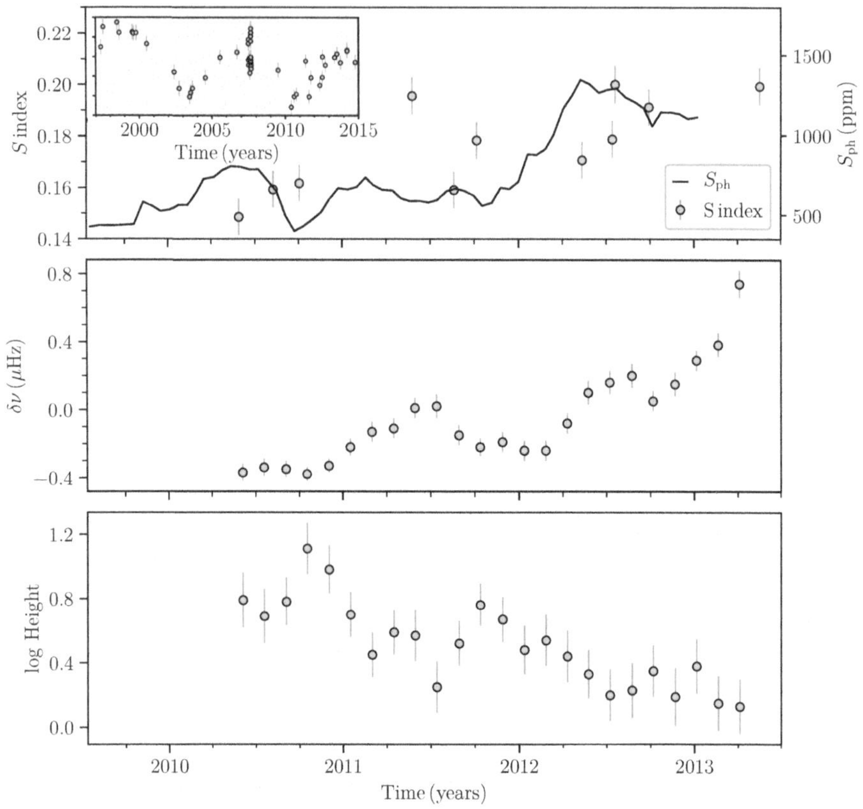

Fig. 3 *Top:* Chromospheric (grey symbols; ground-based spectroscopy) and photometric (solid line; *Kepler* long-cadence) activity proxies for KIC 8006161 during the *Kepler* observations. The panel at the top left corner displays the complete spectroscopic observations where the white region indicates those shown in the main panel. *Middle:* Variations in the acoustic-mode frequency of KIC 8006161 obtained from *Kepler* short-cadence data. *Bottom:* Logarithm of the mode heights. Chromospheric emission and seismic data from Santos et al. [73] and Karoff et al. [43]

While KIC 8006161 has well determined mode properties, spot modulation of the light curve which allows to measure S_{ph}, and independent spectroscopic observations spanning two decades, the same is not verified for the majority of the *Kepler* targets. Based on contemporaneous data, [44] compared the chromospheric emission from NOT observations with the frequency shifts measured from *Kepler* data for 20 targets. With exception of a few targets, the authors found a low agreement between the two properties. However, the disagreement can be explained in part by the poor sampling of the ground-based observations.

Santos et al. [74] investigated the relation between the amplitude of the observed frequency shifts and other stellar parameters for an ensemble of 75 *Kepler* solar-type stars, taking solely the highest signal-to-noise targets. The amplitude of the frequency shifts increases with effective temperature which is in agreement with the theoretical predictions for activity-related frequency shifts by Metcalfe et al. [52]. Frequency shifts decrease with stellar age and surface rotation period. This is in agreement with the observed frequency shifts being connected to activity, as stars become less active as they evolve and spin-down (for example [83, 85, 77, 82, 55]). Interestingly, the relation between frequency shifts and metallicity separate into two sequences: frequency shifts for young stars increase with metallicity, which supports the interpretation by Karoff et al. [43] for KIC 8006161, while the frequency shifts for old stars decrease with metallicity. It is worth noting, however, that the frequency shifts measured from *Kepler* data (length of about 4 years) might be a lower limit of the frequency shifts associated to full stellar cycle. Finally, although only available for 30 targets, the chromospheric activity index is found to be strongly correlated with the frequency-shift amplitude. Therefore, the ensemble analysis indicates that the frequency shifts measured for solar-type stars observed by *Kepler* are indeed activity related. Activity-related frequency shifts rise, however, a concern. The stellar parameters determined through asteroseismology, such as age, mass, and radius, can be biased if the detected mode frequencies have been affected by magnetic activity. This was investigated by Pérez Hernández et al. [57] who found that the bias is generally small but, in the worst case scenario, it can reach about 10% in age.

The fact that acoustic-mode properties are sensitive to magnetic activity allows us to learn about stellar magnetism through asteroseismology. However, as seen for the Sun and other solar-type stars magnetic activity suppresses the amplitude of the acoustic modes which already have small amplitudes. This hampers the detection of acoustic oscillations and consequently activity-related variations of the mode properties. Chaplin et al. [15] analyzed 1-month short-cadence *Kepler* data and were able to detect acoustic modes for about 25% of their target sample of $\sim$ 2000 solar-type stars. The detection fraction decreases with photometric variability suggesting that the low fraction of detection is related to activity. Using a cleaner sample, [51] revisited the impact of magnetic activity on the detection of acoustic oscillations. The results suggest that high activity is not the main contribution to the low detection fraction. Nevertheless, about 30% of the stars with non-detections are more active than the Sun at solar maximum. In fact, the authors concluded that above $S_{\mathrm{ph}} = 2000\,\mathrm{ppm}$, the probability of detecting modes is smaller than 2%. This also means that we are limited to seismic studies, including the search for activity-related variations, for stars with relatively low activity.

4 Future of Seismic Signatures of Magnetic Activity

The properties of the acoustic modes are sensitive to the properties of the medium where the acoustic waves propagate, including magnetic properties. Therefore, one can use helioseismology/asteroseismology to learn about solar/stellar magnetism.

Helioseismology has contributed to a better understanding of the solar structure and dynamics, as well as the solar cycle and its impact on the former. However, many unknowns still remain. Concerning the topic addressed in these proceedings, we still do not understand well the nature of activity-related frequency shifts. In particular, it is still not well established the relative importance of the strong and weak components of the magnetic field, as well as their indirect and direct effects on oscillations. Nevertheless, the most recent study [16] finds that the strong component of the magnetic field plays the main role. Furthermore, it is known that magnetic features suppress acoustic oscillations, but the processes behind are not well understood. Quasi-biennial variations are detected in different activity proxies, including frequency shifts, but their source is still undetermined. The weak solar cycle 24, which was proceeded by a very long and quiet solar minimum, and the departure between surface activity and frequency shifts, together with evidence from solar-type stars [81, 53], may suggest that the nature of the solar dynamo is changing and the Sun may be in transition. Future helioseismic studies are, thus important, to help us to better understand the solar magnetism.

Another way to better understand the solar magnetic cycle is by placing the Sun in the broader context of solar-type stars. The *Kepler* high-precision long-term observations constitute an unique opportunity to study stellar magnetism through asteroseismology. We now have temporal frequency shifts measured for about 100 solar-type stars. Using *Kepler* long and short cadence data, frequency shifts are seen to vary in phase in mode heights, characteristic timescale of the granulation, and/or the photometric activity index $S_{\rm ph}$ [73, 67, 45]. Whether the granulation properties are affected by activity is still a matter of debate in the literature [73, 25, 45, 56, 60, 61, 54, 48, 39, 42]. Nevertheless, frequency shifts varying in phase with the $S_{\rm ph}$ and in anti-phase with mode heights is a good indication for activity-related frequency shifts. Unfortunately, the light curves of about 40 stars do not exhibit rotational modulation and, thus, $S_{\rm ph}$ cannot be measured. In addition, the data source is the same. Therefore, independent data and measurements of activity, such as the chromospheric activity index measured from ground-based spectroscopic data, are preferable to confirm activity cycles. *Kepler* targets are, however, typically faint, hence not suitable for ground-based surveys. The existing ground-based data for *Kepler* targets are sparse, which limits the investigation of the relation between the chromospheric emission and frequency shifts [44]. Although, at the present moment, the relation between the observed frequency shifts and activity cannot be definitely established for most of the *Kepler* targets, ensemble analysis support the hypothesis that the observed frequency shifts are indeed caused by stellar activity [74].

Kepler is long gone, but the legacy of its unprecedented long-term high-precision photometric data still holds on. In fact, all the compelling results on seismic signatures of magnetic activity for *Kepler* targets were obtained after *Kepler* main mission was concluded. Thus, it is possible that *Kepler* data still have secrets to unveil and future analyses may provide further constraints on stellar magnetic properties.

While *Kepler* targets are too faint, TESS (Transiting Exoplanet Survey Satellite, [63]) observes bright stars. This is an advantage in comparison with *Kepler* and allows for ground-based follow-up and independent constraints on stellar activity.

However, one of the main disadvantages of TESS relative to *Kepler* is that the light curves obtained during its main mission will be significantly shorter, which may restrict our ability to measure frequency shifts, as well as surface rotation from spot modulation and consequently the photometric activity proxy $S_{\rm ph}$. That said, TESS is observing the *Kepler* field, thus it will be possible to revisit *Kepler* targets and determined whether the acoustic frequencies have changed and maybe obtain new constraints on $S_{\rm ph}$ for stars that did not show spot modulation during the *Kepler* observations. TESS is likely to be operational for decades. This opens the possibility for mission extensions (currently the first extension has been approved). Therefore, if TESS performs long-term short-cadence observations or observes targets that were previously observed, in principle one will be able to measure temporal frequency shifts with TESS. Using synthetic TESS data and a new scaling relation to estimate activity-related frequency shifts for different spectral types and evolutionary stages, [47] discusses the prospects to detect seismic signatures of magnetic activity for a TESS extended mission. They expect that frequency shifts will be possible to measure for a large number of stars.

The launch of future mission PLATO (PLAnetary Transits and Oscillations of stars, [59]) is planned for 2026. PLATO will obtain relatively long time-series, which are suitable for seismic studies and, in particular, to search for temporal variations in the seismic properties that may be related to activity. Furthermore, PLATO targets are suitable for ground-based follow-up which will provide independent observations.

Acknowledgments The legacy of the scientific work by Michael J. Thompson extends to the impact of magnetic fields/cycles on the properties of the acoustic modes of the Sun and solar-type stars, which is reviewed in these proceedings. Coincidentally, ARGS and MJT first met in the conference "Asteroseismology of stellar activity cycles: From the Sun to the Stars". ARGS acknowledges the support by the National Aeronautics and Space Administration (NASA) under Grant No. NNX17AF27G.

References

1. Baglin, A., Auvergne, M., Barge, P., Deleuil, M., Catala, C., Michel, E., et al. (2006). In M. Fridlund, A. Baglin, J. Lochard, & L. Conroy - ESA (Eds.), *The CoRoT mission pre-launch status – Stellar seismology and planet finding* (vol. 33)
2. Baliunas, S. L., Donahue, R. A., Soon, W. H., Hone, J. H., Frazer, J., Woodard-Eklund, L., et al. (1995). *The Astrophysical Journal, 438*, 269.
3. Basu, S., Broomhall, A.-M., Chaplin, W. J., Elsworth, Y. (2012). *The Astrophysical Journal, 758*, 43.
4. Borucki, W. J., Koch, D., Basri, G., Batalha, N., Brown, T., Cadwell, D., et al. (2010). *Science, 327*, 977.
5. Broomhall, A.-M. (2017). *Solar Physics, 292*, 67.
6. Broomhall, A.-M., & Nakariakov, V. M. (2015). *Solar Physics, 290*, 3095.
7. Broomhall, A.-M., Chaplin, W. J., Elsworth, Y., Fletcher, S. T. , & New, R. (2005). *Astronomy and Astrophysics, 503*, 241.

8. Broomhall, A.-M., Chaplin, W. J., Elsworth, Y., & Simoniello, R. (2012). *Monthly Notices of the Royal Astronomical Society, 420*, 1405.
9. Broomhall, A.-M., Chatterjee, P., Howe, R., Nortan, A. A., Thompson, M. J. (2014). *Space Science Reviews, 186*, 191.
10. Chaplin, W. J., Elsworth, Y., Isaak, G. R., Lines, R., McLeod, C. P., Miller, B. A., et al. (1998). *Monthly Notices of the Royal Astronomical Society, 300*, 1077.
11. Chaplin, W. J., Elsworth, Y., Isaak, G. R., Miller, B. A., & New, R. (2000). *Monthly Notices of the Royal Astronomical Society, 313*,32.
12. Chaplin, W. J., Appourchaux, T., Elsworth, Y., Isaak, G. R., & New, R. (2001). *Monthly Notices of the Royal Astronomical Society, 324*, 910.
13. Chaplin, W. J., Elsworth, Y., Isaak, G. R., Miller, B. A., New, R. (2004). *Monthly Notices of the Royal Astronomical Society, 352*, 1102.
14. Chaplin, W. J., Elsworth, Y., Miller, B. A., Verner, G. A., & New, R. (2007). *The Astrophysical Journal, 659*, 1749.
15. Chaplin, W. J., Bedding, T. R., Bonanno, A., Broomhall, A.-M., García, R. A., Hekker, S., et al. (2011). *The Astrophysical Journal Letters, 732*, L5.
16. Chaplin, W. J., Howe, R., Basu, S., Elsworth, Y., Milbourne, T. W., Haywood, R. D., et al. (2019). *Monthly Notices of the Royal Astronomical Society, 489*, L86.
17. Christensen-Dalsgaard, J., Berthomieu, G. (1991). *Solar interior and atmosphere* (pp. 401). Tucson: University of Arizona Press.
18. Dziembowski, W. A., & Goode, P. R. (2005). *The Astrophysical Journal, 625*, 548.
19. Elsworth, Y., Howe, R., Isaak, G. R., McLeod, C. P., & New, R. (1990). *Nature, 345*, 322.
20. Elsworth, Y., Howe, R., Isaak, G. R., McLeod, C. P., Miller, B. A., New, R., et al. (1994). *The Astrophysical Journal, 434*, 801.
21. Fletcher, S. T., Broomhall, A.-M., Salabert, D., Sarbani, B., Chaplin, W. J., Elsworth, Y., et al. (2010). *The Astrophysical Journal, 718*, L19.
22. Fröhlich, C., Romero, J., Roth, H., Wehrli, C., Andersen, B. N., Appourchaux, T., et al. (1995). *Solar Physics, 162*, 101.
23. Fröhlich, C., Andersen, B. N., Appourchaux, T., Berthomieu, G., Crommelynck, D. A., Domingo V., et al. (1997). *Solar Physics, 170*, 1.
24. García, R. A., & Ballot, J. (2019). *Living Reviews in Solar Physics, 16*, 4.
25. García, R. A., Mathur, S., Salabert, D., Ballot, J., Régulo, C., Metcalfe, T. S., et al. (2010). *Science, 329*, 1032.
26. Goldreich, P., & Keeley, D. A. (1977). *The Astrophysical Journal, 212*, 243.
27. Hindman, B., Haber, D., Toomre, J., & Bogart, R. (2000). *Solar Physics, 192*, 363.
28. Houdek, G., Chaplin, W. J., Appourchaux, T., Christensen-Dalsgaard, J., Däppen W., Elsworth, Y., et al. (2001). *Monthly Notices of the Royal Astronomical Society, 327*, 483.
29. Howe, R., & Komm, R., & Hill, F. (1999). *The Astrophysical Journal, 524*, 1084.
30. Howe, R., Komm, R. W., & Hill, F. (2002). *The Astrophysical Journal, 580*, 1172.
31. Howe, R., Davies, G. R., Chaplin, W. J., Elsworth, Y. P., & Hale, S. J. (2015). *Monthly Notices of the Royal Astronomical Society, 454*, 4120.
32. Howe, R., Davies, G. R., Chaplin, W. J., Elsworth, Y. (2017). *Monthly Notices of the Royal Astronomical Society, 470*, 1935.
33. Isaacson, H., & Fischer, D. (2010). *The Astrophysical Journal, 725*, 875.
34. Jain, R., & Haber, D. (2002). *Astronomy and Astrophysics, 387*, 1092.
35. Jain, K., Tripathy, S. C., & Hill, F. (2009). *The Astrophysical Journal, 695*, 1567.
36. Jain, K., Tripathy, S. C., & Hill, F. (2011). *The Astrophysical Journal, 739*, 6.
37. Jiménez-Reyes, S. J., Régulo, C., Pallé, P. L., Roca Cortes, T. (1998). *Astronomy and Astrophysics, 329*, 1119.
38. Jiménez, A., Roca Cortés, T., Jiménez-Reyes, S. J. (2002). *Solar Physics, 209*, 247.
39. Karoff, C. (2012). *Monthly Notices of the Royal Astronomical Society, 421*, 3170.
40. Karoff, C., Metcalfe, T. S., Chaplin, W. J., Elsworth, Y., Kjeldsen, H., Arentoft, T. et al. (2009). *Monthly Notices of the Royal Astronomical Society, 399*, 914.

41. Karoff, C., Metcalfe, T. S., Chaplin, W. J., Frandsen, S., Grundahl, F., Kjeldsen, H., et al. (2013). *Monthly Notices of the Royal Astronomical Society, 433*, 3227.
42. Karoff, C., Campante, T. L., Ballot, J., Kallinger, T., Gruberbauer, M., García, R. A., et al. (2013). *The Astrophysical Journal, 767*, 34.
43. Karoff, C., Metcalfe, T. S., Santos, A. R. G., Montet, B. T., Isaacson, H., Witzke, V., et al. (2018). *The Astrophysical Journal, 852*, 46.
44. Karoff, C., Metcalfe, T. S., Montet, B. T., Jannsen, N. E., Santos, A. R. G., Nielsen, M. B., et al. (2019). *Monthly Notices of the Royal Astronomical Society, 485*, 5096.
45. Kiefer, R., Schad, A., Davies, G., & Roth, M. (2017). *Astronomy and Astrophysics, 598*, A77.
46. Kiefer, R., Komm, R., Hill, F., Broomhall, A.-M., & Roth, M. (2018). *Solar Physics, 293*,151.
47. Kiefer, R., Broomhall, A.-M., & Ball, W. H. (2019). *Frontiers in Astronomy and Space Sciences, 6*, 52.
48. Lefebvre, S., García, R. A., Jiménez-Reyes, S. J., Turck-Chièze, S., Mathur, S. (2008). *Astronomy and Astrophysics, 490*, 1143.
49. Libbrecht, K. G., & Woodard, M. F. (1990). *Nature, 345*, 779.
50. Mathur, S., García, R. A., Ballot, J., Ceillier, T., Salabert, D., et al. (2014). *Astronomy and Astrophysics, 562*, A124.
51. Mathur, S., García, R. A., Bugnet, L., Santos, A. R. G., Santiago, N., Beck, P. G. (2019). *Frontiers in Astronomy and Space Sciences, 6*, 46.
52. Metcalfe, T. S., Dziembowski, W. A., Judge, P. G., & Snow, M. (2007). *Monthly Notices of the Royal Astronomical Society, 379*, L16.
53. Metcalfe, T. S., Egeland, R., van Saders, J. L. (2016). *The Astrophysical Journal Letters, 826*, L2.
54. Muller, R., Hanslmeier, A., & Saldaña-Muñoz, M. (2007). *Astronomy and Astrophysics, 475*, 717.
55. Noyes, R. W., Hartmann, L. W., Baliunas, S. L., Duncan, D. K., Vaughan, A. H. (1984). *The Astrophysical Journal, 279*, 763.
56. Pallé, P. L., Jiménez, A., Pérez Hernández, F., Roca Cortés, T., & Sanchez, L. (1995). *The Astrophysical Journal, 441*, 952.
57. Pérez Hernández, F., García, R. A., Mathur, S., Santos, A. R. G., & Régulo, C. (2019). *Frontiers in Astronomy and Space Sciences, 6*, 41.
58. Raskin, G., van Winckel, H., Hensberge, H., Jorissen, A., Lehmann H., Waelkens, C., et al. (2011). *Astronomy and Astrophysics, 526*, A69.
59. Rauer, H., Catala, C., Aerts, C., Appourchaux, T., Benz, W., Brandeker, A., et al. (2014). *Experimental Astronomy, 38*, 249.
60. Régulo, C., Roca Cortés, T., & Vázquez Ramió, H. (2002). *Solar variability: From core to outer frontiers – The 10th European solar physics meeting* (vol. 506, pp. 889). Noordwijk: A. Wilson, ESA Publications Division
61. Régulo, C., Vázquez Ramió, H., & Roca Cortés, T. (2005). *Astronomy and Astrophysics, 443*, 1013.
62. Régulo, C., García, R. A., & Ballot, J. (2016). *Astronomy and Astrophysics, 589*, A103.
63. Ricker, G. R., Winn, J. N., Vanderspek, R., et al. (2014). *Proceedings of the SPIE. Space telescopes and instrumentation: optical, infrared, and millimeter wave* (vol. 9143, pp. 15)
64. Salabert, D., & Jiménez-Reyes, S. J. (2006). *The Astrophysical Journal, 650*, 451.
65. Salabert, D., Régulo, C., Ballot, J., García, R. A., & Mathur, S. (2011). *Astronomy and Astrophysics, 530*, A127.
66. Salabert, D., García, R. A., & Turck-Chièze, S. (2015). *Astronomy and Astrophysics, 578*, A137.
67. Salabert, D., Régulo, C., García, R. A., Beck, P. G., Ballot, J., Creevey, O. L., et al. (2016). *Astronomy and Astrophysics, 589*, A118.
68. Salabert, D., García, R. A., Jiménez, A., Bertello, L., Corsaro, E., Pallé, P. L. (2017). *Astronomy and Astrophysics, 608*, A87.
69. Salabert, D., Régulo, C., Pérez Hernández, F., García, R. A. (2018). *Astronomy and Astrophysics, 611*, A84.

70. Santos, A. R. G., Cunha, M. S., & Lima, J. J. G. (2012). *Astronomische Nachrichten, 333*, 1032.
71. Santos, A. R. G., Cunha, M. S., Avelino, P. P., Chaplin, W. J., Campante, T. L. (2016). *Monthly Notices of the Royal Astronomical Society, 461*, 224.
72. Santos, A. R. G., Cunha, M. S., Avelino, P. P., Chaplin, W. J., Campante, T. L. (2017). *Monthly Notices of the Royal Astronomical Society, 464*, 4408.
73. Santos, A. R. G., Campante, T. L., Chaplin, W. J., Cunha, M. S., Lund, M. N., Kiefer, R., et al. (2018). *The Astrophysical Journal Supplement Series, 237*, 17.
74. Santos, A. R. G., Campante, T. L., Chaplin, W. J., Cunha, M. S., van Saders, J. L., Karoff, C., et al. (2019). *The Astrophysical Journal, 883*, 65.
75. Simoniello, R., Finsterle, W., Salabert, D., García, R. A., Turck-Chièze, S., Jiménez, A., et al. (2012). *Astronomy and Astrophysics, 539*, A135.
76. Simoniello, R., Jain, K., Tripathy, S. C., Turck-Chièze, S., Baldner, C., Finsterle, W., et al. (2013). *The Astrophysical Journal, 765*, 100.
77. Skumanich, A. (1972). *The Astrophysical Journal, 171*, 565.
78. Thomas, A. E. L., Chaplin, W. J., Davies, G. R., Howe, R., Santos, A. R. G., Elsworth, Y., et al. (2019). *Monthly Notices of the Royal Astronomical Society, 485*, 3857.
79. Tripathy, S. C., Hill, F., Jain, K., & Leibacher, J. W. (2007). *Solar Physics, 243*, 105.
80. Tripathy, S. C., Jain, K., Hill, F., & Leibacher, J. W. (2010). *The Astrophysical Journal Letters, 711*, L84.
81. van Saders, J. L., Ceillier, T., Metcalfe, T. S., Silva Aguirre, V., Pinsonneault, M. H., García, R. A., et al. (2016). *Nature, 529*, 181.
82. Vaughan, A. H., Baliunas, S. L., Middelkoop, F., Hartmann, L. W., Mihalas, D., Noyes, R. W., et al. (1981). *The Astrophysical Journal, 250*, 276.
83. Wilson, O. C. (1963). *The Astrophysical Journal, 138*, 832.
84. Wilson, O. C. (1978). *The Astrophysical Journal, 226*, 379.
85. Wilson, O. C., Skumanich, A. (1964). *The Astrophysical Journal, 140*, 1401.
86. Woodard, M. F., & Noyes, R. W. (1985). *Nature, 318*, 449.

Confinement of Magnetic Fields Below the Solar Convection Zone

Nicholas Brummell

Abstract Confining the primordial dipole magnetic field below the convection zone affords the possibility of a serendipitous solution that fits many of the observed qualities of the solar interior rotation profile. Gough and McIntyre [6] pointed this out and derived a laminar balanced model that operates on the slow Eddington-Sweet timescale for this purpose. An interesting question is, what is the role of turbulence? Does the presence of turbulence help or hinder? In the spirit of Michael Thompson's love of tachocline dynamics, we survey some recent work on these issues.

1 Introduction

The Gough and McIntyre model [6] (hereinafter GM) offers a theoretical explanation of tachocline dynamics that is consistent with the whole internal solar rotation profile. The rotation profile is made up of a solid-body-rotating radiative interior coupled to the differentially-rotating convection zone via a surprisingly thin transition layer known as the tachocline. The GM model is based on the presence of magnetic fields, in contrast to other models that are purely hydrodynamic (see e.g. [11], hereinafter SZ). The advantage of the GM model is that, provided that the large-scale interior primordial field invoked remains confined below the convection zone (i.e. field lines do not cross into the convection zone), then it explains not only the thinness of the tachocline, but also the solid-body rotation of the radiative interior, due to rigidization there [3]. The tachocline problem in this context then can be recast as one of confinement of the magnetic field.

In the balanced boundary layer model of GM, the radiative spreading (or "burrowing": see SZ) of differential rotation and meridional flows (induced via

N. Brummell (✉)
University of California Santa Cruz, Santa Cruz, CA, USA
e-mail: brummell@ucsc.edu

M. J. P. F. G. Monteiro et al. (eds.), *Dynamics of the Sun and Stars*, Astrophysics and Space Science Proceedings 57,
https://doi.org/10.1007/978-3-030-55336-4_31

"gyroscopic pumping") counteracts the spread of a dipole located in the radiative interior. The timescale for the operation of this laminar process is the slow Eddington-Sweet timescale, $t_{ES} = (N/2\Omega)^2(R^2/\kappa)$, where N is the buoyancy frequency, Ω is the rotation rate, R is the radius and κ is the thermal diffusivity.

It is also well-established that turbulent transport of magnetic field can lead to confinement. Whilst this can come in many guises (diamagnetic pumping; a γ mean-field effect; topological pumping; flux expulsion; etc), the transport is often down a gradient of turbulent intensity and aided by asymmetries in the flow (see e.g. [2]). These very conditions occur naturally at the interface between the radiative and convection zones and therefore some sort of turbulent expulsion of large-scale, mostly-horizontal magnetic field from the convection zone might be anticipated. Note that the timescale for the operation of this "turbulent pumping" is the advective turnover timescale, which is much faster than t_{ES}.

Questions about the role of turbulence in the formation of the tachocline and its dynamics therefore naturally arise. Does the presence of turbulence affect the laminar mechanism proposed by GM? Could fast turbulent transport confine an interior primordial dipole instead of the slow GM mechanism? Do both fast and slow mechanisms play a role?

2 Slow Confinement in Cartesian Models

Two papers by Wood and Brummell [13, 14] examined the first of these questions. These papers use numerical simulations of a rotating, compressible fluid in a Cartesian box in a two-layer system where a convective layer is maintained overlying a convectively-stable layer (as previously in [2]). A volume forcing function is applied that drives a sinusoidal horizontal mean shear flow in the simulated convection zone, mimicking differential rotation in latitude. The first paper examined the gyroscopic pumping of meridional flows and their subsequent burrowing into the stable region in the presence of the convective turbulence and waves. The main result was that, despite the presence of turbulence and waves, the differential rotation and the meridional circulations generated did indeed burrow, provided that the viscous time scale was longer than the Eddington-Sweet time scale (as was predicted by earlier models, e.g. [5, 4]). The second paper added a magnetic field in the stably-stratified region and looked for confinement of that field by the GM mechanism. Confinement and the formation of a tachocline with the balances predicted by GM was found, despite the presence of convective turbulence and waves, when the viscous forces did not dominate the burrowing mechanism. Figure 1 shows the results from one of the two scenarios examined, where the interior field is purely horizontal (mimicking a region at mid-latitudes in the solar problem). Panel (*a*) shows the (unsteady) SZ burrowing (to the bottom of the box) of both differential rotation and meridional circulation when conditions are appropriate and there is no interior magnetic field. Panels (*b*) and (*c*) show two equivalent cases but with interior magnetic fields of

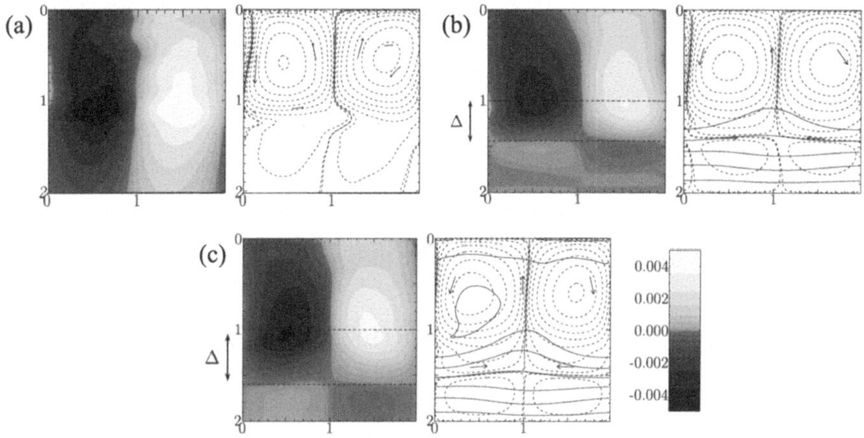

Fig. 1 The GM mechanism in action. Time- and azimuthally- averaged flows and fields. Shaded panels (left) are differential-rotation; right panels show meridional flow (dashed contours) and magnetic field (solid contours). Tachoclines formed are shown by horizontal dotted lines

differing strengths. Balanced (steady) tachoclines of varying widths are created and the accompanying magnetic fields are confined below the convection zone.

Under the conditions of the model, that GM mechanism appears to function even in the presence of moderate turbulence and waves. The model however is not self-consistent in the sense that the differential rotation is forced and not generated by the turbulence itself. A fully self-consistent model requires the presence of rotation in a spherical geometry to provide the Reynolds stresses that drive the differential rotation.

3 Fast Confinement in Spherical Shell Models

The Ph.D. thesis of Lydia Korre, and papers generated from that thesis [7, 8, 9, 10], attempt to follow up on the self-consistency issue raised above, by moving from Cartesian (compressible) two-layer simulations to spherical Boussinesq two-shell simulations using the PARODY code [1]. Before attempting the far more complex simulations involving both fast and slow processes, Korre and collaborators examined the issue of potential fast confinement alone. This required the spherical geometry but no rotation, so that differential rotation (and subsequently meridional flows and the slow dynamics) are not generated. After examining pure convection [8] and then two-shell overshooting convection [9], Chapter 4 of Korre's thesis [7] (and papers soon to be forthcoming [10]) address the confinement of an initial dipole field by the overshooting convective system. Figure 2a shows the idea schematically: a spatially-localized dipole is initially placed in the stably-stratified interior ($0.2 \leq r \leq 0.7$) below a zone ($0.7 \leq r \leq 1.0$) that is driven at a Rayleigh number, Ra, such

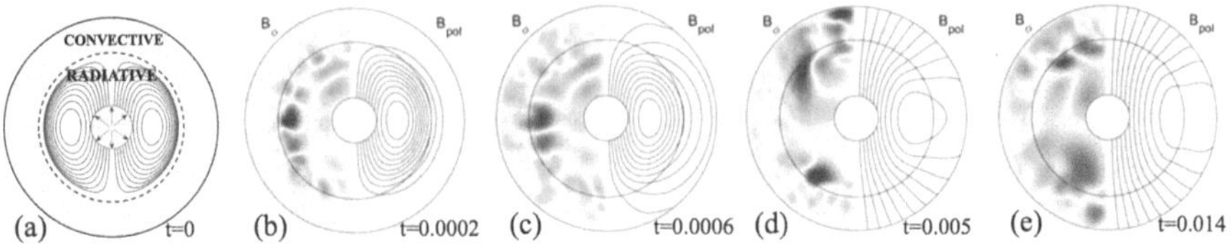

Fig. 2 Time evolution of the dipole field (right hemisphere) at $Ra = 10^7$, $Pm = 0.1$ (plus the azimuthal average of the toroidal field: left hemisphere)

that it is fairly strongly convectively unstable ($10^6 \leq Ra \leq 10^9$ at Prandtl number $Pr = 0.1$). The dipole is added to an initially nonlinearly-saturated convecting (and overshooting) simulation. We will discuss here simulations at magnetic Prandtl number $Pm = 0.1$ and $Pm = 1.0$ but, due to lack of space, for other detailed parameters, equations and boundary conditions, see [7, 10].

For $Pm = 0.1$, none of the Rayleigh numbers simulated are sufficiently turbulent to generate convective dynamo action. Figure 2b–e exhibits a typical example ($Ra = 10^7$) of the evolution of the magnetic field. The plots show intensity plots of the axisymmetric component of the toroidal field B_ϕ and contours of the dipole component of the poloidal field $B_{\text{pol}} = (B_r, B_\theta)$. Despite the fairly vigorous convection occurring in the outer shell, the dipole component of the field is clearly not confined to the stable region, as the field lines are transported outwards, very quickly extending and opening up into the convection zone. A more quantitative measurement of confinement (or lack thereof) is shown in Fig. 3 where the simulated evolution of the dipole component of the field is compared to a calculated purely-diffusive transport that would occur in the absence of convection. Figure 3a–c shows the spherically-averaged dipole field for both the simulated (solid) and diffusive (dashed) versions as a function of radius for three times. These plots demonstrate that the dipole component of the simulations is spread into the convection zone rapidly and then removed faster everywhere than would be expected by pure diffusion alone. This suggests that not only is the convection not confining the field below the convection zone, but further it is acting as a turbulent diffusion to destroy the field faster than microscopic diffusion. The latter idea can be confirmed by replacing the constant magnetic diffusivity in the purely diffusive model with a radially-dependent diffusivity (as show in Fig. 3d) representing stronger (turbulent) diffusion in the convection zone. Even such a simple model as this (hyperbolic tangent) step function gives a much better fit to the actual dynamics (Fig. 3e). The upcoming paper [10] incorporates more physical models.

Raising the magnetic Prandtl number to $Pm = 1.0$ decreases the critical magnetic Reynolds number for the onset of dynamo action in the convective region. At this Pm, $Ra \geq 10^7$ are convective dynamos (but $Ra = 10^6$ is not). Figure 4 contrasts snapshots of the full velocity and magnetic fields for the two cases, $Pm = 0.1, 1.0$ at $Ra = 10^7$. Whilst the velocity fields can be seen to be active in the

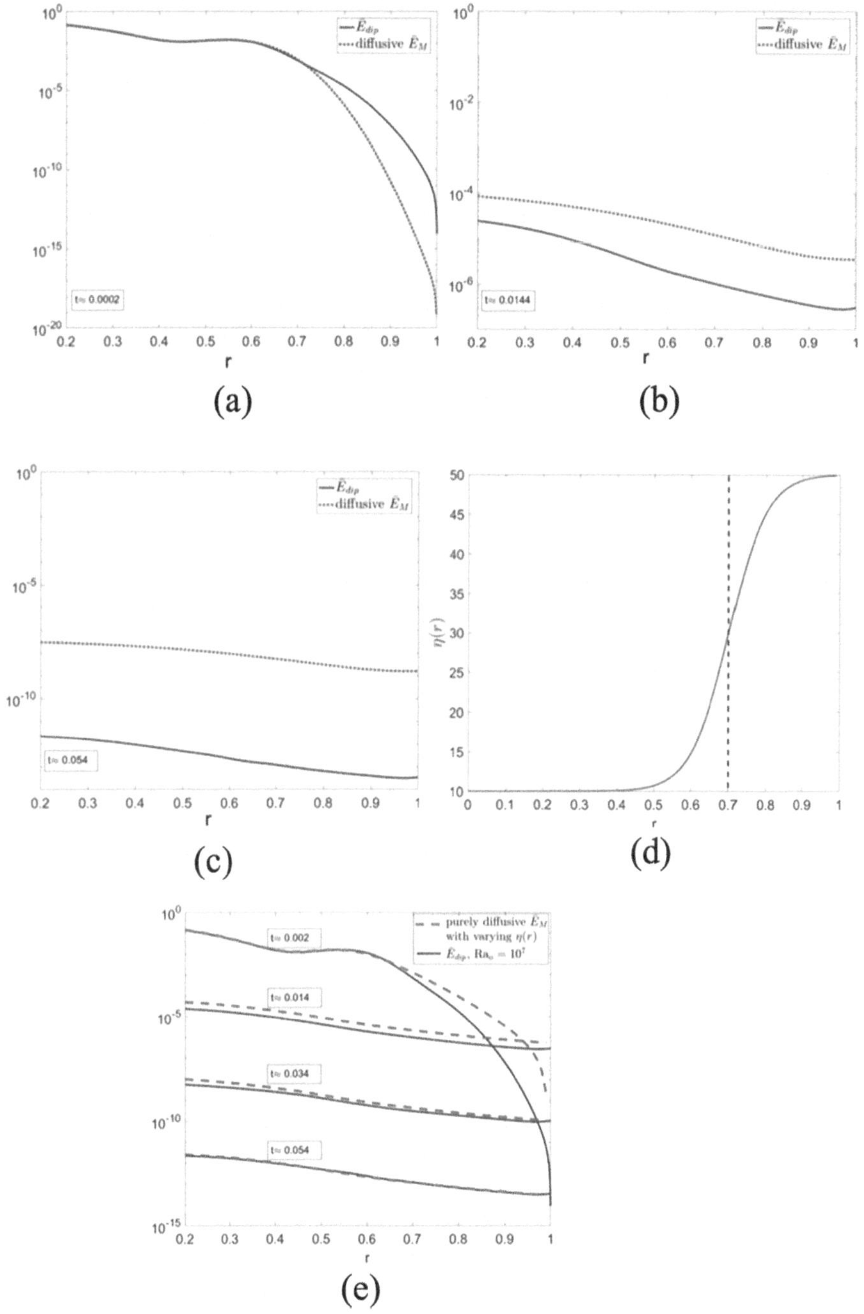

Fig. 3 (**a**)–(**c**) Time evolution of the spherical average of the dipolar field as a function of r compared to the evolution of a purely diffusive solution. (**d**) A model for spatially-varying magnetic diffusivity. (**e**) A diffusive solution using the model of (**d**), compared to the simulation results

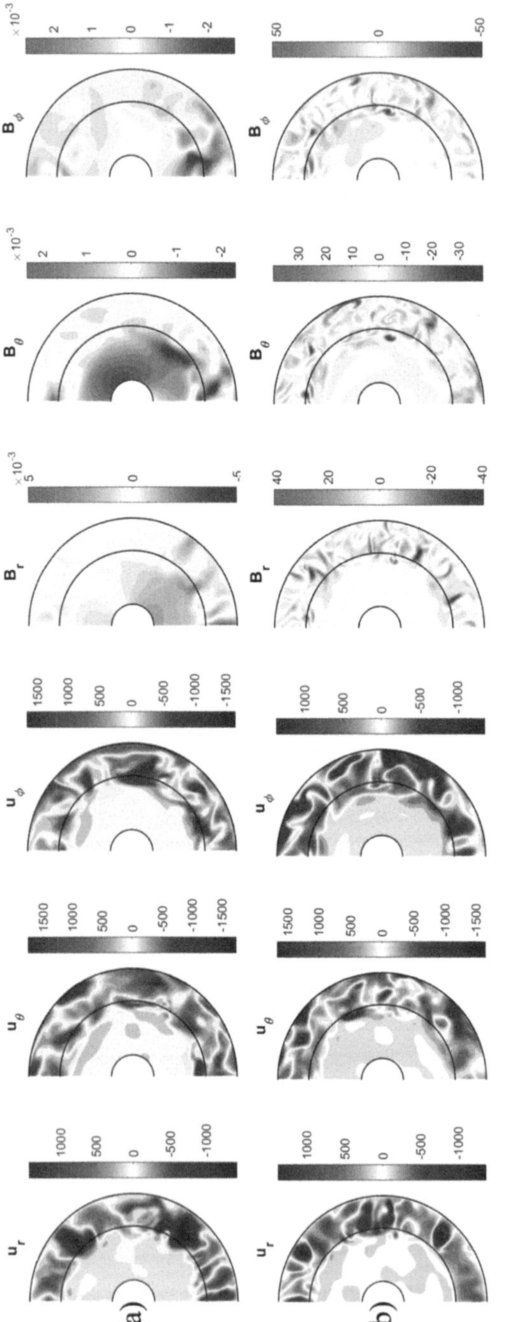

Fig. 4 Snapshots of all the variables from solutions at $Ra = 10^7$ for **(a)** $Pm = 0.1$ and **(b)** $Pm = 1.0$

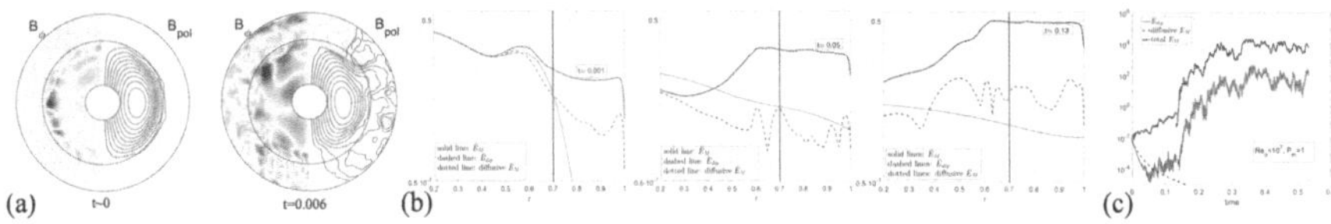

Fig. 5 Results for $Ra = 10^7$, $Pm = 1.0$. (**a**) The evolution of the dipole. (**b**) The spherically-averaged dipole and total energies as a function of radius, compared to a diffusive solution, at various times. (**c**) The time evolution of the volume-averaged total and dipole magnetic energy

convection zone and overshooting significantly in both cases, small-scale (velocity-scale) magnetic field is created and maintained in the convecting and overshooting regions in the higher Pm case in panel (b). Figure 5a shows the evolution of the dipole component as before, demonstrating that in these dynamo cases confinement of the original (primordial) dipole is still not achieved. Figure 5b shows three time snapshots of the spherically-averaged dipolar component as compared to a purely diffusive solution (with constant diffusion) again, and shows that the dipole is destroyed faster than diffusively initially, as before by the action of convection dragging the field into the convection zone and tangling it rapidly for turbulent diffusion. However, in this case, the dipole energy appears to be somewhat more elevated in the convection zone (at least) at later times, and clearly the total magnetic energy is higher, thanks to the presence of the dynamo. The effect of this dynamo action is perhaps more obviously seen in Fig. 5c, which shows the volume-averaged total magnetic energy (upper solid line) and also the energy in the dipole component alone (lower solid line), versus time, as compared to what would be found in a purely diffusive case (dashed). The total magnetic energy clearly climbs exponentially and then saturates at a finite value as is expected of a dynamo. The dipole component initially decays faster than diffusively but then clearly begins to grow at roughly the same rate as the total magnetic energy. This appears to be an indication that the dynamo, despite lacking any symmetry breaking (e.g. rotation), still has a significant dipole component solely as a result of nonlinear triadic interactions. The presence of significant dynamo field at all scales in the convection and overshoot zones means that there is a spatially-decaying tail of field in the deeper interior. The ramifications of the presence of this field for theories of the tachocline might be significant, and would depend on the strength of the dynamo components that could be generated.

4 Conclusions

Progress has been made in understanding tachocline dynamics. The slow GM mechanism has been demonstrated to be viable, but only so far in simple, Cartesian models that lack self-consistent generation of the differential rotation [14]. Fast turbulent confinement has also been demonstrated in simple Cartesian models [2], but was not found in the Boussinesq spherical models described here [7]. The lack of turbulent confinement could be due to the need for higher degrees of turbulence, or higher degrees of asymmetry, ingredients typically thought to be beneficial. In particular, the asymmetry for these cases is only provided by the spherical geometry, the boundary conditions and the layered setup of the system. Compressible (or anelastic) fluids have considerably more natural asymmetry than the Boussinesq fluids studied here, so perhaps that is an obvious avenue for improved turbulent confinement. Ultimately, one wishes to study a rotating spherical shell allowing both slow and fast mechanisms with spontaneously and self-consistently generated differential rotation. Some previous studies, e.g. [12], have attempted this and found no confinement. The work in [14] points to the lack of slow confinement being due to the simulations being viscously dominated, and perhaps the lack of fast confinement may be due to some of the reasons cited above. This needs to be revisited now.

An interesting new wrinkle is the question of what influence a dynamo field in the convection and overshooting zones has on any of the theoretical ideas for tachocline dynamics. For example, GM assumes that there is a field-free region in the tachocline above the tachopause boundary layer. Can the tail of a dynamo field be sufficiently strong to interfere with this? Similarly, purely hydrodynamic arguments regarding shear instabilities in the tachocline might be strongly affected by the presence of significant magnetic field. Ultimately, there is still much work to be done on tachocline dynamics and I believe Michael Thompson would be excited about this!

Acknowledgments The material for this paper is drawn from work with co-authors Pascale Garaud, Lydia Korre and Toby Wood.

References

1. Aubert, J., Aurnou, J., & Wicht, J. (2008). *Geophysical Journal International, 172*, 945.
2. Brummell, N. H., Clune, T. C., & Toomre, J. (2002). *The Astrophysical Journal, 570*, 825.
3. Ferrara, V. C. A. (1937). *Monthly Notices of the Royal Astronomical Society, 97*, 458.
4. Garaud, P., & Acevedo-Arreguin, L. A. (2009). *The Astrophysical Journal 704*, 1.
5. Garaud, P., & Brummell, N. H. (2008). *The Astrophysical Journal, 674*, 498.
6. Gough, D. O. & McIntyre, M. E. (1998). *Nature, 394*, 755.
7. Korre, L. (2018). University of California Santa Cruz Ph.D. thesis
8. Korre, L., Brummell, N. H., & Garaud, P. (2017). *Physical Review E, 96*, 033104.
9. Korre, L., Garaud, P., & Brummell, N. H. (2019). *Monthly Notices of the Royal Astronomical Society, 484*, 1220.
10. Korre, L., Garaud, P., Brummell, N. H., & Guervilly, C. (2020, submitted). *Monthly Notices of the Royal Astronomical Society*.
11. Spiegel, E. A. & Zahn, J.-P. (1992). *The Astrophysical Journal, 256*, 106.
12. Strugarek, A., Brun, A. S., & Zahn, J.-P. (2011). *Astronomy and Astrophysics, 532*, A34.
13. Wood, T. S., & Brummell, N. H. (2012). *The Astrophysical Journal, 755*, 99.
14. Wood, T. S., & Brummell, N. H. (2018). *The Astrophysical Journal, 853*, 97.

Asteroseismic Study of KIC 11145123: Its Structure and Rotation

Yoshiki Hatta, Takashi Sekii, Masao Takata, and Donald W. Kurtz

Abstract KIC 11145123 is one of the Kepler targets that has been actively studied asteroseismically. Its well-resolved frequency splittings for p, g, and mixed modes have enabled us to infer the 1-dimensional surface-to-core rotation of the star (Kurtz et al., Mon Not R Astron Soc 444:102, 2014) and to measure the asphericity of the star sensed by the modes (Gizon et al., Sci Adv 2:e1601777, 2016). These have been the first such attempts for main-sequence stars other than the Sun. In this paper, we report further detailed asteroseismic analyses of KIC 11145123, focusing on inferring 2-dimensional rotation profile and on fine-tuning the equilibrium model of the star. Our main results are: (1) the convective core might be rotating six times faster than the other parts of the star, and (2) adopting diffusion "weaker" than the usual settings used in the standard 1-d stellar evolution calculations leads to a better agreement with the observed g-mode period spacing pattern. These two results suggest that the rotational velocity shear might be causing some extra mixing around the boundary between convective core and radiative region above, and the extra mixing weakens, in effect, the diffusion process there. This resembles a mechanism thought to be at work around the solar tachocline.

Y. Hatta (✉)
Department of Astronomical Science, School of Physical Sciences, SOKENDAI, Mitaka, Tokyo, Japan
e-mail: yoshiki.hatta@nao.ac.jp

T. Sekii
National Astronomical Observatory of Japan, Mitaka, Tokyo, Japan
e-mail: takashi.sekii@nao.ac.jp

M. Takata
Department of Astronomy, School of Science, The University of Tokyo, Tokyo, Japan
e-mail: takata@astron.s.u-tokyo.ac.jp

D. W. Kurtz
Jeremiah Horrocks Institute, University of Central Lancashire, Preston, UK
e-mail: dwkurtz@uclan.ac.uk

M. J. P. F. G. Monteiro et al. (eds.), *Dynamics of the Sun and Stars*, Astrophysics and Space Science Proceedings 57,
https://doi.org/10.1007/978-3-030-55336-4_32

1 Introduction

Since the launch of the Kepler space telescope [6], asteroseismology has greatly developed thanks to the unprecedented precision and the long duration achieved by the series of observations. This has led to numerous important studies in stellar physics, such as measuring stellar global parameters based on $\nu_{\max}$ [3], determining stellar evolutionary stages by g-mode period spacings ΔPg [2], and inferring internal rotation profiles [7].

One of the remarkable Kepler targets is KIC 11145123, which used to be categorized as a main-sequence A-type star based on photometry [5]. This star is not only pulsating with p, g, and mixed modes, but also exhibits well-resolved frequency splitting for p, g, and mixed modes. The number of such stars in the Kepler Input Catalog is very small, and, as far as we know, only a few of them are main-sequence stars [11, 14]. The star KIC 11145123 is one of them.

The first intensive asteroseismic study of the star was done by Kurtz et al. [7]. They measured the surface-to-core rotation based on the observed p- and g-mode frequency splitting, and model-independently showed that the envelope of the star is rotating slightly faster than the deep radiative region. When we assume local conservation of the angular momentum in a star, it is usually expected that the core of the star rotates faster than the envelope does, because the core (envelope) shrinks (expands) as the star evolves. Their finding thus sheds new light on the theory of angular momentum transfer inside stars.

There are several scenarios for the origin of the peculiar rotational profile of KIC 11145123. One possibility is that the star has experienced some interactions with another star, probably in a binary system, and has accreted angular momentum, resulting in the spin-up of the envelope. Interestingly, a spectroscopic study of the star conducted with Subaru/HDS [15] concluded that the star is spectroscopically a blue straggler star [13], a kind of star thought to be born via interactions such as mergers or collisions with other stars.

We have set two main goals. The first is to infer two-dimensional internal rotation of the star, and to obtain insight for latitudinal angular momentum transfer. This can be a simple extension of the work by Kurtz et al. [7], and is featured in Sect. 2. The second goal is to refine the equilibrium model assuming that the star has experienced some interaction with another star during its evolution. We have, as the first step toward the second goal, fine-tuned the model by fitting the modulated pattern of the observed g-mode period spacing as shown in Sect. 3. We conclude with some remarks on relationships between the inferred internal rotation and the structure of the fine-tuned model in Sect. 4.

2 Two-Dimensional Rotation of the Star

In this section, we first show the latitudinal dependence of the internal rotation of the star. Then, we show an indication for the fast-core rotation of the star. For inferring the (two-dimensional) rotation profile of the star, we have carried out inversion with the Optimally Localized Averaging (OLA) method [1]. Examples of rotational splitting kernels used in our analysis are shown in Fig. 1. The equilibrium model is calculated via MESA [9], ver 9793, based on which linear adiabatic oscillation is computed via GYRE [17]. We have used 15 g modes, 2 p modes, and 3 mixed modes, and have selected several target points where the splitting kernels are expected to have sensitivity. For further details and discussions, see [4].

2.1 *Latitudinally Differential Rotation*

The averaging kernels for a pair of target points, one in the high latitude region and another in the low latitude region, are shown with the corresponding estimates of the rotation rate in Fig. 2, clearly showing that we can successfully extract the information separately. We found that the high latitude region rotates slightly faster than the low latitude region does, so called "anti-solar-like rotation." We have also computed a3 coefficients [10] for the observed frequency splittings of the three modes with $l = 2$, and found that the averaged value is negative with more than 2σ significance, leading to the same conclusion.

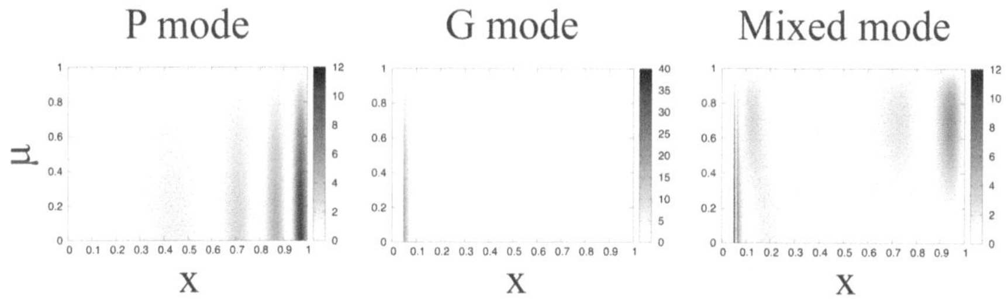

Fig. 1 Examples of rotational splitting kernels: for a p mode $(n, l, m) = (3, 1, 1)$, a g mode $(n, l, m) = (-33, 1, 1)$, and a mixed mode $(n, l, m) = (-1, 2, 1)$, from left to right. The horizontal axis (x) is the fractional radius and the vertical axis (μ) is cosine of colatitude. The mixed-mode kernel has sensitivity around the high latitude region, showing the possibility that we can infer the rotation rate there

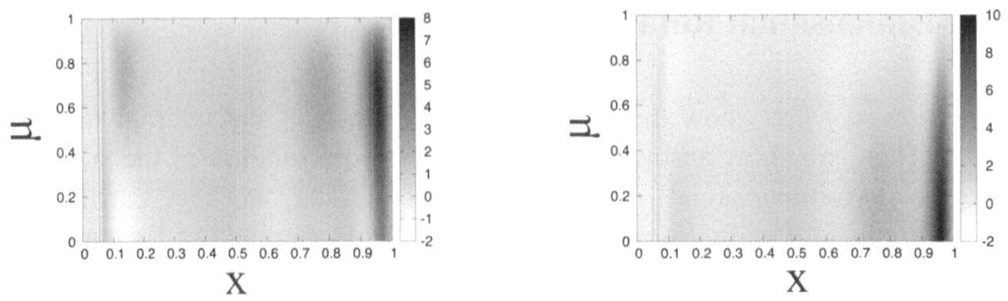

$\hat{\Omega}(0.95, 0.70) = (0.9564 \pm 0.0006)\,\Omega_{100}$ $\hat{\Omega}(0.95, 0.00) = (0.9511 \pm 0.0005)\,\Omega_{100}$

Fig. 2 Averaging kernels for a target point in the high latitude region (left) and in the low latitude region (right). The corresponding estimates of the rotation rate, in units of $\Omega_{100} = 2\pi \times 0.01\ \mathrm{d}^{-1}$, are presented below each panel

2.2 *Hint of a Fast Core Rotation*

Another interesting result we have found is that the convective core might be rotating 5–6 times faster than the other regions of the star. When we attempted to infer the rotation rate in the deep radiative region, we had two possible estimates depending on a choice of the trade-off parameter. According to the inversion coefficients, one estimate is mostly determined by g-mode rotational shifts, and the result thus obtained is consistent with that of the previous study [7] that the envelope rotates faster than the deep radiative region (slower estimate). On the other hand, the other estimate is mostly determined by mixed-mode rotational shifts, and suggests the opposite trend of the rotational profile (faster estimate). We closely looked at the averaging kernels for the two estimates and noticed that the averaging kernel for the faster estimate has a small sensitivity inside the convective core where g modes cannot propagate (Fig. 3). Then, we can explain the difference between the two estimates if the convective core is rotating much faster than the radiative

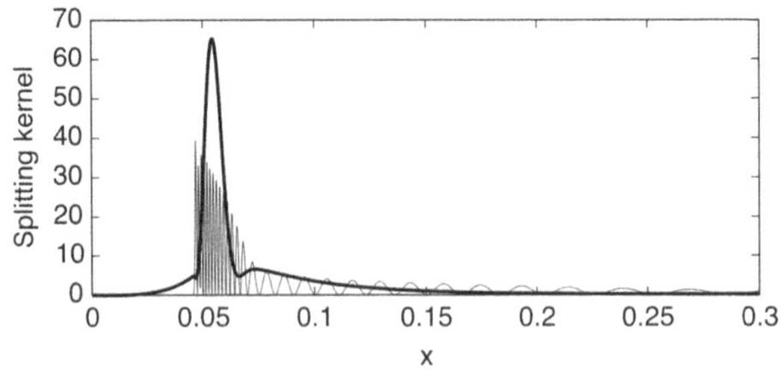

Fig. 3 Slices of rotational splitting kernels at $\mu = 0$ for a g mode (thin curve) and a mixed mode with $(n, l, m) = (2, 2, 2)$ (thick curve). The g-mode kernel has no sensitivity below $x = 0.05$ because there is a convective region and g modes cannot propagate through a convectively unstable region. The mixed-mode kernel, on the other hand, has a sensitivity there, enabling us to extract some information on rotation rate inside the convective core

region above, affecting the results even with such a small sensitivity of mixed-mode splitting kernels inside the convective core.

In order to quantify the fast rotation of the core, we carried out three-zone modeling of the internal rotation, and found that the convective core might be rotating 5–6 times faster than the other regions of the star. Note that we have also confirmed the result of [7]; the results can be roughly summarized as $\Omega_{\mathrm{cov,core}} \gg \Omega_{\mathrm{env}} > \Omega_{\mathrm{rad,deep}}$.

3 Refining the Equilibrium Model of the Star Based on the Observed Period Spacing Pattern

There have been several attempts to calculate an equilibrium model of KIC 11145123 [7, 15, 12]. However, those models do not reproduce the observed frequencies well, especially for the p modes. As introduced in Sect. 1, the star might have experienced some interaction with another star. But all the models calculated in the previous studies are based on an assumption of a single star evolution, thus there is room for improving the model taking some interactions during its evolution into account.

In this section, as a first step toward the non-standard modeling of the star, we fine-tuned the model made by Takada-Hidai et al. [15] focusing on the modulated pattern of the observed g-mode period spacings ΔPg (Fig. 4). The spacing ΔPg is defined as the difference between two g-mode periods with the consecutive radial order and with the same spherical degree. For high-order g modes, it is generally considered that ΔPg approaches a constant value determined by the integration of the Brunt-Väisälä frequency throughout the g-mode cavity [16]. Though many observations have supported the relation predicted based on the asymptotic approx-

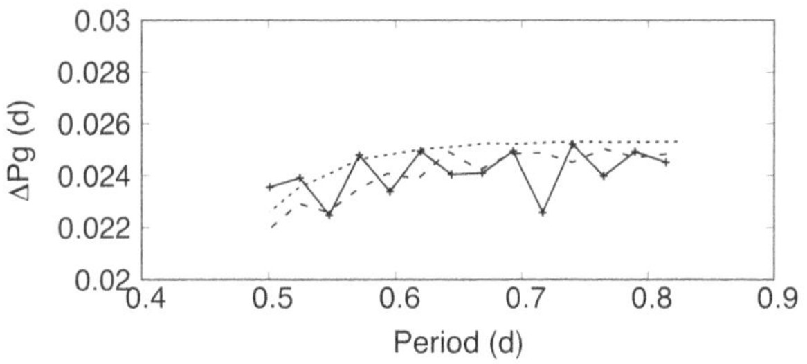

Fig. 4 Modeled g-mode period spacing patterns (dotted and dashed lines) are compared with the observed one (solid line). The dotted line is calculated based on [15] (and reproduced by us); the dashed line is calculated based on the model with weaker diffusion setting. There is a clear difference in the behavior of the modeled g-mode period spacing patterns; the weaker diffusion model can seemingly mimic the modulated pattern of the observed period spacing, which is thought to be caused by a steep feature in the Brunt-Väisälä frequency in the weaker diffusion model

imation theory, some deviations from the constant values certainly exist in ΔPg patterns as well. These deviations are usually interpreted as a manifestation of sharp features in the Brunt-Väisälä frequency distribution [8].

In the case of the previous model of [15], the Brunt-Väisälä frequency distribution is well smoothed due to diffusion process adopted in the model, leading to the smooth ΔPg pattern (dotted line in Fig. 4). We then weakened the diffusion process in our model calculation by 5 orders of magnitude, intentionally made the distribution of the Brunt-Väisälä frequency steeper than the previous one, and succeeded in mimicking the amplitude of the oscillatory pattern of the observed ΔPg (dashed line in Fig. 4).

4 Conclusions

We have found from the detailed asteroseismic analyses for KIC 11145123 that: (1) the star is exhibiting latitudinally differential rotation with more than 2σ significance and it is anti-solar-like rotation, (2) the convective core of the star might be rotating much faster than the other parts of the star, and (3) the diffusion process should be much weaker (by 5 orders of magnitude) in the deep region of the star than usual settings in MESA [9]. The question is if there is any relationship among these results.

One possible relationship can be found between the second result and the third result: The diffusion process in the deep region of the star is weakened due to extra mixing caused by the velocity shear which exists at the boundary between the convective core and the radiative region above. This scenario is reminiscent of what is thought to be at work at the solar tachocline, motivating us toward further studies of, for instance, coupling between rotational velocity shear and mixing processes inside stars.

Acknowledgments The authors would like to thank NASA and Kepler team for their invaluable data. H. Saio and H. Shibahashi are thanked for their insightful and constructive advices. YH thanks SOKENDAI for the financial support. YH and TS thank late Michael Thompson for delightful discussions about this interesting star.

References

1. Backus, G. E., & Gilbert, J. F. (1967). *Geophysical Journal, 13*, 247.
2. Beck, P. G., Montalban, J., Kallinger, T., De Ridder, J., Aerts, C., García, R. A., et al. (2012). *Nature, 481*, 55.
3. Chaplin, W. J., & Miglio, A. (2013). *Annual Review of Astronomy and Astrophysics, 51*, 353–392.
4. Hatta, Y., Sekii, T., Takata, M., & Kurtz, D. W. (2019). *The Astrophysical Journal, 871*, 135.

5. Huber, D., Aguirre, V. S., Matthews, J. M., Pinsonneault, M. H., Gaidos, E., García, R. A., et al. (2014). *The Astrophysical Journal Supplement Series, 211*, 2.
6. Koch, D. G., Borucki W. J., Basri, G., et al. (2010). *The Astrophysical Journal Letters, 713*, L79.
7. Kurtz, D. W., Saio, H., Takata, M., Shibahashi, H., Murphy, S. J., & Sekii, T. (2014). *Monthly Notices of the Royal Astronomical Society, 444*, 102.
8. Miglio, A., Montalbán, J., Noels, A., & Eggenberger, P. (2008). *Monthly Notices of the Royal Astronomical Society, 386*, 1487–1502.
9. Paxton, B., Marchant, P., Schwab, J., Bauer, E. B., Bildsten, L., Cantiello, M., et al. (2015). *The Astrophysical Journal Supplement Series, 220*, 15.
10. Ritzwoller, M. H., & Lavely, E. M. (1991). *The Astrophysical Journal, 369*, 557.
11. Saio, H., Kurtz, D. W., Takata, M., Shibahashi, H., Murphy, S. J., Sekii, T., & Bedding, T. R. (2015). *Monthly Notices of the Royal Astronomical Society, 447*, 3264.
12. Sánchez Arias, J. P., Córsico, A. H., & Althaus, L. G. (2017). *Astronomy and Astrophysics, 597*, A29.
13. Sandage, A. R. (1953). *Astronomical Journal, 58*, 61.
14. Schmid, V. S., & Aerts, C. (2016). *Astronomy and Astrophysics, 592*, A116.
15. Takada-Hidai, M., Kurtz, D. W., Shibahashi, H., Murphy, S. J., Takata, M., Saio, H., et al. (2017). *Monthly Notices of the Royal Astronomical Society, 470*, 4908.
16. Tassoul, M. (1980). *The Astrophysical Journal Supplement Series, 43*, 469.
17. Townsend, R. H. D., & Teitler, A. (2013). *Monthly Notices of the Royal Astronomical Society, 435*, 3406–3418.

The Impact of a Fossil Magnetic Field on Dipolar Mixed-Mode Frequencies in Sub- and Red-Giant Stars

L. Bugnet, V. Prat, S. Mathis, R. A. García, S. Mathur, K. Augustson, C. Neiner, and M. J. Thompson

Abstract The recent discovery of low-amplitude dipolar oscillation mixed modes in massive red giants indicates the presence of a missing physical process inside their cores. Stars more massive than $\sim 1.3\ M_\odot$ are known to develop a convective core during the main-sequence: the dynamo process triggered by this convection could be the origin of a strong magnetic field inside the core of the star, trapped when it becomes stably stratified and for the rest of its evolution. The presence of highly magnetized white dwarfs strengthens the hypothesis of buried fossil magnetic fields inside the core of evolved low-mass stars. If such a fossil field exists, it should affect the mixed modes of red giants as they are sensitive to processes affecting the deepest layers of these stars. The impact of a magnetic field on dipolar oscillations modes was one of Pr. Michael J. Thompson's research topics during the 90s when preparing the helioseismic SoHO space mission. As the detection of gravity modes in the Sun is still controversial, the investigation of the solar oscillation modes did not provide any hint of the existence of a magnetic field in the solar radiative core. Today we have access to the core of evolved stars thanks to the asteroseismic observation of mixed modes from CoRoT, *Kepler*, K2 and TESS missions. The idea of applying and generalizing the work done for the Sun came from discussions with Pr. Michael Thompson in early 2018 before we lost him. Following the path we drew together, we theoretically investigate the effect of a stable axisymmetric mixed poloidal and

L. Bugnet (✉) · V. Prat · S. Mathis · R. A. García · K. Augustson
AIM, CEA, CNRS, Université Paris-Saclay, Université Paris Diderot, Gif-sur-Yvette, France
e-mail: lisa.bugnet@cea.fr

S. Mathur
Instituto de Astrofísica de Canarias, La Laguna, Tenerife, Spain

C. Neiner
LESIA, Paris Observatory, CNRS, PSL University, Sorbonne Université, Univ. Paris Diderot, Meudon, France

M. J. Thompson
National Center for Atmospheric Research, Boulder, CO, USA

M. J. P. F. G. Monteiro et al. (eds.), *Dynamics of the Sun and Stars*, Astrophysics and Space Science Proceedings 57,
https://doi.org/10.1007/978-3-030-55336-4_33

toroidal magnetic field, aligned with the rotation axis of the star, on the mixed modes frequencies of a typical evolved low-mass star. This enables us to estimate the magnetic perturbations to the eigenfrequencies of mixed dipolar modes, depending on the magnetic field strength and the evolutionary state of the star. We conclude that strong magnetic fields of $\sim$ 1MG should perturb the mixed-mode frequency pattern enough for its effects to be detectable inside current asteroseismic data.

1 Introduction: Understanding Red Giant Dynamics

Amongst key observations of red-giant (RG) interiors that stand out are their interestingly low core rotation rate (e.g., [1, 20, 14]) that is only about 10 times higher than the surface rate. However, state-of-the-art stellar evolution models predict that the core should rotate about 100 times faster than the stellar surface according to current transport models (e.g., [9, 6, 18, 4]). This highlights an strong bottleneck concerning the understanding of the strength of the angular momentum transport inside evolved Solar-like stars. This is complemented by the surprising discovery of two populations of RG where the character of their oscillations depends upon their masses (e.g., [13, 25, 21]). On the one hand, the lower mass (1 $M_\odot < M_\star < 1.3\ M_\odot$) range of stars exhibit solar-like oscillations with normal amplitude. On the other hand, nearly 50% of the more massive group ($1.5\ M_\odot < M_\star < 2.5\ M_\odot$) exhibit some mixed modes with very low amplitudes. This emphasises that some oscillation modes are weakened and sometimes they disappear in the core of a large fraction of RG. Once again, the current simplified theory needs to be modified to account for the origin of these weakened modes.

The idea of a strong stable buried magnetic field altering the oscillation modes of red giants (RG) came from [11] who invoked a conversion from gravity waves towards magneto-gravity waves that would remain trapped inside the core of the star. This theory is controversial [21], but has been systematically studied and improved [5, 17, 16], revealing that magnetism is still a very good candidate for both angular momentum transport [12, 10] and mode suppression. It was shown in [2] that the relaxation of a stochastic field in a radiative zone leads to the formation of a stable, fossil, mixed poloidal and toroidal magnetic field. The energy minimization principle employed in [7] yields analytical approximations to such a mixed field solution. We use this configuration as provided in [8] to model a fossil magnetic field inside RG being the remnant of the field generated by the convective core dynamo during the main sequence (MS) for $M_\star > 1.3\ M_\odot$ (see Fig. 1). We then theoretically investigate the effect of this stable fossil mixed poloidal and toroidal field buried inside the core of RG on their mixed-mode frequencies. In this study, we focus on the axisymmetric case, where the magnetic axis is aligned with the rotation axis of the star. The convective envelope is also represented by the vortex symbols.

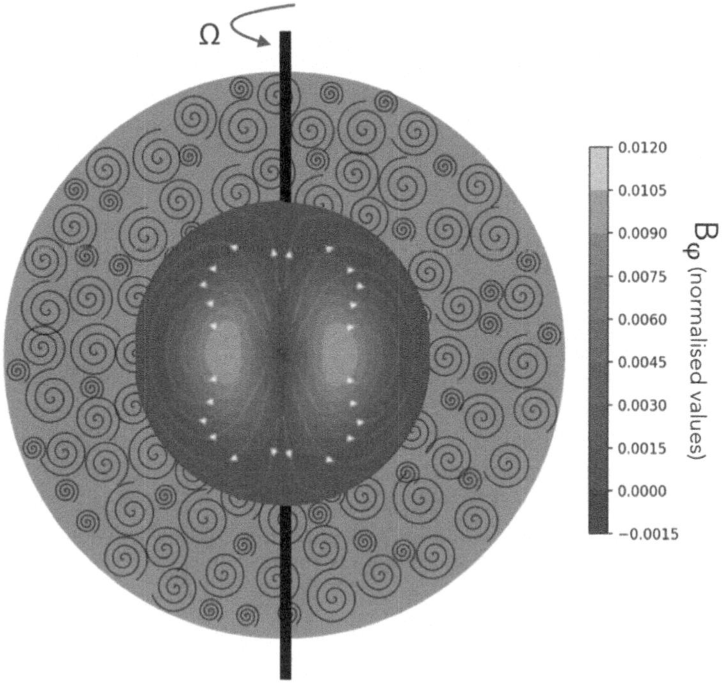

Fig. 1 Representation of a trapped magnetic field inside an RG core. The magnetic field has a poloidal (white lines) and a toroidal (grey scale) components normalised by the strength B_0 of the field computed by using the semi-analytic description of [7, 8]

2 Perturbative Methods

We use the linearised equations for stellar oscillations [27], perturbed by both rotation and magnetism to evaluate the first-order frequency perturbation $\delta\omega$ through

$$\delta\omega = -\frac{\langle\vec{\xi}_0, \delta\vec{F}_\ell(\vec{\xi}_0)/\rho\rangle + \langle\vec{\xi}_0, \vec{F}_c(\vec{\xi}_0)\rangle}{2\omega_0\langle\vec{\xi}_0, \vec{\xi}_0\rangle}, \tag{1}$$

with $\delta\vec{F}_\ell$ being the perturbed volumetric Lorentz force, $\vec{F}_c$ the Coriolis acceleration, $\vec{\xi}_0$ the unperturbed eigenfunction, ρ the density profile of the star, and ω_0 the unperturbed mixed-mode frequencies. For each of the terms composing Eq. 1 (see [23] for more details), we numerically estimate the mass mode by following the dominant components

$$\langle\vec{\xi}_0, \vec{F}_c(\vec{\xi}_0)\rangle \simeq 8\pi m\Omega \int_0^R \rho r^2 |\xi_h|^2 \mathrm{d}r \int_0^\pi Y_\ell^m \partial_\theta Y_\ell^{m*} \frac{\cos\theta}{\sin\theta} \mathrm{d}\theta, \tag{2}$$

$$\langle \vec{\xi_0}, \delta \vec{F_\ell}(\vec{\xi_0}) \rangle \simeq 2\pi B_0^2 \int_0^R r\xi_h^* b_r (r\xi_h b_r)'' \mathrm{d}r \times \int_0^\pi \left(\left(\frac{mY_\ell^m}{\sin\theta} \right)^2 + (\partial_\theta Y_\ell^m)^2 \right) \cos^2\theta \sin\theta \mathrm{d}\theta, \quad (3)$$

and the mode mass

$$\langle \vec{\xi_0}, \vec{\xi_0} \rangle \simeq 2\pi \int_0^R \rho r^2 |\xi_h|^2 \mathrm{d}r \int_0^\pi \left(\left(\frac{mY_\ell^m}{\sin\theta} \right)^2 + (\partial_\theta Y_\ell^m)^2 \right) \sin\theta \mathrm{d}\theta, \quad (4)$$

with B_0 the magnetic field strength, $\xi_h(r)$ the radial eigenfunction describing the horizontal displacement on the spherical harmonics ($Y_\ell^m(\theta, \varphi)$ where ℓ is the degree, m the azimuthal degree, and (r, θ, φ) are the usual spherical coordinates), $b_r(r)$ the radial function describing the radial component of the magnetic field on the $\ell = 1$ dipolar spherical harmonics, Ω the angular velocity. The unperturbed eigenfunctions of mixed modes ($\vec{\xi_0}$) are evaluated by using the stellar pulsation code GYRE [26] along with the stellar evolution model MESA [22].

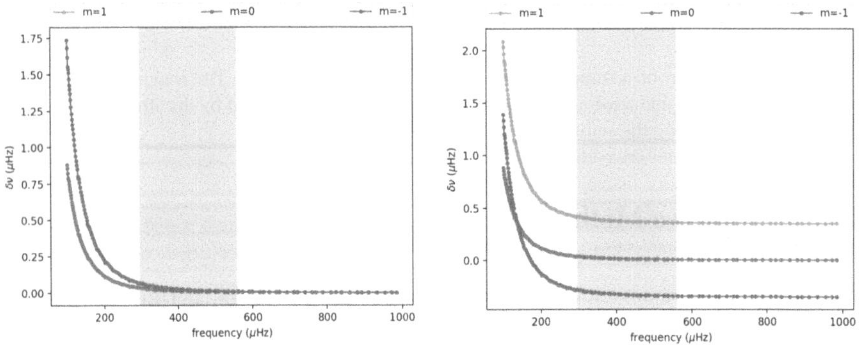

Fig. 2 *Left:* Mode splittings ($\delta\nu$) induced by magnetism only for a $M_\star = 1.5\, M_\odot$ early RG with $\nu_{max} = 425\,\mu$Hz as a function of mode frequencies. The light gray area represents the range of frequencies $[\nu_{max} - 5\Delta\nu : \nu_{max} + 5\Delta\nu]$ for which mixed modes should be visible in real data. Each of them represents a simulated mixed mode with the lower line representing $m = 0$ and the upper line $m = 1$ and $m = -1$ modes, which overlap as a result of the m dependency in Eq. 3. *Right:* Same as left panel with rotation splitting corresponding to a uniform rotation of $0.7\,\mu$Hz added

3 Mixed-Mode Splittings in RG

Figure 2 shows the strength of the frequency perturbation of mixed modes due to the magnetic field only (left panel) for a field strength of $B_0 = 1$ MG. We observe that one individual $\ell = 1$ mixed mode splits into two modes: the m =0 component and the m =1 and m =-1 components (overlapping as described by Eq. 3). In the frequency region where we expect to see mixed modes in the power spectrum density (light gray area), magnetic splittings are of the order of 0.1 μHz. This is a small effect by comparison with the rotational frequency splitting due to a uniform rotation of the star of 0.7μHz as shown on the right panel. Nevertheless, the effect of such a magnetic field can be large enough to be observable with current instruments.

The dependence of the magnetic splittings on the evolution of the star and on the magnetic field strength is represented in Fig. 3. The right panel indicates typical data frequency boundaries for the observation of magnetic splittings to be compared to the left panel on which values of magnetic splitting are represented: if the combination of magnetic field strength and of the evolution leads to a magnetic splitting below the *Kepler* resolution, then the effect of magnetism on mixed-mode frequencies would not be visible in *Kepler* data. Above the typical $\ell = 0$ linewidth the magnetic effect is considered as being too large for our perturbative analysis to be valid (see [3, 19]). The region in between the two black lines approximatively shows the ideal conditions for the detection of the magnetic effect in *Kepler* data. This region would be narrowed to the area between the TESS 1 year white line and

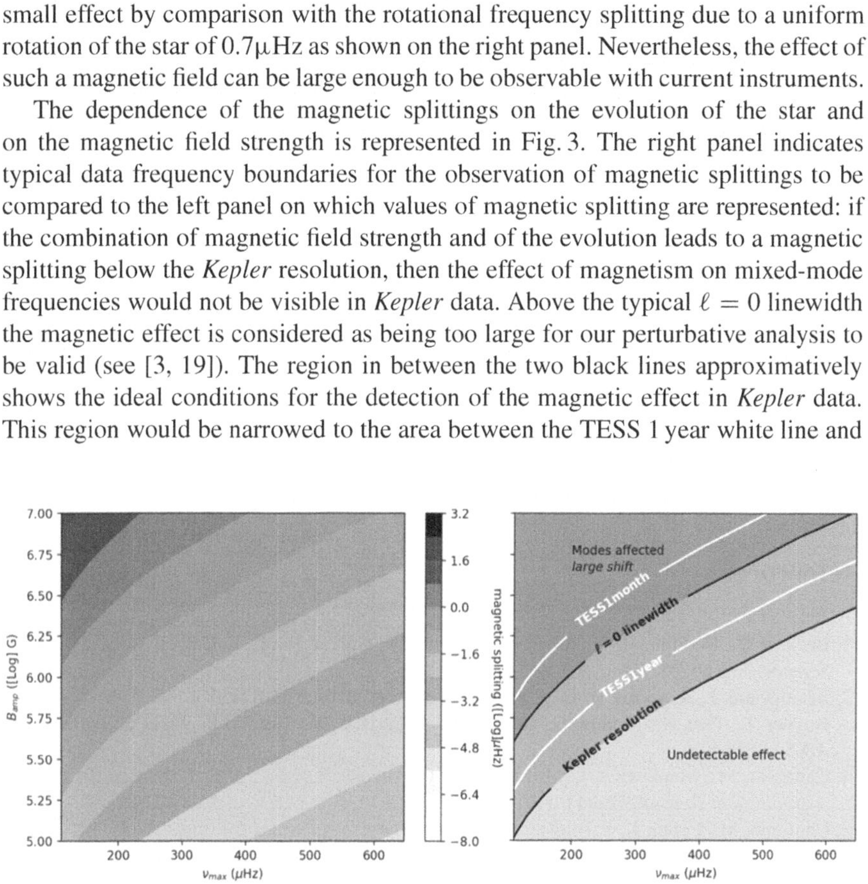

Fig. 3 *Left:* Mode splitting induced by magnetism only for a $M_\star$ = 1.5 $M_\odot$ giant as a function of the frequency of maximum power ν_{max} and the magnetic field strength. The mode splitting is calculated at each evolutionary stage by taking the splitting corresponding to the central mixed mode (the closest to ν_{max}). *Right:* Same diagram as the left panel simplified to compare with typical frequencies: the "*Kepler* resolution" line corresponds to the 7.9 nHz resolution for the 4 years *Kepler* data, the "TESS 1 year" line corresponds to the 32 nHz resolution for 1 year of TESS data, the "TESS 1 month" line corresponds to the 380 nHz resolution for 1 month of TESS data, and the "$\ell = 0$" line corresponds to the typical radial mode linewidths [28] for sub-giants, as being an upper limit for mixed-mode linewidth

the $\ell=0$ linewidth black line if considering TESS 1 year data, and we do not expect to be able to detect any magnetic effect on mixed-mode frequencies when looking at only 1 month TESS data.

4 Conclusion and Perspectives

We study the axisymmetric case for which a buried stable fossil magnetic field, with poloidal and toroidal components, is aligned with the rotation axis of the star. We find that a typical field strength of about 1MG results in a detectable frequency shift of the mixed modes inside RG's power spectrum density in addition to the already measured rotational splittings. The study will be extended towards fossil dipolar fields inclined with respect to the rotational axis of the star, and towards non-fossil field topologies such as those derived in [24, 15, 12], in order to explore the variety of possible seismic signatures that would allow us to probe the deep magnetism of evolved low-mass stars and its consequences for the transport of angular momentum.

Acknowledgments The authors of this work acknowledge the support received from the PLATO CNES grant, the National Aeronautics and Space Administration under Grant NNX15AF13G, by the National Science Foundation grant AST-1411685, the Ramon y Cajal fellowship number RYC-2015-17697, the ERC SPIRE grant (647383), and the Fundation L'Oréal-Unesco-Académie des sciences.

References

1. Beck, P. G., Bedding, T. R., Mosser, B., Stello, D., Garcia, R. A., Kallinger, T., et al. (2011). *Science, 332*(6026), 205.
2. Braithwaite, J., & Spruit, H. C. (2004). *Nature, 431*(7010), 819–821.
3. Bugnet, L., Prat, V., Mathis, S., Astoul, A., Augustson, K., Garcia, R. A., et al. (submitted) *A&A*.
4. Cantiello, M., Mankovich, C., Bildsten, L., Christensen-Dalsgaard, J., & Paxton, B. (2014). *Astrophysical Journal, 788*(1), 1–7.
5. Cantiello, M., Fuller, J., & Bildsten, L. (2016). *The Astrophysical Journal, 824*(1), 14.
6. Ceillier, T., Eggenberger, P., García, R. A., & Mathis, S. (2013). *Astronomy and Astrophysics, 555*, 1–8.
7. Duez, V., & Mathis, S. (2010). *Astronomy and Astrophysics, 517*(8), 1–13.
8. Duez, V., Braithwaite, J., & Mathis, S. (2010). *Astrophysical Journal Letters, 724*(1 PART 2), 34–38.
9. Eggenberger, P., Montalbán, J., & Miglio, A. (2012). *Astronomy and Astrophysics, 544*, 1–4.
10. Eggenberger, P., Deheuvels, S., Miglio, A., Ekström, S., Georgy, C., Meynet, G., et al. (2019). *Astronomy & Astrophysics, 621*, A66.
11. Fuller, J., Cantiello, M., Stello, D., Garcia, R. A., & Bildsten, L. (2015). *Science, 350*(6259), 423–426.
12. Fuller, J., Piro, A. L., & Jermyn, A. S. (2019). *Monthly Notices of the Royal Astronomical Society, 485*(3), 3661–3680.

13. García, R. A., Pérez Hernández, F., Benomar, O., Silva Aguirre, V., Ballot, J., Davies, G. R., et al. (2014). *Astronomy and Astrophysics, 563*, 1–17.
14. Gehan, C., Mosser, B., Michel, E., Samadi, R., & Kallinger, T. (2018). *Astronomy and Astrophysics, 616*, 1–12.
15. Jouve, L., Gastine, T., & Lignières, F. (2014). *Astronomy and Astrophysics, JGL*(12), 12.
16. Lecoanet, D., Vasil, G. M., Fuller, J., Cantiello, M., & Burns, K. J. (2017). *Monthly Notices of the Royal Astronomical Society, 466*(2), 2181–2193.
17. Loi, S. T., & Papaloizou, J. C. B. (2017). *Monthly Notices of the Royal Astronomical Society, 3225*, 3212–3225.
18. Marques, J. P., Goupil, M. J., Lebreton, Y., Talon, S., Palacios, A., Belkacem, K., et al. (2013). *Astronomy and Astrophysics, 549*, A74.
19. Mathis, S., Bugnet, L., Prat, V., Augustson, K., Mathur, S., & Garcia, R.A. (submitted) *A&A*.
20. Mosser, B., Elsworth, Y., Hekker, S., Huber, D., Kallinger, T., Mathur, S., et al. (2012). *Astronomy and Astrophysics, 537*, 1–15.
21. Mosser, B., Belkacem, K., Pinçon, C., Takata, M., Vrard, M., Barban, C., et al. (2017). *Astronomy and Astrophysics, 598*, 1–12.
22. Paxton, B., Bildsten, L., Dotter, A., Herwig, F., Lesaffre, P., & Timmes, F. (2011). *Astrophysical Journal, Supplement Series, 192*(1), 1–110.
23. Prat, V., Mathis, S., Buysschaert, B., Van Beeck, J., Bowman, D. M., Aerts, C., et al. (2019). *Astronomy and Astrophysics, 64*, 1–9.
24. Spruit, H. C. (1999). *Astronomy and Astrophysics, 349*(1), 189–202.
25. Stello, D., Cantiello, M., Fuller, J., Garcia, R. A., & Huber, D. (2016). *Publications of the Astronomical Society of Australia, 33*, 1–6.
26. Townsend, R. H., & Teitler, S. A. (2013). *Monthly Notices of the Royal Astronomical Society, 435*(4), 3406–3418.
27. Unno, W., Osaki, Y., Ando, H., Saio, H., & Shibahashi, H. (1989). *Nonradial oscillations of stars*. Tokyo: University of Tokyo Press.
28. Vrard, M., Mosser, B., & Barban, C. (2017). *EPJ Web of Conferences, 160*, 2–4.

Asteroseismic Stellar Modelling: Systematics from the Treatment of the Initial Helium Abundance

Nuno Moedas, Benard Nsamba, and Miguel T. Clara

Abstract Despite the fact that the initial helium abundance is an essential ingredient in modelling solar-type stars, its abundance in these stars remains a poorly constrained observational property. This is because the effective temperature in these stars is not high enough to allow helium ionization, not allowing any conclusions on its abundance when spectroscopic techniques are employed. To this end, stellar modellers resort to estimating the initial helium abundance via a semi-empirical helium-to-heavy element ratio, anchored to the standard Big Bang nucleosynthesis value. Depending on the choice of solar composition used in stellar model computations, the helium-to-heavy element ratio, $(\Delta Y/\Delta Z)$ is found to vary between 1 and 3. In this study, we use the *Kepler* LEGACY stellar sample, for which precise seismic data is available, and explore the systematic uncertainties on the inferred stellar parameters (radius, mass, and age) arising from adopting different values of $\Delta Y/\Delta Z$, specifically, 1.4 and 2.0. The stellar grid constructed with a higher $\Delta Y/\Delta Z$ value yields lower radius and mass estimates. We found systematic uncertainties of 1.1%, 2.6%, and 13.1% on radius, mass, and ages, respectively.

N. Moedas · M. T. Clara
Instituto de Astrofísica e Ciências do Espaço, Universidade do Porto, Porto, Portugal

Departamento de Física e Astronomia, Faculdade de Ciências da Universidade do Porto, Porto, Portugal

B. Nsamba (✉)
Max-Planck-Institut für Astrophysik, Garching, Germany

Instituto de Astrofísica e Ciências do Espaço, Universidade do Porto, Porto, Portugal
e-mail: nsamba@mpa-garching.mpg.de

M. J. P. F. G. Monteiro et al. (eds.), *Dynamics of the Sun and Stars*, Astrophysics and Space Science Proceedings 57,
https://doi.org/10.1007/978-3-030-55336-4_34

1 Introduction

Chemical abundances are some of the most essential ingredients in stellar modelling, complementing our understanding of the formation, structure, and evolution of stars. Solar abundances are commonly adopted in stellar evolution codes, e.g., Modules for Experiments in Stellar Astrophysics (MESA; [20]), among others, however, some significant discrepancies exist. Among the different element abundances, helium abundance measurements in solar-type stars are one of the most poorly constrained ingredients in stellar modelling. This is because the temperature required to excite an atomic transition of the helium exceeds 20,000K, a value higher than the characteristic effective temperature of the solar-type stars (e.g., [6]).

To overcome the helium abundance problem when constructing stellar models, a common solution is use the "Galactic chemical evolution law" in that the iron content ([Fe/H]) is transformed into fractional abundances (i.e., hydrogen mass fraction, X, helium mass fraction, Y, and heavy elements mass fraction, Z) via the helium enrichment ratio ($\Delta Y/\Delta Z$) using the expression

$$\frac{\Delta Y}{\Delta Z} = \frac{Y - Y_0}{Z - Z_0} \tag{1}$$

set to the BNN values of Z_0= 0.0 and Y_0= 0.2484 [4]. [13] reported $\Delta Y/\Delta Z$ = 2.1 ± 0.4 using observations of nearby K dwarfs and a set of isochrones. Similar results were found by Casagrande et al. [3] using a set of Padova isochrones and observations of nearby K dwarfs (i.e., $\Delta Y/\Delta Z = 2.1 \pm 0.9$). [1] published $\Delta Y/\Delta Z$ = 1.6 obtained using metal-poor H II regions, Magellanic cloud H II regions and M17 abundances, while taking into account the effects of temperature fluctuations. Interestingly, when using only galaxy H II region S206 and M17, [1] determines $\Delta Y/\Delta Z = 1.41 \pm 0.62$, a value reported to be consistent with that from standard chemical evolution models. Depending on the choice of solar composition, [25] reported the initial helium abundance of the Sun to be in agreement with a slope of $1.7 \leq \Delta Y/\Delta Z \leq 2.2$. In general, acceptable values for the helium enrichment ratio tend to vary between 1 and 3.

Lebreton et al. [15] reported a scatter of $\sim 5\%$ in mass arising from the treatment of initial helium mass fraction. Using synthetic data of about 10,000 artificial stars, [30] found a systematic bias of 2.3% and 1.1% in mass and radius, respectively, arising from a variation of ±1 in $\Delta Y/\Delta Z$. Further, [31] reported a systematic bias in age to be about one-fourth of the statistical error in the first 30% of the evolution, while its negligible for more evolved stars. The treatment of the initial helium mass fraction in stellar models is therefore a substantial source of systematic uncertainties on stellar properties (such as mass, radius, and age) derived through forward modelling techniques.

In this article, we take advantage of the available stars with multi-year *Kepler* photometry [16] and with parallax measurements from the Gaia mission. We complement all these constraints with spectroscopic constraints (i.e., effective temperature and metallicity) and quantify systematic uncertainties on stellar parameters (mainly radius, mass, and age) arising from the variation in $\Delta Y/\Delta Z$ values used Eq. 1.

2 Target Sample

Our sample consists of 66 *Kepler* LEGACY stars [26, 16] with at least 12 months of short cadence data ($\Delta t = 58.89$ s). Figure 1 shows the location of our sample on a $\Delta\nu$ (large separation)–$T_{\rm eff}$ diagram. The spectroscopic parameters (metallicity, [Fe/H], and effective temperature, $T_{\rm eff}$) for each star were obtained from [26] and references therein. [2] used *Gaia* Data Release 2 (*Gaia* DR2) parallaxes as inputs in the stellar classification code "isoclassify" [10] to derive stellar radii for majority of the *Kepler* stars. Adopting the stellar radii and $T_{\rm eff}$ measurements in the Stefan-Boltzmann relation, we derived the stellar luminosities for the majority of our sample. For the stars in our sample that were not analysed by Berger et al. [2], we obtained their luminosities using the expression [21]

$$\log(L/{\rm L}_\odot) = 4.0 + 0.4M_{\rm bol,\odot} - 2.0\log(\pi[{\rm mas}]) - 0.4(V - A_V + BC_V), \qquad (2)$$

where $M_{\rm bol,\odot}$ is the bolometric magnitude of the Sun with a value of 4.73 mag, π[mas] is the parallax, V is the magnitude in the V band obtained from [9], A_V is

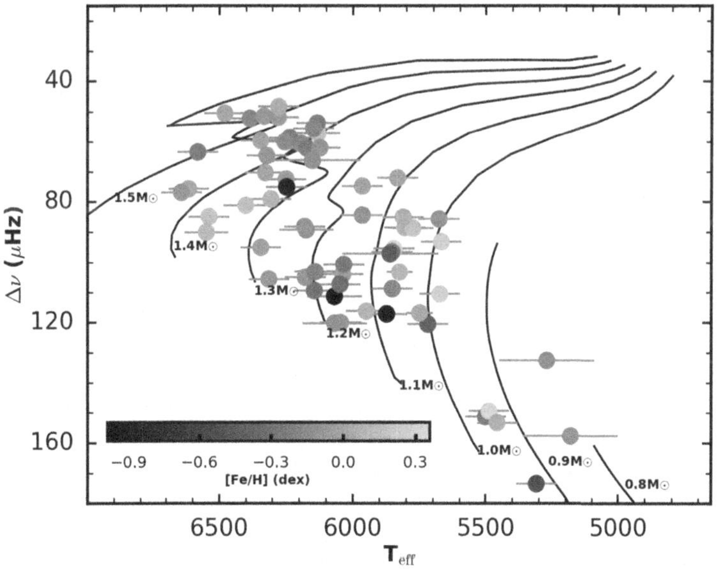

Fig. 1 $\Delta\nu$–$T_{\rm eff}$ diagram. Each circle represents a target star colour coded according to its metallicity. The black lines represent stellar evolutionary tracks, ranging in mass from 0.8M$_\odot$ to 1.5M$_\odot$, with $Z = 0.02$ and a mixing length parameter ($\alpha_{\rm MLT}$) of 1.8 constructed using MESA

Table 1 Stellar grid constituents

Grid	Mass ($M_\odot$)	Diffusion	Overshoot	$\Delta Y / \Delta Z$
A	0.7–1.1	Yes	No	1.4
	1.2–1.6	No	Yes	1.4
B	0.7–1.1	Yes	No	2.0
	1.2–1.6	No	Yes	2.0

the extinction in the V band taken from [17], and BC_V is the bolometric correction calculated using the polynomial expression from [28]. We note that for the binary system 16 Cyg, we used the luminosities presented by Metcalfe et al. [18].

2.1 Stellar Models

We built two stellar grids (namely A and B) varying mainly in the treatment of initial helium mass fraction (Y) using MESA version 9793. The evolution tracks were only terminated when models reach: (1) a stellar age of $\sim$ 16 Gyr and (2) a point along the evolutionary tracks where the $\log(\rho_c) = 4.5$ (ρ_c is the central density). We note that only models from the Zero Age Main-Sequence (ZAMS) to the termination point were stored.

Table 1 summarizes the different grid constituents. The evolution tracks vary in mass M $\in$ [0.7–1.6] $M_\odot$ in steps of 0.05 $M_\odot$, Z $\in$ [0.004–0.04] in steps of 0.002, and $\alpha_{\rm mlt} \in$ [1.0–3.0] in steps of 0.4. Atomic diffusion is known to be an efficient transport process in low-mass stars and was included in our low-mass models (see Table 1) based the description of [27]. For models with M $\in$ [1.2–1.6] $M_\odot$, we include convective core overshoot by adopting the exponential diffusive overshoot procedure as implemented in MESA based on [8]. The overshoot parameter was set to vary in the range [0.0–0.03] in steps of 0.005. We note models in the mass range [1.1–1.15] $M_\odot$ lie in the transition region where models may develop convective cores [19]. For models within this mass range with a convective core, both diffusion and core overshoot were included.

The general input physics used for the grids include nuclear reaction rates obtained from Joint Institute for Nuclear Astrophysics Reaction Library (JINA REACLIB; [4]) version 2.2 with specific rates for $^{12}C(\alpha, \gamma)^{16}O$ and $^{14}N(p, \gamma)^{15}O$ described by Kunz et al. [14] and [12], respectively. At high temperatures, OPAL tables [11] were used to cater for opacities while tables from [5] were used at lower temperatures. Both grids A and B used the 2005 updated version of the OPAL equation of state [24]. The surface boundary of stellar models is described using the standard Grey-Eddington atmosphere. Both grids used metallicity mixtures from [7].We note that GYRE [29] was used to calculate model oscillation frequencies for spherical degrees $\ell = 0, 1, 2, 3$.

In order to generate a set of models that best-fit the seismic and classical constraints of our sample stars, we employ the Asteroseismic Inference on a Massive

Scale (AIMS; [23]) yielding the mean and standard deviation of the posterior probability density functions of different stellar parameters.

3 Discussion

The top left panel of Fig. 2 shows that grid A yields optimal solutions with higher masses compared to grid B, with systematic uncertainties of $\sim 2.6\%$ and a bias of $\sim 2\%$. A similar trend can be seen in the bottom panel of Fig. 2, with grid A yielding higher radii compared to grid B, with systematic uncertainties of $\sim 1.1\%$ and a bias of $\sim 0.6\%$. Based on Eq. 1, for a given value of Z, grid B (i.e., with $\Delta Y/\Delta Z = 2$) has models with higher initial helium mass fraction (Y) compared to models in grid A (i.e., with $\Delta Y/\Delta Z = 1.4$). This implies that models in grid B have a higher mean molecular weight which, in turn, increases the energy production rate resulting into an increase in the energy flux at the surface—increasing the model luminosity.

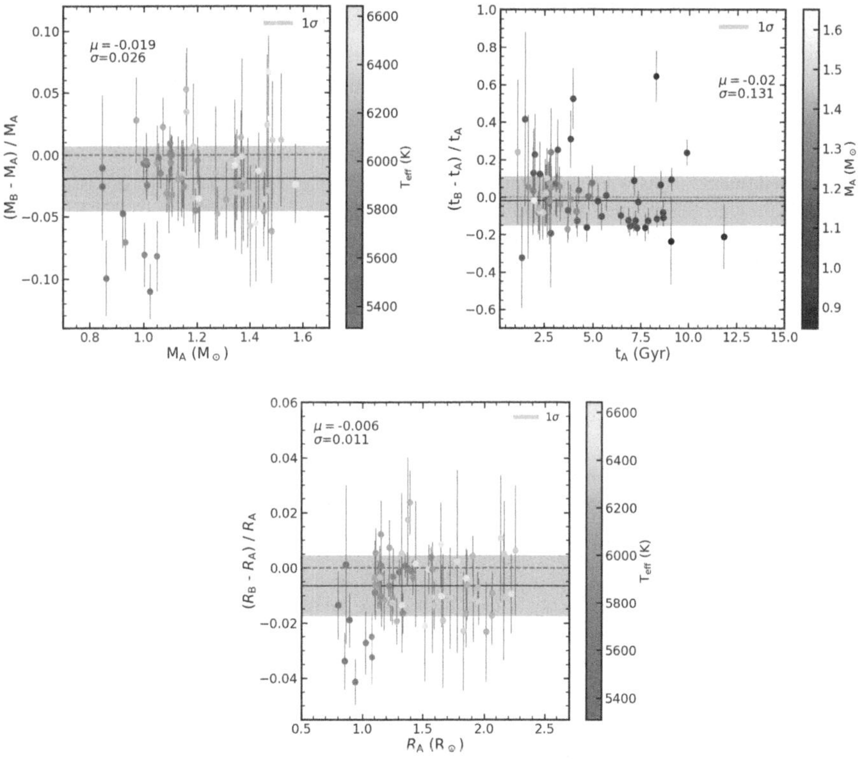

Fig. 2 Fractional differences in stellar mass, age, and radius resulting from the treatment of the initial helium mass fraction (abscissa values from grid A). The black line represents the bias (μ), the scatter (σ) is represented by the blue region, and the red line is the null offset

We stress that stellar luminosities of our target stars are included as part of the surface constraints in our optimisation process. That being said, in order to have best-fit models from grid A (i.e., with lower $\Delta Y/\Delta Z$ ratio) which satisfy the observed stellar luminosities, they should have higher masses and radius compared to those from grid B as shown in the top left and bottom panels of Fig. 2, respectively. The top right panel of Fig. 2 shows a relatively good agreement in the stellar ages from both grid A and B, with a bias (μ) of 2% and systematic uncertainties of 13.1%.

4 Summary

This article highlights our preliminary findings on the systematic uncertainties arising from the variation in the treatment of initial helium abundance on the inferred stellar parameters.

An in depth study on the treatment of initial helium abundance is being addressed in Nsamba et al. (in prep), including a comprehensive comparison to findings of [30] based on synthetic stellar data, as well as assessing if the scatter arising from the differences in the treatment of initial helium abundance in stellar grids still satisfy the stellar parameter accuracy requirements expected for precise exoplanet characterisation for the future ESA's PLATO space mission [22].

Acknowledgments This work was supported by Fundação para a Ciência e a Tecnologia (FCT, Portugal) through national funds (UID/FIS/04434/2013) and by FEDER through COMPETE2020 (POCI-01-0145-FEDER-007672). BN acknowledges support from the project "CIAAUP-21/2019-CTTC", Alexander Humboldt Fundation and travel support from the workshop "Dynamics of the Sun & Stars: Honoring the Life & Work".

References

1. Balser, D. S. (2006). *Astronomical Journal, 132*, 2326–2332.
2. Berger, A. T., Huber. D., Gaidos, E., van Saders, J. L. (2018). arXiv:1805.00231
3. Casagrande, L., Flynn, C., Portinari, L., Girardi, L., Jimenez, R. (2007). *Monthly Notices of the Royal Astronomical Society, 382*, 1516–1540.
4. Cyburt, R. H., Fields, B. D., Olive, K. A. (2003). *Physics Letters B, 567*, 227–234.
5. Ferguson, J. W., Alexander, D. R., Allard, F., Barman, T., Bodnarik, J. G., Hauschildt, P. H., et al. (2005). *The Astrophysical Journal, 623*, 585–596.
6. Gennaro, M., Prada Moroni, P. G., Degl'Innocenti, S. (2010). *Astronomy and Astrophysics, 518*, A13.
7. Grevesse, N. and Sauval, A. J. (1998). *Space Science Reviews, 85*, 161–174.
8. Herwig, F. (2000). *Astronomy and Astrophysics, 360*, 952–968.
9. Høg, E., Fabricius, C., Makarov, V. V., Urban, S., Corbin, T., Wycoff, G. (2000). *Astronomy and Astrophysics, 355*, L27–L30.
10. Huber, D., Zinn, J., Bojsen-Hansen, M., Pinsonneault, M., Sahlholdt, C., Serenelli, A., et al. (2017). *The Astrophysical Journal, 844*, 102.

11. Iglesias, C. A., & Rogers, F. J. (1996). *The Astrophysical Journal, 464*, 943.
12. Imbriani, G., Costantini, H., Formicola, A., Vomiero, A., Angulo, C., Bemmerer, D. et al. (2005). *The European Physical Journal, 25*, 455–466.
13. Jimenez, R., Flynn, C., MacDonald, J., Gibson, B. K. (2003). *Science, 299*, 1552–1555.
14. Kunz, R., Fey, M., Jaeger, M., Mayer, A., Hammer, J. W., Staudt, G. (2002). *The Astrophysical Journal, 567*, 643–650.
15. Lebreton, Y. & Goupil, M. J. (2014). *Astronomy and Astrophysics, 569*, A21.
16. Lund, M. N., Silva Aguirre, V., Davies, G. R., Chaplin, W. J., Christensen-Dalsgaard, J., Houdek, G., et al. (2017). *The Astrophysical Journal, 835*, 172.
17. Mathur, S., Huber, D., Batalha, N. M., Ciardi, D. R., Bastien, F. A., Bieryla, A., et al. (2017). *The Astrophysical Journal Supplement Series, 229*, 2.
18. Metcalfe, T. S., Chaplin, W. J., Appourchaux, T., García, R. A., Basu, S., Brandão, I., et al. (2012). *The Astrophysical Journal Letters, 748*, L10.
19. Nsamba, B., Campante, T. L., Monteiro, M. J. P. F. G., Cunha, M. S., Sousa, S. G. (2018). *Monthly Notices of the Royal Astronomical Society, 479*, L55–L59.
20. Paxton, B., Schwab, J., Bauer, E. B., Bildsten, L., Blinnikov, S., Duffell, P., et al. (2018). *The Astrophysical Journal Supplement Series, 234*, 34.
21. Pijpers, F. P. (2003). *Astronomy and Astrophysics, 400*, 241–248.
22. Rauer, H., Catala, C., Aerts, C., Appourchaux, T., Benz, W., Brandeker, A., et al. (2014). *Experimental Astronomy, 38*, 249–330.
23. Rendle, B. M., Buldgen, G., Miglio, A., Reese, D., Noels, A., Davies, G. R., et al. (2019). *Monthly Notices of the Royal Astronomical Society, 484*, 771–786.
24. Rogers, F. J. & Nayfonov, A. (2002). *The Astrophysical Journal, 576*, 1064–1074.
25. Serenelli, A. M. & Basu, S. (2010). *The Astrophysical Journal, 719*, 865–872.
26. Silva Aguirre, V., Lund, M. N., Antia, H. M., Ball, W. H., Basu, S., Christensen-Dalsgaard, J., et al. (2017). *The Astrophysical Journal, 835*, 173.
27. Thoul, A. A., Bahcall, J. N., Loeb, A. (1994). *The Astrophysical Journal, 421*, 828.
28. Torres, G. (2010). *The Astronomical Journal, 140*, 1158–1162.
29. Townsend, R. H. D., Teitler, S. A. (2013). *Monthly Notices of the Royal Astronomical Society, 435*, 3406–3418.
30. Valle, G., Dell'Omodarme, M., Prada Moroni, P. G., Degl'Innocenti, S. (2014). *Astronomy and Astrophysics, 561*, A125.
31. Valle, G., Dell'Omodarme, M., Prada Moroni, P. G., Degl'Innocenti, S. (2015). *Astronomy and Astrophysics, 575*, A12.

Direct Travel Time of X-ray Class Solar Storms

Alan S. Hoback

Abstract Plasma from solar events such as coronal mass ejections and solar flares can impact the Earth's magnetosphere and cause disruptions of electrical systems on the ground. Warning systems exist to help prepare people for possible disruptions. Generally, it is thought that there would be a few days' warning. However, larger flares on the scale of the Carrington Event have much faster travel times. For X-ray class flares above X of 17 with data available, all have travel times to one astronomical unit of 30 h or less. Since those are more likely to have the greatest impact on the operation of electrical systems, their travel time is more important than the average event.

1 Plasma Speed

The possibility of an event with an impact similar to the Carrington Event is hard to predict, but helioseismology is increasing understanding of the physics of the Sun [4]. However, warning of impending disturbance on Earth is currently only from watching for solar events as they happen. Knowing the speed of plasma to 1 AU will set limits on preparation time.

The escape velocity at the corona of the Sun is about 220 km/s, so only plasma exceeding that is of concern. There are many factors that influence how the initial speed of plasma changes over the distance from the Sun to 1 AU [1, 3]. The net effect of all of the factors can cause the plasma to accelerate or decelerate. The major known factors are gravity, solar wind, expansion of the plasma bubble and work from magnetic field lines. The trend is that the plasma accelerates until it reaches about 0.7 AU.

A. S. Hoback (✉)
University of Detroit Mercy, Detroit, MI, USA
e-mail: hobackas@udmercy.edu

M. J. P. F. G. Monteiro et al. (eds.), *Dynamics of the Sun and Stars*, Astrophysics and Space Science Proceedings 57,
https://doi.org/10.1007/978-3-030-55336-4_35

Table 1 Top solar storms since 1976

Rating	Date	Initial speed (km/s)
X45	Nov. 4, 2003	2657 [1]
X20	Apr. 2, 2001	2300 [3]
X20	Aug. 16, 1989	1377 [2]
X17.2	Oct. 28, 2003	2459 [1]
X14.4	Apr. 15, 2001	2773 [3]

Previous researchers have shown relationships between size of a Coronal Mass Ejection (CME) and the speed of the solar energetic particles contained in it [2]. Speed calculation is not always possible because sometimes satellite data may not be available. The calculation of speed is done through photographic analysis.

Data for the October to November "Halloween" storm of 2003 is available for events above M1 [1]. Space doesn't allow for showing the result, but a correlation coefficient of 0.602 was found between log of the X-ray rating and the velocity of the ejection. This confirms that larger solar events have a faster plasma ejection speed.

Additionally, the top eight solar storms since 1976 were investigated, and available data is shown in Table 1. There was incomplete data for the events on Sept. 7, 2005, March 6, 1989 and July 11, 1978, so those were left off. There is a clear trend that all of the top storms have fast speeds that result in travel times of 17–30 h.

In conclusion, solar storm events as large as the Carrington event have been commonly observed, but they have not hit the Earth in the last century. There is a positive correlation between the size of the solar flare and its velocity, so the powerful storms most commonly reach one astronomical unit in one day. A Carrington-sized event headed towards Earth would likely only give one day of warning versus average-sized storms that give a few days of warning. Solar storms that are that large but not directly aimed at Earth would provide a longer warning time, but that is not as important because the longer route also weakens the effect on the Earth.

References

1. Gopalswamy, N., Yashiro, S., Liu, Y., Michalek, G., Vourlidas, A., Kaiser, M. L., et al. (2005). Coronal mass ejections and other extreme characteristics of the 2003 October–November solar eruptions. *Journal of Geophysical Research: Space Physics, 110*(A9), 1–18.
2. Kahler, S. W. (2001). The correlation between solar energetic particle peak intensities and speeds of coronal mass ejections: Effects of ambient particle intensities and energy spectra. *Journal of Geophysical Research: Space Physics, 106*(A10), 20947–20955.
3. Sun, W., Dryer, M., Fry, C. D., Deehr, C. S., Smith, Z., Akasofu, S.-I., et al. (2002). Evaluation of solar type II radio burst estimates of initial solar wind shock speed using a kinematic model of the solar wind on the April 2001 solar event swarm. *Geophysical Research Letters, 29*(8), 12-1.
4. Thompson, M. J., Christensen-Dalsgaard, J., Miesch, M.S., & Toomre, J. (2003). The internal rotation of the Sun. *Annual Review of Astronomy and Astrophysics, 41*(1), 599–643.

On the Limits of Seismic Inversions for Radial Differential Rotation of Solar-Type Stars

Ângela R. G. Santos, Savita Mathur, Rafael A. García, and Michael J. Thompson

Abstract Seismic data contains information on stellar internal rotation, which plays an important role on dynamo models. Due to the uncertainties on the observations and stellar models, determining internal rotation of main-sequence solar-type stars has been challenging. Here, we use artificial rotational splittings for two-zone profiles to explore the limitations to constrain internal rotation profiles.

To test the limitations of the inversions for radial differential rotation we use artificial rotational splittings for the solar-analog KIC 8006161. Using its AMP (Asteroseismic Modeling Portal [5]) stellar model and ADIPLS [1] through Modules for Experiments in Stellar Astrophysics [6, and references therein], we obtain the corresponding rotational kernels, which describe the mode sensitivity to the rotation. We consider a grid of two-zone rotational profiles with different core-envelope rotation ratio Ω_C/Ω_E (Ω_E fixed at the surface rotation value [3]) and transition/step radius r_t (see Fig. 1). We then retrieve the artificial splittings only for $l = 1, 2$ modes within the observed radial orders [4] by forward modelling. In

Â. R. G. Santos (✉)
Space Science Institute, Boulder, CO, USA
e-mail: asantos@spacescience.org

S. Mathur
Instituto de Astrofisica de Canarias, Santa Cruz de Tenerife, Spain

Universidad de La Laguna , Santa Cruz de Tenerife, Spain

R. A. García
IRFU, CEA, Université Paris-Saclay , Gif-sur-Yvette, France

Université Paris Diderot, AIM, CEA, CNRS , Gif-sur-Yvette, France

M. J. Thompson
High Altitude Observatory, National Center for Atmospheric Research, Boulder, CO, USA

M. J. P. F. G. Monteiro et al. (eds.), *Dynamics of the Sun and Stars*, Astrophysics and Space Science Proceedings 57,
https://doi.org/10.1007/978-3-030-55336-4_36

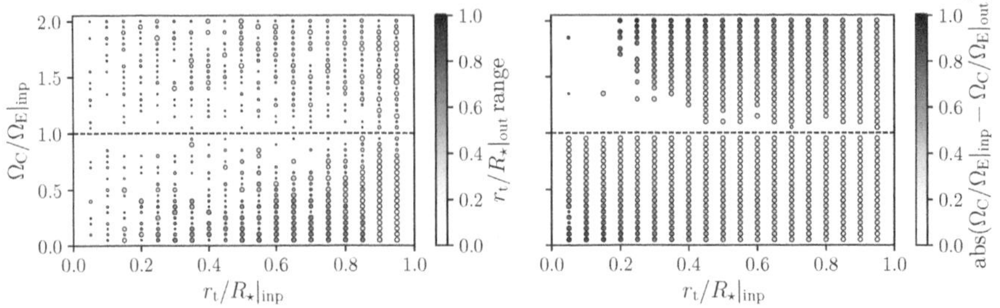

Fig. 1 Results for 30% noise level for the grid of models with different r_t and Ω_C/Ω_E. The left panel summarizes the results for r_t while the right-hand panel concerns Ω_C/Ω_E. The size of the symbols indicates the percentage of realizations for which the input r_t and sign of differential rotation are retrieved. The absence of symbol indicates that we were successful for 0 realizations. The color code indicates the $r_t/R_\star|_{out}$ range and the difference between the input and output Ω_C/Ω_E

addition to the no noise case, the artificial splittings are then degraded by randomly varying the input values according to a Gaussian distribution for different noise levels. The solution for the rotational profile is obtained using the inversion method developed by Christensen-Dalsgaard et al. [2]. For each noise level, the procedure above is repeated for 500 realizations.

We test the inversion results (denoted as "out") in terms of the retrieved step location r_t and sign of differential rotation. Under the same conditions ($r_t/R_\star|_{inp}$; $\Omega_C/\Omega_E|_{inp}$; noise—where "inp" denotes input), we compute the χ^2 as a function of $r_t/R_\star|_{out}$ and normalize it to the maximum value. The range of possible values $r_t/R_\star|_{out}$ corresponds to the solutions with normalized χ^2 smaller than 5%. In terms of r_t, the solution is considered successful if $r_t/R_\star|_{out}$ corresponding to the minimum χ^2 is within 10% of $r_t/R_\star|_{inp}$. In terms of Ω_C/Ω_E, the solution is considered successful if the sign of differential rotation of the input model is correctly found.

We find that at 0% noise, we successfully recover $r_t/R_\star|_{inp}$ and the correct sign of differential rotation for the full grid of models. At fixed $r_t/R_\star|_{inp}$, $r_t/R_\star|_{out}$ range is independent on $\Omega_C/\Omega_E|_{inp}$ and is smaller when r_t is the furthest of $0.5R_\star$. $\Omega_C/\Omega_E|_{inp}$ is better constrained when the step of the input model happens near the surface. As the noise level increases, the rate of successful detection for both r_t and Ω_C/Ω_E decreases in particular for fast cores. Also, the best set of solutions is found near the surface. Thus, the results indicate that the best targets for seismic inversions for internal rotation may be those with a slow core relative to the envelope rotation, similar to the solar case, and transition between the two zones near the surface.

Acknowledgments This project started with M. J. Thompson in the summer of 2017. Support: Visitor Program at HAO; the Nonprofit Adopt a Star program at WDRC; NASA No. NNX17AF27G; RYC-2015-17697; PLATO and GOLF CNES grants.

References

1. Christensen-Dalsgaard, J. (2008). *Astrophysics and Space Science, 316*, 113.
2. Christensen-Dalsgaard, J., Schou, J., & Thompson, M. J. (1990). *Monthly Notices of the Royal Astronomical Society, 242*, 353.
3. García, R. A., Ceillier, T., Salabert, D., Mathur, S., Van Saders, J. L., Pinsonneault, M., et al. (2014). *Astronomy and Astrophysics, 572*, A34.
4. Lund, M. N., Aguirre, V. S., Davies, G. R., Chaplin, W. J., Christensen-Dalsgaard, J., Houdek, G., et al. (2017). *The Astrophysical Journal, 835*, 172.
5. Metcalfe, T. S., Creevey, O. L., & Christensen-Dalsgaard, J. (2009). *The Astrophysical Journal, 699*, 373.
6. Paxton, B., Schwab, J., Bauer, E. B., Bildsten, L., Blinnikov, S., Duffell, P., et al. (2018). *The Astrophysical Journal Supplement Series*, 234, 34.

The Physical Origin of the Luminosity Maximum of the RGB-Bump

S. Hekker, G. C. Angelou, Y. Elsworth, and S. Basu

Abstract In their post-main-sequence evolution, low-mass stars go through a subgiant phase and subsequently evolve up the red-giant branch (RGB). Upon ascending the RGB these stars go through a short phase of contraction, the red-giant-branch bump (RGBB), before continuing their ascent. The RGBB is well-known both in stellar evolution models as well as in observations of for instance clusters, where the RGBB is visible as an over density of the number of stars observed. Despite, the clear detections of the RGBB, the underlying physical reason for the RGBB to occur is still an enigma. Here, we show that during the bump the full star is contracting, i.e. the mirror—the interface between contracting and expanding layers—has disappeared.

S. Hekker (✉)
Max Planck Institute for Solar System Research, Göttingen, Denmark

Stellar Astrophysics Centre , Aarhus, Denmark
e-mail: saskia.hekker@h-its.org

G. C. Angelou
Max Plank Institute for Astrophysic, Garching, Germany

Y. Elsworth
University of Birmingham, Birmingham, UK

Stellar Astrophysics Centre , Aarhus, Denmark

S. Basu
Yale University, New Haven, CT, USA

M. J. P. F. G. Monteiro et al. (eds.), *Dynamics of the Sun and Stars*, Astrophysics and Space Science Proceedings 57,
https://doi.org/10.1007/978-3-030-55336-4_37

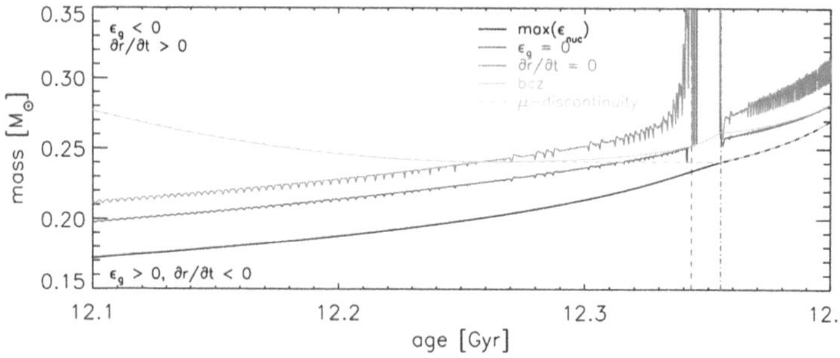

Fig. 1 The evolution around the RGBB of a 1 $M_\odot$ stellar model computed using MESA [2] showing the location of the base of the convection zone (bcz), the stationary point ($\partial r/\partial t = 0$), the pivot ($\epsilon_g = 0$), the peak of the burning ($\max(\epsilon_{nuc})$) and the mean molecular weight discontinuity (μ-discontinuity) as a function of mass ordinate. The vertical dashed and dashed-dotted lines indicate the luminosity maximum and luminosity minimum of the bump. Figure adapted from Hekker et al. [1]

1 Mirror and Bump

Following [1] we define the 'stationary point' in star as the location(s) below which shells move inwards and above which shells move outwards (or vice versa), i.e. $\partial r/\partial t = 0$ with r radius and t time (see Fig. 1). Furthermore, we define a 'pivot', which indicates the location(s) where shells change from compressing (increasing density) to expanding (decreasing density), i.e. $\epsilon_g = 0$ (see Fig. 1), where ϵ_g is the 'gravothermal' energy generation rate defined as:

$$\epsilon_g = -T\frac{\partial s}{\partial t} = \epsilon_{\text{internal}} + \epsilon_{\text{compression}}, \tag{1}$$

where s is specific entropy, T is temperature, $\epsilon_{\text{internal}}$ the rate of change of the internal energy per unit mass and $\epsilon_{\text{compression}}$ the rate at which work is being done per unit mass to compress matter. A mirror is present when ϵ_g changes sign.

Figure 2 shows that ϵ_g is positive at the base of the convection zone between the luminosity maximum and luminosity minimum in the RGBB. As shown by Hekker et al. [1] this results in a positive value of ϵ_g throughout the star and no mirror is present as evident from (temporal changes of) the specific entropy. So the star is fully contracting. For further analysis of this phenomenon we refer the reader to [1].

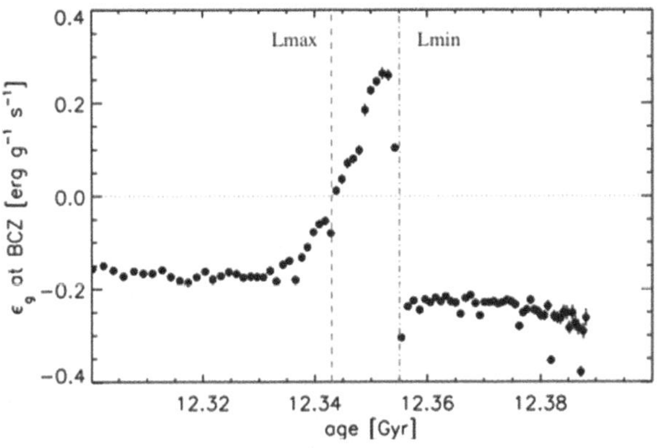

Fig. 2 For the same stellar model as shown in Fig. 1, the value of ϵ_g at the base of the convection zone as a function of age. The vertical dashed and dashed-dotted lines indicate the stellar ages of the luminosity maximum (Lmax) and luminosity minimum (Lmin) of the bump, respectively. The horizontal dotted line indicates zero. Figure adapted from Hekker et al.[1]

References

1. Hekker, S., Angelou, G. C., Elsworth, Y., & Basu, S. (2020). *Monthly Notices of the Royal Astronomical Society*, *492*(4), 5940–5948.
2. Paxton, B., Smolec, R., Schwab, J., Gautschy, A., Bildsten, L., Cantiello, M., et al. (2019). *The Astrophysical Journal Supplement Series*, *243*, 10.

A New Utility to Study Strong Chemical Gradients in Stellar Interiors

Stefano Garcia, Margarida S. Cunha, and Mathieu Vrard

Abstract Asteroseismology uses stellar pulsations to probe the stellar interior. Those pulsations are sensitive to rapid variations on the stellar structure known as glitches. Here we present a program to easily add and remove buoyancy glitches in an existing stellar model.

In radiative regions inside the star, the buoyancy force allows propagation of gravity waves. The asymptotic analysis of the pulsation equations predicts regularly-spaced periods of such waves. These periods get deflected when a sharp variation in the buoyancy frequency, often associated with strong chemical gradients, is present. These features are known as buoyancy glitches and their study is one of the few methods allowing us to probe the physical conditions in a localized region inside a star. Theoretical works (e.g., [3, 4]) provide the analytical bases to interpret the signatures of glitches of arbitrary amplitude in seismic observables. However, testing the analytical predictions is not straightforward, because finding a model with a glitch with specific properties (e.g., position, amplitude and width) implies, in the best case scenario, to explore the space of stellar parameters until the desired glitch is attained. This approach has the following disadvantages: (1) no warranty to generate the pursued glitch, (2) time consuming, and (3) preclusion of a direct comparison of the effect of different glitches on the seismic data. We have developed here a Python utility that allows to artificially add/remove a Gaussian-like glitch into/from the buoyancy profile N of an already existing stellar model, in that way solving points (1–3). The modification to the buoyancy frequency is done in a

S. Garcia (✉)
Instituut voor Sterrenkunde, Leuven, Belgium
e-mail: stefanorenzo.garciacastaneda@kuleuven.be

M. S. Cunha · M. Vrard
Instituto de Astrofísica e Ciências do Espaço, Universidade do Porto, CAUP, Porto, Portugal
e-mail: mcunha@astro.up.pt; mathieu.vrard@astro.up.pt

M. J. P. F. G. Monteiro et al. (eds.), *Dynamics of the Sun and Stars*, Astrophysics and Space Science Proceedings 57,
https://doi.org/10.1007/978-3-030-55336-4_38

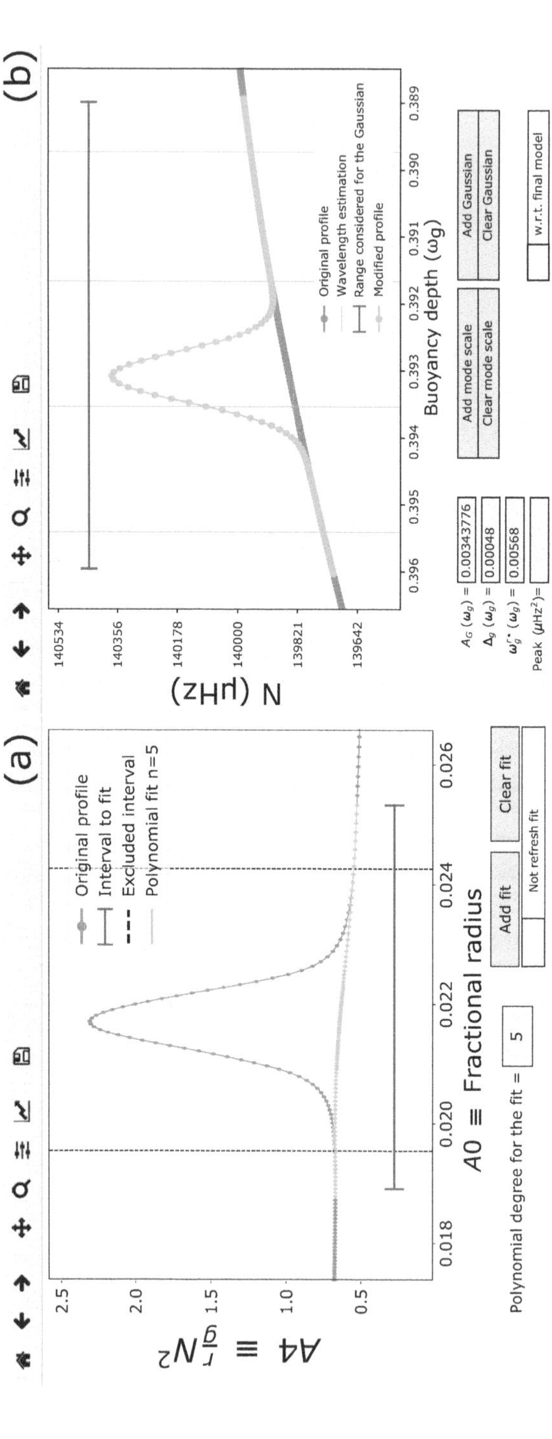

Fig. 1 Program's GUI. Panel (**a**) shows the interactive mode for removing a glitch from a stellar model. The axes *A4* and *A0* refer to the variables in the AMDL format. Panel (**b**) shows the interactive mode for adding a Gaussian-like glitch. Figures have been adapted for better readability

consistent way by a correspondent modification of the first adiabatic exponent as described in [1, Section 2.3]. The program counts with an interactive GUI as well as a command line mode and is available as a Github repository at https://github.com/stefano-rgc/glitch_add_and_remove. Further documentation and examples can be found in the repository. To this article's date, the program can operate only with stellar models stored as unformatted AMDL files, which is the format the Aarhus adiabatic pulsation code (ADIPLS, [2]) uses.

Figure 1 presents an example of the interactive GUI. Panel (a) shows a glitch in the N profile been replaced with a polynomial fit. Lagrange multipliers are used to ensure a smooth match between the fit and the original model. Panel (b) shows the addition of a Gaussian-like glitch to the N profile. The amplitude, width and, position of the glitch can be set interactively with the cursor or in the text boxes below and follow the parametrization from [4, equation (13)] within it. If there is an eigenmode file, the program offers to load it and overplots estimations of its wavelength (orange vertical lines in Fig. 1). The option of remeshing the stellar model is also available and allows to introduce glitches with a width similar to the mesh resolution of the stellar model. Glitch's parameters can be set with respect to either the perturbed or unperturbed model.

Acknowledgments This work was supported by FCT through Portuguese funds (PTDC/FIS-AST/30389/2017; UID/FIS/04434/2019) and by FEDER through COMPETE2020 (POCI-01-0145-FEDER030389).

References

1. Ball, W. H., Themeßl, N., & Hekker, S. (2018). *Monthly Notices of the Royal Astronomical Society, 478*, 4697.
2. Christensen-Dalsgaard, J. (2008). *Astrophysics and Space Science, 316*, 113–120.
3. Cunha, M. S., Stello, D., Avelino, P. P., Christensen-Dalsgaard, J., & Townsend, R. H. D. (2015). *Astrophysical Journal, 805*, 127.
4. Cunha, M. S., Avelino, P. P., Christensen-Dalsgaard, J., Stello, D., Vrard, M., Jiang, C., et al. (2019). *Monthly Notices of the Royal Astronomical Society, 490*, 909.

16 Cygni A: A Testbed for Stellar Core Physics

Catarina I. S. A. Rocha, Cristiano J. G. N. Pereira, Margarida S. Cunha, Mário J. P. F. G. Monteiro, Bernard Nsamba, and Tiago L. Campante

Abstract We present a novel method aimed at reducing the number of stellar models accepted by the Forward Modelling down to the ones that better represent the core of a specific star. The method is illustrated on the bright, solar-like pulsator 16 Cygni A. We show that by comparing the observed frequency ratios for this benchmark star to the ones derived from stellar models, we are able to constrain further the fraction of hydrogen in the core, establishing its precise evolutionary state.

We performed a forward modelling of the bright, benchmark solar-like oscillator 16 Cyg A [5] using the observations listed in Table 1 and a grid of stellar models based on MESA [7], with a mass, M, heavy element mass fraction, Z, and mixing length, $\alpha_{\rm mlt}$, varying as follows: $M \in [0.7,1.1]$ in steps of $0.05\,{\rm M}_\odot$, $Z \in [0.004,0.04]$ in steps of 0.002, and $\alpha_{\rm mlt} \in [1.0,3.0]$ in steps of 0.4. The initial helium abundance was determined using the expression $Y = (\delta Y/\delta Z)\,Z + Y_0$, with $Y_0 = 0.2484$ when $Z = 0.0$ [2] and $\delta Y/\delta Z = 1.4$. Diffusion of hydrogen and gravitational settling of heavy elements were implemented according to [11]. A summary of the stellar grid input physics is also given in [6]. The adiabatic oscillation frequencies were calculated using GYRE [12] for l = 0, 1, 2, and 3. A two-term surface correction [1] was applied. Using AIMS [8], stellar parameters and their uncertainties were obtained from posterior probability distributions. The inferred stellar parameters are shown in Table 1.

C. I. S. A. Rocha (✉) · C. J. G. N. Pereira · M. S. Cunha · B. Nsamba
Instituto de Astrofísica e Ciências do Espaço, Universidade do Porto, CAUP, Porto, Portugal
e-mail: up201505702@g.uporto.pt

M. J. P. F. G. Monteiro · T. L. Campante
Instituto de Astrofísica e Ciências do Espaço, Universidade do Porto, CAUP, Porto, Portugal

Departamento de Física e Astronomia, Faculdade de Ciências, Universidade do Porto, Porto, Portugal

M. J. P. F. G. Monteiro et al. (eds.), *Dynamics of the Sun and Stars*, Astrophysics and Space Science Proceedings 57,
https://doi.org/10.1007/978-3-030-55336-4_39

Table 1 Classical and seismic constraints used in the modelling and derived stellar parameters

Observations			Outputs		
$T_{\rm eff}$ (K)	[Fe/H] (dex)	Frequencies [4]	Mass ($M_\odot$)	Radius ($R_\odot$)	Age (Gyr)
5825 ± 50	0.10 ± 0.03	$l \le 3$, 17 radial orders	1.08 ± 0.01	1.226 ± 0.004	6.52 ± 0.27

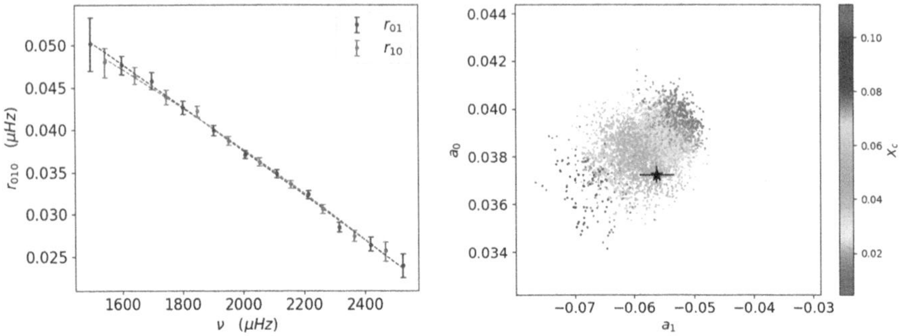

Fig. 1 Left: Ratios r_{01} and r_{10} for 16 Cyg A. The dashed lines show the second-order polynomial fits to r_{01}(blue) and r_{10}(red). Right: Location of models and star (with 3σ error bar) in the (a_1, a_0) plane for the r_{10} fit, where a_0 and a_1 are the independent term and linear coefficient of the fitted second order polynomial, respectively. Colours indicate the hydrogen abundance in the core

The ratios, r_{01} and r_{10}, are a combination of frequencies proposed by [9] that are insensitive to the surface layers. To down select the set of models accepted by forward modelling, we follow [3] and perform a second order polynomial fit to r_{01} and r_{10}, extracting the first two polynomial coefficients a_0 and a_1. Figure 1 shows the ratios and the second order polynomial fits. We compared the pair (a_1, a_0) from the fit to the observations with the coefficients obtained from the fit to the model ratios.

Figure 1 confirms that only a selection of models satisfies the observed ratios. We see in the (a_1, a_0) plane a clear separation between models with different core hydrogen abundances (X_c) (see also [10]) that allows us to gain new information about the evolution state of the star.

Acknowledgments This work was supported by the following grants: POCI-01-0145-FEDER-030389 (FEDER). TC acknowledges funds from the EU Horizon 2020 R&I programme (Marie Skłodowska-Curie grant agreement No. 792848, PULSATION). BN acknowledges support from the project "CIAAUP-21/2019-CTTC", Alexander Humboldt Foundation and travel support from the workshop "Dynamics of the Sun & Stars: Honoring the Life & Work".

References

1. Ball, W. H., & Gizon, L. (2014). *The American Academy of Pediatrics, 568*, A123.
2. Cyburt, R. H., Fields, B. D., & Olive, K. A. (2003). *Physics Letters B, 567*, 227.

3. Deheuvels, S., Brandão, I., Silva Aguirre, V., Ballot, J., Michel, E., Cunha, M. S., et al. (2016). *The American Academy of Pediatrics, 589*, A93.
4. Lund, M. N., Silva Aguirre, V., Davies, G. R., Chaplin, W. J., Christensen-Dalsgaard, J., Houdek, G., et al. (2017). *The Astrophysical Journal, 835*, 172.
5. Metcalfe, T. S., Chaplin, W. J., Appourchaux, T., García, R. A., Basu, S., Brandão, I., et al. (2012). *The Astrophysical Journal Letters, 748*, L10.
6. Nsamba, B., Campante, T. L., Monteiro, M. J. P. F. G., Cunha, M. S., Rendle, B. M., Reese, D. R., et al. (2018). *Monthly Notices of the Royal Astronomical Society, 477*, 5052.
7. Paxton, B., Marchant, P., Schwab, J., Bauer, E. B., Bildsten, L., Cantiello, M., et al. (2015). *The Astrophysical Journal Supplement, 220*, 15.
8. Rendle, B. M., Buldgen, G., Miglio, A., Reese, D., Noels, A., Davies, G. R., et al. (2019). *Monthly Notices of the Royal Astronomical Society, 484*, 771.
9. Roxburgh, I. W., & Vorontsov, S. V. (2003). *The American Academy of Pediatrics, 411*, 215.
10. Silva Aguirre, V., Ballot, J., Serenelli, A. M., & Weiss, A. (2011). *The American Academy of Pediatrics, 529*, A63.
11. Thoul, A. A., Bahcall, J. N., & Loeb, A. (1994). *The Astrophysical Journal, 421*, 828.
12. Townsend, R. H. D., & Teitler, S. A. (2013). *Monthly Notices of the Royal Astronomical Society, 435*, 3406.

Exploring the Origins of Intense Magnetism in Early M-Dwarf Stars

Connor Bice and Juri Toomre

Abstract We present the results of new global 3D MHD simulations of early M-Dwarf stars, exploring the influence that a tachocline of shear has on their instantaneous and long-term magnetic activity.

Despite their standing as the smallest, dimmest stars on the main sequence, many M-dwarfs have been found to be extraordinarily magnetically active. Large-scale surveys of these stars have found that where only about 10% of the early (more massive) M-dwarfs demonstrate chromospheric markers for strong magnetism, this percentage rises sharply later than M3 (0.35 $M_\odot$) to nearly 90% of the less massive M-dwarfs [5]. For these active low-mass stars, flaring events are extremely common, often with typical quiescent intervals of only hours. The largest superflares, however, seem to be more common on the more massive active G-, K-, and M-dwarfs [4].

Within the Sun, helioseismology has revealed a layer of remarkable rotational shear separating the convection zone (CZ) and radiative zone (RZ), which has come to be called the tachocline, and is thought to be a significant player in the Sun's dynamo. While early M-dwarfs have solar-like internal structures, those less massive than 0.35 $M_\odot$ become fully convective and may no longer contain a tachocline.

We present here a summary of a set of global 3D anelastic MHD simulations of M2 (0.4 $M_\odot$) stars, the methods and results of which are discussed in more detail in forthcoming papers [1, 2]. Our models are computed using the open-source code

This work funded by NASA ATP grant NNX17AG22G.

C. Bice (✉) · J. Toomre
JILA & Department of Astrophysical and Planetary Sciences, University of Colorado Boulder, Boulder, CO, USA
e-mail: connor.bice@colorado.edu

M. J. P. F. G. Monteiro et al. (eds.), *Dynamics of the Sun and Stars*, Astrophysics and Space Science Proceedings 57,
https://doi.org/10.1007/978-3-030-55336-4_40

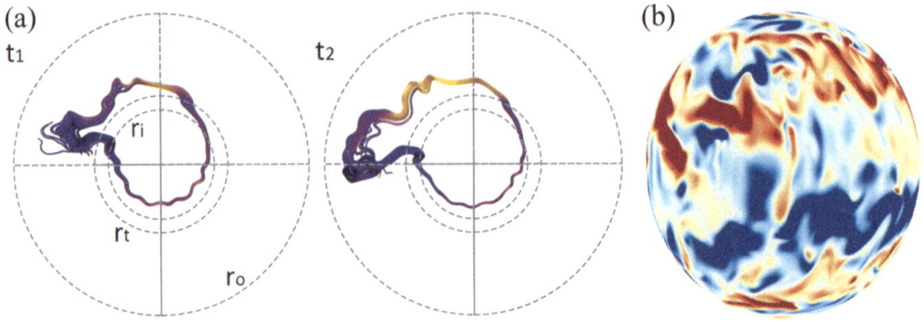

Fig. 1 (**a**) Field-line tracings of a loop rising from the tachocline at two instants separated by roughly 24 days. Colors denote field amplitude, ranging from <2 kG (black) to >30 kG (yellow). (**b**) Snapshot of mid-depth B_ϕ in orthographic projection. Colors saturate at ±12 kG

Rayleigh [3], and survey a range of rotation rates and magnetic Prandtl numbers. To explore the influence the tachocline has on these stars magnetic fields, we consider pairs of models with otherwise identical parameters. Each pair consists of a model whose computational domain ends in an impenetrable boundary at the base of the CZ, and one which permits overshoot into the underlying RZ and tachocline.

The magnetism we achieve in our models demonstrates a rich time-dependence. Global fields migrate through symmetric, antisymmetric, and asymmetric states as a result of decadal-scale polarity reversals in each hemisphere. A snapshot of an example antisymmetric state is shown in Fig. 1b. Applying Fourier analysis to the fields in the bulk of the CZ, we find that in our more rapidly rotating models, the inclusion of a tachocline tends to increase the duration of the cycle and reduce the signal of high frequency noise. In one pair, the presence of a tachocline eliminates the magnetic cycle altogether, with the tachocline case persisting in a time-steady configuration until its symmetry is disrupted by other instabilities.

The instability responsible for ending the time-steady magnetic field configuration in that model is magnetic buoyancy. The fields within the tachocline of one hemisphere reach amplitudes on the order of 30 kG, and begin to rise. A time-lapse of their progress through the CZ is presented in Fig. 1a. The magnetic forces exerted by this self-consistently generated rising loop are sufficient to disrupt the equilibrium of the global scale fields, causing the entire wreath of strong field embedded in the tachocline to rise after it, with a magnetic energy content on the order of 10^{37} ergs. Although the boundary conditions of our models do not permit this structure to breach the surface, it is decidedly suggestive. The largest early M-dwarf superflares have energy budgets of approximately 10^{35} ergs, and so the magnetic energy being transported toward the surface of the star in our model would be sufficient to power dozens of superflares over the course of several years with even a modest efficiency. The dearth of superflares on otherwise active late M-dwarfs may then be related to their lack of a stable reservoir such as the tachocline within which to build up their fields.

Because magnetic activity is closely tied to rotation rate, the differing active fractions across spectral type M can be understood in terms of spin-down rate. Although early and late M-dwarfs share a spin-down mechanism in the torques from their magnetized stellar winds, this process seems to occur much more gradually on the less massive stars, and so they stay active longer. In our pairs of models, those which contained tachoclines exhibited significantly stronger near-surface magnetic fields, which were organized on larger spatial scales. These conditions both translate to greater wind torques, and so our simulations are consistent with the explanation that the absence of a tachocline in the later M-dwarfs is at least partially responsible for their markedly different magnetic population in the modern universe.

References

1. Bice, C. P., & Toomre, J. (2020). *The Astrophysical Journal, 893*, 107–125.
2. Bice, C. P., & Toomre, J. (2020). In prep.
3. Featherstone, N. A., & Hindman, B. W. (2016). *The Astrophysical Journal, 818*, 32–45.
4. Kowalski, A. F., Hawley, S. L., Hilton, E. J., Becker, A. C., West, A. A., & Bochanski, J. J. et al. (2009). *The Astrophysical Journal, 138*, 633–648.
5. West, A. A., Hawley, S. L., Bochanski, J. J., Covey, K. R., Reid, I. N., & Dhital, S. et al. (2008). *The Astrophysical Journal, 135*, 785–795.

Part V
The Future

A Future Path for Solar Synoptic Ground-Based Observations

Markus Roth

Abstract This contribution provides a status overview on the work on the Solar Physics Research Integrated Network Group (SPRING), which is a study for a new ground-based network for future synoptic observations of the Sun. The planning of this started together with Michael J. Thompson. He strongly pushed for its realization. Several steps were already completed: The science requirements were defined, and a technical feasibility concept was completed in 2017. Based on this, work is ongoing towards a preliminary design.

1 The Need for Synoptic Observations of the Sun

The Sun changes on different time scales. Studying the respective related processes reveals valuable insights into the physics of our host star. The ambition of solar physics is to understand all these phenomena. This requires observing the Sun as continuously as possible in order to obtain observations without temporal aliasing and with statistical significance also for the rare events. Obtaining observations of the Sun is based on two approaches. One approach uses large-aperture telescopes as they provide spatially highly resolved observations of small areas on the Sun (small field-of-view). The complementary approach is employing smaller telescopes that provide time-resolved measurements of physical characteristics of the Sun with a scope of the entire solar disk and the nearby solar space environment. This second type of observations is referred to as "synoptic solar observations" as they provide recurrent measurements of various observables resulting in records over long times.

Such continuous observations can be carried out either in space by a single observatory or on ground in form of a network of observatories. In the case of a ground-based network, the telescopes need to be geographically distributed such

M. Roth (✉)
Leibniz-Institut für Sonnenphysik (KIS), Freiburg, Germany
e-mail: mroth@leibniz-kis.de

M. J. P. F. G. Monteiro et al. (eds.), *Dynamics of the Sun and Stars*, Astrophysics and Space Science Proceedings 57,
https://doi.org/10.1007/978-3-030-55336-4_41

that gaps from the night time, bad weather and instrumental interruptions are minimized. Ideally the individual observatories are equipped with nearly-identical observing instruments to ease the data merging and analysis.

The success of this strategy has been clearly demonstrated, e.g. by the UK BiSON (Birmingham Solar Oscillations Network) and US GONG (Global Oscillation Network Group) networks, and others (see [10] for an overview). These are based on small-aperture telescopes with a large field-of-view. They provide important and valuable observations of the Sun and given the fact that they can be operated for a long time, their recorded time series complement the observations taken at the large telescopes with only limited field-of-view.

Both ground-based and space observations have their advantages and disadvantages. The advantage of a ground-based network versus an observatory in space is the possibility of easy maintenance and upgrading of the instrumentation, and the larger data volumes that it can provide, i.e. the greater number of observables that can be recorded and handled. The first of these advantages could, in principle, result in an indefinite operation. GONG for example, which was originally designed to obtain high-quality imaging helioseismic data, is still doing so today and obtains now in addition, after instrumental upgrades, full-disk magnetograms of the Sun, and intensity images in the Hα spectral line. These data products have become a major data source for space weather forecasts from the USA, the National Oceanic and Atmospheric Agency Space Weather Prediction Center (NOAA SWPC), and the UK Meteorological Bureau.

The second of these advantages affects the number of science questions that can be addressed. If equipped with modern instrumentation, such a network will be able to provide those highly demanded synoptic observations that integrates a large fraction of scientific topics of solar physics.

The Solar Physics Research Integrated Network Group (SPRING) is a concept for such a new network of ground-based telescopes. The scientific objectives are manifold and cover topics like: the physical origins of the solar activity cycle; the interaction of the p-mode oscillations and the solar magnetic field; the formation, growth, decay, and disappearance of active regions; the connections of the solar magnetic field from the interior to the corona; the mechanisms of coronal mass ejections (CMEs), erupting filaments, flares, and other phenomena that can affect terrestrial technology and society; the variations in solar irradiance that may affect terrestrial climate; and solar-stellar connections.

Michael J. Thompson was a driving part of the team promoting SPRING, working on its scientific requirements, and studying the technical feasibility. This paper summarizes the current status of the developments of SPRING that Michael helped to achieve. Moreover it gives an outlook on the further steps that we now need to take without Michael and on which we will strongly miss his contributions.

2 Paths Gone: European Helio- and Asteroseismology Network (HELAS)

Large undertakings like SPRING require good collaborations with reliable partners. Often it takes years until a new research infrastructure starts its operation. The excellent and always reliable collobaration with Michael started with the foundation of the European Helio- and Asteroseismology Network (HELAS), which received funding by the Sixth European Framework Programme (FP6) from 2006 to 2010. The objectives of this project were to coordinate the activities in European helio- and asteroseismology. At that time the preparations for the new space missions CoRoT, Kepler, and SDO were already ongoing and support in the form of organizing workshops and developing tools for this research area could be provided by HELAS. Michael worked at the University in Sheffield and he was the first with whom I discussed the plans for the HELAS proposal in October 2004. He strongly supported and encouraged me to follow that idea. When the HELAS project received funding, Michael took a leading role and chaired the developments in global helioseismology and built synergies with asteroseismology.

HELAS turned out to be great success thanks to the engagement and positive spirit of all involved. This success could also be measured by the number of publications, tools, and data products that were made publicly available, and the numerous meetings that were organized to strengthen the research field. Actually, Michael organized the first out of four international conferences, i.e. "SOHO 18, GONG 2006, HELAS I—Beyond the spherical Sun" on August 7–11, 2006 in Sheffield. And of course, he was present at all the other three international conferences in Göttingen, Paris, and Lanzarote, too.

HELAS laid the foundation for further successful work and collaborations with Michael. These could be continued when he moved to Boulder, CO, as director of the High Altitude Observatory (HAO). One of the follow-up projects was "Exploitation of Space Data for Innovative Helio- and Asteroseismology" (SpaceInn, http://www.spaceinn.eu), which received funding under the Seventh European Framework Programme (FP7) from 2013 to 2016, and where Michael participated with HAO as one of the associated partners. Highlights of the developments are the new virtual infrastructure that was built for helio- and asteroseismology, i.e. the *Seismic Plus Portal* (http://voparis-spaceinn.obspm.fr/seismic-plus/, operated by the Paris Observatory), and many tools and data sets that were developed or made available for the research community.

These projects triggered also thinking about the future of the research area, which was found to strongly depend on having suitable instrumentation for continuing monitoring the Sun. One of the thoughts brought up a new ground-based network for helioseismology that replaces GONG and that serves a larger community. This idea had its origin in particular at the National Solar Observatory (NSO) and in parallel at the Kiepenheuer-Institut für Sonnenphysik (KIS), and was further discussed with the collaboration partners in Europe at a workshop in Tatranska Lomnica in Slovakia. Michael supported this idea when he became director at HAO.

The need for synoptic observations of the Sun was discussed at a workshop held at the International Space Science Institute (ISSI) on helioseismology with Michael as an organizer. The results of these discussions are summarized by [2].

The required synoptic observations of the Sun must enable a long-term monitoring of the solar magnetic field in order to understand the solar dynamo. Observations that allow to follow the evolution of the magnetic field at the poles and in active regions are crucial for such studies. The latter is relevant for space weather studies. In addition synoptic observations of the Sun must allow to continue monitoring velocity fields on the solar surface. This includes to enable helioseismology to probe the subsurface flows and their variation with the solar cycle to establish the relationship to the processes involved in the solar dynamo. An open question of solar physics is how magnetic flux emerges and forms active regions on the solar surface. Here studies of the smaller scale flows around active regions are required as they could carry relevant information for space weather predictions. In addition, as the upcoming next generation of 4-meter-telescopes will allow observations of small fractions of the solar surface in high spatial resolution, a ground-based network of synoptic telescopes is needed to complement the view on the full solar disk and to provide context information on the large-scale effects, e.g. flares and filament eruption, or small-scale events such as flux emergence.

3 The Solar Physics Research Integrated Network Group—SPRING

A first study in this direction was carried out as a joint research activity chaired by KIS within the High-resolution Solar Physics Network (SOLARNET) funded under the European Community's Seventh Framework Programme (FP7) from 2013 to 2017. Several international partners worked together on the science requirements for a new ground-based network and to study the technical feasibility.

3.1 Science Requirements

The kick-starting workshop to discuss the science requirements for a new ground-based network was organized by Michael on April 22–24, 2013 in Boulder, USA. The results of this "Synoptic Network Workshop" are summarized by [7].

At that meeting it was made clear that most scientific challenges of solar physics can only be answered with a combined approach of both high-resolution and synoptic observations. The scientific questions driving the development of solar physics in general are the questions and major puzzles on

1. **How is the variable solar magnetic field generated, maintained and dissipated?**
   To answer this question, one needs to discriminate between solar dynamo models, which can only be done by synoptic observations, as it requires long-term monitoring. However, for determining the role of induction effects near the surface in the creation of the global field, only high-resolution observations can provide the required measurements. Understanding the solar dynamo requires an understanding of the large-scale flows inside the Sun, i.e., the differential rotation and meridional flow, and their role in the angular momentum transport inside the Sun, which is, again, only accessible using synoptic facilities with long observational baselines in time and a full-disk view.
2. **How are the corona and solar wind maintained and what determines their properties?**
   Here, both synoptic and high-resolution observations are needed to identify the large-scale and small-scale contributions to the coronal heating, and to study the short- and long-term temporal behaviour of the outer layers of the Sun.
3. **What triggers transient events and how to forecast them?**
   This links back to the first question on the variable solar magnetic field and making transient events predictable. The effects of the Sun on space weather and space climate can only be understood by understanding the triggering mechanisms of energetic events, which are thought to be in the interaction of interior flows and magnetic fields. This interaction requires synoptic observations, while the identification and characterization of magnetic reconnection processes needed for the establishment of reliable space weather prediction call for both high-resolution and synoptic observations.
4. **How does solar magnetism influence the internal structure and radiative output of the Sun?**
   This fundamental question links to astrophysics in general when comparing the Sun with stars of similar and different magnetic activity. The synoptic facilities provide the required observations.

In short, solar physics is able to answer key questions of physics and plasma astrophysics, providing important services to protect humans and their technological assets (such as satellites, aircrafts in high-latitude flight paths, power grids and power lines) from hazards originating on the Sun, which are collectively known as space weather, if the right observations are available. Moreover, studies of Earth's climate require knowledge of long-term evolution of the Sun's activity and surface magnetism. The main sources of such information are archives of "synoptic solar observations".

In the following meeting in Germany held on November 25–28, 2013 in Titisee, four working groups were formed to develop the science requirements for "Synoptic magnetic fields", "Solar Seismology", "Transient events", and "Solar Awareness". A further meeting organised in Tatranska Lomnica, Slovakia continued discussing the ongoing studies of the groups, which were finally concluded in Boulder again at a workshop organized by the National Solar Observatory on May 15–17, 2016.

The results of this is the Science Requirement Document for SPRING [11], which is available at http://science-media.org/paper/251.

A relevant requirement for all four science topics are observations of the various features on the Sun in multiple lines and at higher spatial, spectral and temporal resolution. In this way the expectations for improvements in the studies of the Sun's magnetic field are that observations will allow to

- determine the three-dimensional topology of active region magnetic fields,
- improve the coronal field extrapolations due to the force-free behaviours in the upper layers of the solar atmosphere,
- provide continously ground-based vector magnetometry for near real-time space weather predictions,
- measure changes in the magnetic field and electric currents in the chromosphere related to flares, and
- enable long-term magnetic field records with improved spatio-temporal resolution.

Furthermore, the improvements for helioseismology are that the new ground-based observations result in

- an improved accuracy and precision of the helioseismic mapping in the vicinity of active regions [6],
- a reduction of systematic errors [1],
- a multi-height mapping of the solar atmosphere [13, 4, 3, 9], and
- an understanding of the convective energy transport through the solar atmosphere [8].

In addition many more ideas were presented in the talks given at this workshop.

3.2 Feasibility Study

The authors of [5] evaluated the technical feasibility based on the science requirements. This final proposed instrument concept is available at: http://science-media.org/paper/930. The following is based on this document.

The key idea for the translation of the science requirements into a technical concept is simplicity. A single instrument cannot fulfil all the collected requirements. Even if it could, its complexity would lead to higher costs and to a short mean time between failures. Therefore, the preferred option is based on multiple instruments on a single platform. Representative examples for this concept are

- SOLIS (Synoptic Optical Long-term Investigations of the Sun) operated by NSO, which has a spectro-polarimeter, fulldisk patrol telescope and a spectrometer for integrated sunlight,
- SMART (The Solar Magnetic Activity Research Telescope) at the Hida Observatory in Kyota, Japan, with its fulldisk Hα imager, the fulldisk vector magneto-

graph, a high-resolution magnetograph, an a high-resolution flare patrol in Hα, or

- SFT (Solar Flare Telescope) Mitaka in Japan, with Hα and whitelight fulldisk telescopes, an infrared spectrometer, and a G-band imager.

Given the science requirements for magnetometry and spectro-polarimetry the front-end telescope is a 50 cm Ritchey-Chretien design with a cooled secondary as tip-tilt mirror. Such a telescope is polarization free and could be filled with Helium to reduce internal seeing. The covered wavelength range is 500–1565 nm. The attached spectrograph works in Littrow configuration with an Echelle grating of 408 × 204 mm with 79 lines per mm. The Blaze angle is 63.5°. This results in a spectral resolution of 200,000. The focal length is 1100 mm.

As the spatial scanning with a slit spectrograph requires approximately 20 minutes to obtain a full disk vector magnetogram, a multiple-slit design can be implemented to reduce the cadence.

The second instrument assembly would provide the Doppler measurements in multiple lines independently from the magnetometer. For most of the photospheric lines the required sensitivity of 10 m/s can be reached with an aperture size of 20 cm. For the chromospheric lines one needs to integrate more photons. Such a telescope could either be a downsized SOLIS or a lens in form of an achromatic doublet.

The requirement for the temporal cadence dictates to use tunable filters with high throughput, such as Fabry-Perot interferometers, for three optical wavelength bands: one for the blue (380–550 nm), one for red (550–850 nm), and one for the infrared (850–1565 nm). A demonstration instrument is the Helioseismic Large Region Interferometric Device (HELLRIDE, [12]) operated at the Vacuum Tower Telescope on Tenerife. The dual-etalon instrument has a fast pre-filter positioning system for minimizing the dead time between subsequent spectral scans. With HELLRIDE it was demonstrated that fulldisk polarimetric measurements are possible in a collimated setup.

A decision about the location of the individual network stations has so far not been made. The GONG sites would be available locations if the science requirements can be met there. In particular, this means that a high duty cycle can be achieved and that the seeing conditions are such that a spatial resolution of 1 arcsecond per pixel can be obtained for the whole observing time at one location.

Regarding the duty cycle, GONG has demonstrated that a coverage of 90% and better throughout the year is possible, even in the case when one station is not active. Six network stations seem to be an optimum. Adding more sites increases the duty cycle only marginally. Furthermore, when evaluating the seeing conditions in the recorded GONG data, it becomes obvious that the new observatories need to be located on a tower to avoid ground-layer seeing. Furthermore the integration times need to be short.

Requirements for the data processing are to deliver the data in real-time. This is possible for Doppler velocities, line-of-sight magnetic field and intensity. The full Stokes inversions, however, require longer processing times. Clearly, the data rates will be high. A simple estimate based on observations of five spectral lines at

20 wavelength bins per spectral line with a 2k × 2k 16-bit detector at a temporal cadence of 10 s, and for the four Stokes components would result in 3.1 GB per cycle. This means that each minute 18.7 GB are recorded. With compression this could be reduced to approximately 10 GB per minute. Assuming that each site observes for 12 h, the data volume would be approximately 7 TB per day and per site. This is still handable with today's means of technology: a connection of 1 GB/s can transfer a maximum of approximately 10 TB per day to a data center, where further data processing can be carried out.

A storage buffer to hold data for 10 weeks at a site is suggested, which means that 500 TB of storage at each network node are needed.

3.3 Preliminary Design Study

The next steps for the further development of SPRING are currently undertaken. Within the next round of funding for SOLARNET under the European Union's Horizon 2020 Framework Programme, the studies for SPRING are continued until 2022. The goal is now to translate the technical concept into a preliminary design. The tasks that are currently carried out are designing the mounting and the telescopes, designing and proto-typing the post-focus instruments, and defining the data processing pipelines.

4 Future Path

Michael helped to kick-start this long-term project of building a new ground-based network for solar physics. He leaves us with the first concepts and ideas. The realisation of such a network requires an international effort. Currently, this effort is jointly led by partners in the USA and in Europe. Furthermore, there is a world-wide interest in the concept.

The days after this workshop to honor the life of Michael J. Thompson were used to discuss in Boulder with colleagues from abroad the further science requirements for a new ground-based network in detail. SPRING is a concept one will further build on. In addition one needs to integrate synoptic observations of the solar corona in the network concept in order to complete the view on the Sun. The further steps ahead are now to complete the preliminary design studies for all parts. Then, it is needed to carry out a detailed design study and to prototype the post-focus instruments and data pipelines.

Michael is strongly missed in these recent discussions. But he would have loved to see the strong will of everyone to continue jointly on this path towards building a new network for advanced solar synoptic observations.

Acknowledgments With infinite thanks to Michael J. Thompson for his continuous support over many years.

The HELAS project received funding from the European Community's Sixth Framework Programme (Grant Agreement No. 026138).

The SpaceInn project has received funding from the European Community's Seventh Framework Programme ([FP7/2007-2013]) under grant agreement no. 312844.

The SOLARNET project was supported by the European Commission's FP7 Capacities Programme for the period April 2013 — March 2017 under the Grant Agreement number 312495.

The SOLARNET project has received funding from the European Union's Horizon 2020 research and innovation programme under grant agreement No 824135.

References

1. Baldner, C. S., & Schou, J. (2012). *The Astrophysical Journal Letters, 760*, L1.
2. Elsworth, Y., Broomhall, A.-M., Gosain, S., Roth, M., Jefferies, S. M., & Hill, F. (2015). *Space Science Reviews, 196*, 137.
3. Finsterle, W., Jefferies, S. M, Cacciani, A., Rapex, P., Giebink, C., Knox, A., et al. (2004). *Solar Physics, 220*, 317.
4. Finsterle, W., Jefferies, S. M., Cacciani, A., Rapex, P., & McIntosh, S. W. (2004). *The Astrophysical Journal Letters, 613*, L185.
5. Gosain, S., Roth, M., Hill, F., Staiger, J., Lin, H., Zima, W., et al. (2017). *Science Media Papers*. https://doi.org/10.18147/smn.2017/paper:930
6. Hill, F. (2009). *ASP Conference Series* (Vol. 416, p. 557).
7. Hill, F., Thompson, M. J., & Roth, M. (2013). *Space Weather, 11*, 392–393.
8. Jefferies, S. M., McIntosh, S. W., Armstrong, J. D., Bogdan, T. J., Cacciani, A., & Fleck, B. (2006). *The Astrophysical Journal Letters, 648*, L151.
9. Nagashima, K., Sekii, T., Kosovichev, A. G., Zhao, J., & Tarbell, T. D. (2009). *ASP Conference Series* (Vol. 415, p. 417).
10. Pevtsov, A. (2016). *ASP Conference Series* (Vol. 504, p. 71).
11. Roth, M., Hill, F., Gosain, S., Pevtsov, A.A., Jain, R., Sobotka, M., et al. (2016). *Science Media Papers*. https://doi.org/10.18147/smn.2016/paper:251
12. Staiger, J. (2011). *Astronomy & Astrophysics, 535*, 83.
13. Wisniewska, A., Musielak, Z. E., Staiger, J., & Roth, M. (2016). *The Astrophysical Journal Letters, 819*, L23.

Observational Asteroseismology of Solar-Like Oscillators in the 2020s and Beyond

Daniel Huber

Abstract Building on the foundations in helioseismology and asteroseismology by pioneers such as Michael Thompson, observational progress in asteroseismology of solar-like oscillators has exploded over the past decade thanks to space-based missions such as CoRoT and *Kepler*/K2. Over the coming decade, this revolution is set to continue through facilities that will provide a several orders of magnitude increase of known solar-like oscillators, providing unprecedented possibilities to study stellar physics, exoplanets, and galactic stellar populations. In this contribution I will provide a brief history of the detection of solar-like oscillations, followed by a review of the first results by the *TESS* mission. Finally, I will discuss the prospects and challenges for asteroseismology of solar-like oscillators in the coming decade and beyond using space-based telescopes such WFIRST and PLATO, as well as ground-based radial velocity observations using dedicated networks such as SONG, photometric transient surveys, and 8-m class telescopes.

1 A Brief History of Solar-Like Oscillations

The detection of oscillations in the Sun in the early 1960s [46, 71] launched the field of helioseismology, including revolutionary studies of the internal rotation profile of the Sun through helioseismic inversions by pioneers such as Michael Thompson [e.g. 69, 70]. The spectacular success of helioseismology made it clear that the detection of oscillations in other stars would be extremely valuable for our understanding of stellar structure and evolution. Early efforts to detect oscillations in other stars focused on ground-based radial-velocity observations, with detections in the bright subgiants Procyon [8] and η Boo [43]. Improvements in radial velocity precision enabled the detection of oscillations in other nearby main sequence and

D. Huber (✉)
Institute for Astronomy, University of Hawai'i, Honolulu, HI, USA
e-mail: huberd@hawaii.edu

M. J. P. F. G. Monteiro et al. (eds.), *Dynamics of the Sun and Stars*, Astrophysics and Space Science Proceedings 57,
https://doi.org/10.1007/978-3-030-55336-4_42

subgiant stars such as β Hyi [4, 14], α Cen A [6, 10] and B [13, 44] as well as red giant stars such as ξ Hya [22] and ϵ Oph [18].

Ground-based detections often suffered from aliasing due to nightly data gaps, motivating space-based photometric observations. The Canadian space telescope MOST [Microvariability and Oscillations in Stars, 72, 47] initially yielded a non-detection in Procyon [48] but later confirmed a detection that was consistent with radial velocity observations [31, 39]. MOST also detected oscillations in red giants [3], including evidence for non-radial modes [41]. Additional space-based detections were performed using the startracker of the WIRE (Wide-Field Infrared Explorer) satellite [60, 56, 9, 64], the SMEI (Solar Mass Ejection Imager) experiment [68] and the Hubble Space Telescope [20, 26, 63, 27]. In total, the combined observational efforts prior to 2009 yielded detections in a total of ~20 stars (left panel of Fig. 1).

A major breakthrough, which is now widely recognized as the beginning of the space photometry revolution of asteroseismology, was achieved by the French-led CoRoT (Convection Rotation and Planetary Transits) satellite. CoRoT detected oscillations in a number of main sequence stars [e.g. 1, 50] and several thousands of red giant stars [e.g. 35] (middle panel of Fig. 1). Importantly, CoRoT unambiguously confirmed that red giants oscillate in non-radial modes [19], which opened the door for detailed studies of the interior structure of red giants [see 34, for a recent review].

The *Kepler* space telescope, launched in 2009, completed the revolution of asteroseismology by covering the low-mass H-R diagram with detections. *Kepler* detected oscillations in over 500 main-sequence and subgiant stars [16] and over twenty thousand red giants [36, 65, 73], enabling the study of oscillations across the low-mass H-R diagram (right panel of Fig. 1). The larger number of red giants with detected oscillations is due to the increase of oscillation amplitudes with luminosity

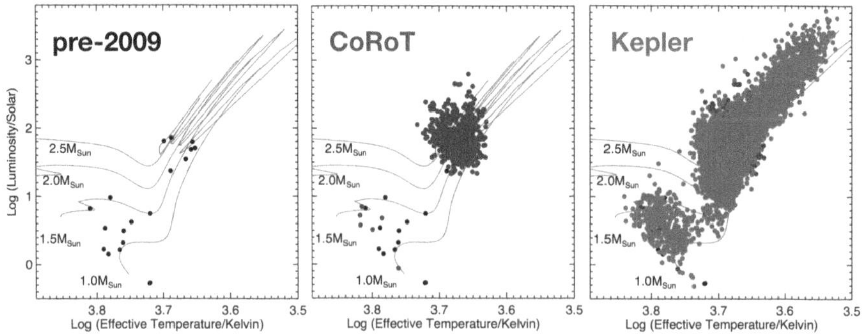

Fig. 1 H-R diagram showing stars with detected solar-like oscillations prior to 2009 (left panel) and after adding detections by the CoRoT (middle panel) and *Kepler* (right panel) missions. Grey lines show solar-metallicity evolutionary tracks for different masses. The space-photometry revolution has increased the number of solar-like oscillators by three orders of magnitude over the past decade

[42] and the fact that the majority of targets were observed with 30-min sampling, setting an upper limit of $\log g \sim 3.5$ dex.

2 First Results from the *TESS* Mission

2.1 Target Selection and Expected Yield

The NASA *TESS* Mission [57] was launched in April 2018 and provides high-precision photometric observations in 24×96 degree fields of view for 27 days, with continuous coverage near the ecliptic poles. *TESS* downloads the entire field of view every 30-min (full-frame images, FFIs), and observes subset of targets in 2-min cadence, which is suitable for the detection of oscillations in solar-type stars (Nyquist frequency of ~4100 μHz).

The target selection for solar-like oscillators was coordinated within the TESS Asteroseismic Science Consortium (TASC). A rank-ordered list was generated in part based on a detection probability calculated using effective temperature, luminosity, apparent *TESS* magnitude and the number of observed sectors following the detection recipe by Chaplin et al. [15], modified for the *TESS* mission. The resulting prioritized targets resulted in the Asteroseismic Target List [ATL, 59].

Figure 2 shows a representative expected yield of solar-like oscillators from *TESS* compared to ground-based observations and the *Kepler* mission. Due to its smaller

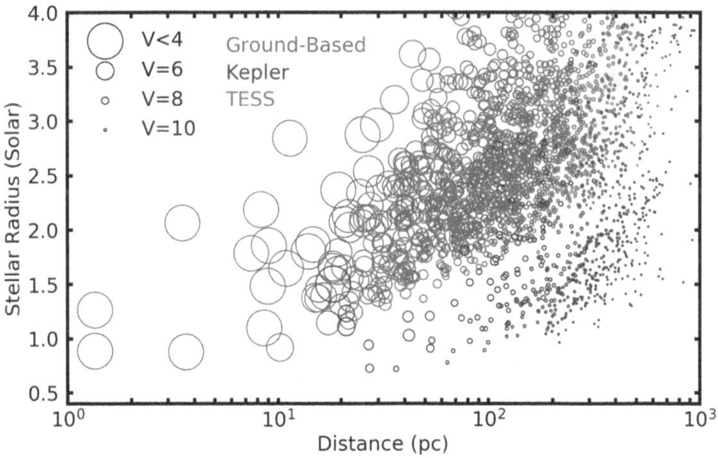

Fig. 2 Stellar radius versus distance for solar-like oscillators detected using ground-based observations (green circles), *Kepler* (blue circles), and a representative expected yield from *TESS* (red circles) based on the *TESS* Asteroseismic Target List [ATL, 59]. Symbol sizes scale with the apparent V-band magnitude as indicated in the plot. The brightest and closest *Kepler* detections are θ Cyg [32] and 16 Cyg A and B [49]. *TESS* is expected to complement the *Kepler* yield by detecting oscillations in bright, evolved stars

aperture, the average *TESS* detection is expected to be ~5 magnitude brighter, more evolved, and closer compared to *Kepler*. *TESS* is thus expected to complement the parameter space explored by *Kepler* which yielded a substantial number of solar-type stars that were relatively faint. Based on preliminary performance the total yield of solar-like oscillators from *TESS* in the prime mission is expected to range between 1000–2000 stars, a 2–4-fold yield increase over the *Kepler* mission.

2.2 Exoplanet Host Stars

The first detections of solar-like oscillations with *TESS* were made for exoplanet host stars, for which light curves were made publicly available early to facilitate ground-based follow-up observations. The first claimed detection of oscillations was made for the solar-type star π Men [23], which hosts the first transiting exoplanet discovered by *TESS* [38]. Subsequent analysis of the light curve showed that the power spectrum noise level is twice as large as the predicted oscillation amplitude,[1] demonstrating that the claimed detection of oscillations by Gandolfi et al. [23] could not have been correct.

The first confirmed detection of oscillations was made in the exoplanet host star HD 221416 (TESS Object of Interest 197, TOI-197), a $V = 8.2$ mag late subgiant star [40]. The power spectrum shows a clear detection of mixed dipole modes (Fig. 4, left). Asteroseismic modeling combined with spectroscopic $T_{\rm eff}$ metallicity and *Gaia* luminosity yielded a precise characterization of the host star radius ($R_\star = 2.943 \pm 0.064 R_\odot$), mass ($M_\star = 1.212 \pm 0.074 M_\odot$) and age ($4.9 \pm 1.1$ Gyr), and demonstrated that it has just started ascending the red-giant branch. The combination of asteroseismology with transit modeling and radial-velocity observations showed that the planet is a "hot Saturn" ($R_{\rm p} = 9.17 \pm 0.33 R_\oplus$) with an orbital period of ~14.3 days, irradiance of $F = 343 \pm 24 F_\oplus$, moderate mass ($M_{\rm p} = 60.5 \pm 5.7 M_\oplus$) and density ($\rho_{\rm p} = 0.431 \pm 0.062\,{\rm g\,cm^{-3}}$). The properties of HD 221416 b showed that the host-star metallicity—planet mass correlation found in sub-Saturns [54] does not extend to larger radii, indicating that planets in the transition between sub-Saturns and Jupiters follow a relatively narrow range of densities. With a density measured to ~15%, HD 221416 b is one of the best characterized Saturn-sized planets to date (Fig. 3).

In addition to stars hosting newly discovered transiting planets, *TESS* has detected oscillations in stars previously known to host planets discovered using the Doppler method [e.g. 12]. *TESS* is expected to yield a significant number of new and known exoplanet hosts that are amenable to asteroseismic characterization [11], including new discoveries of transiting planets around oscillating red-giant branch stars [e.g. 29].

[1] ExoFOP Observing Notes.

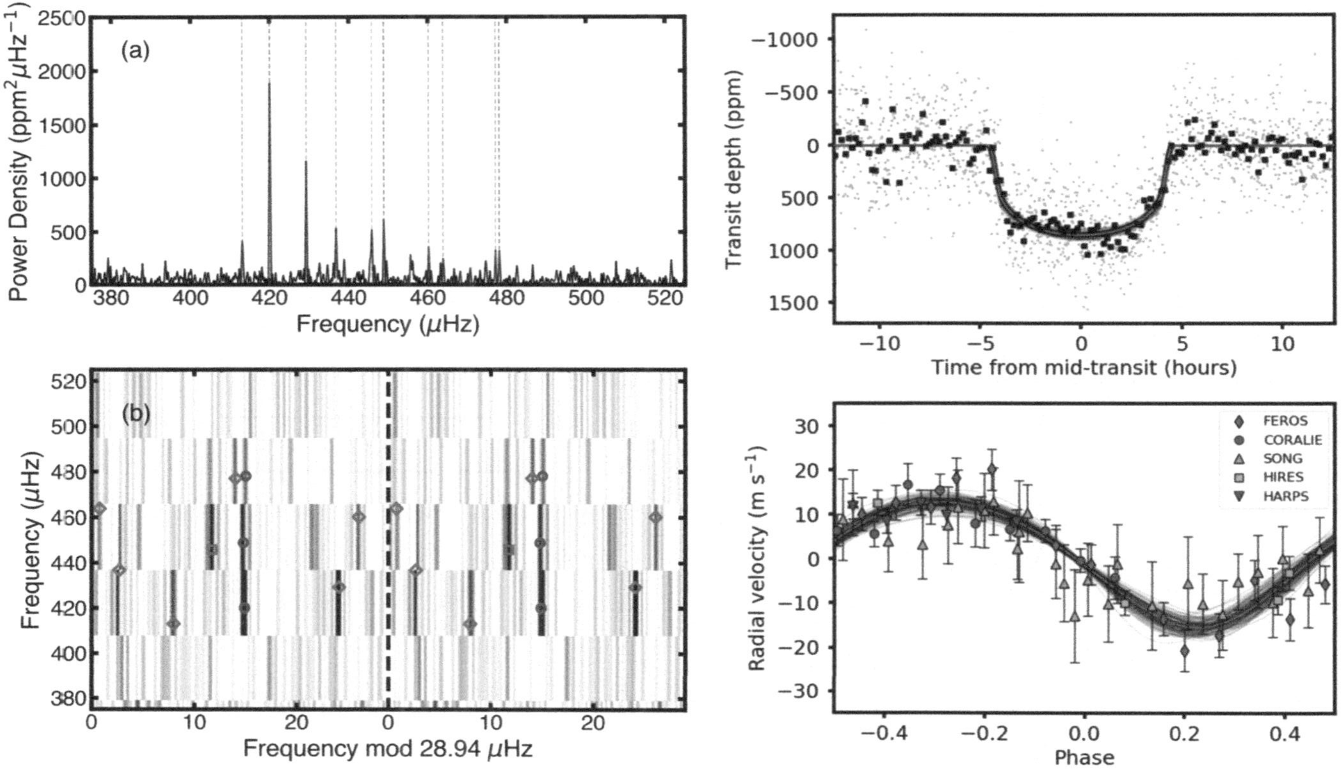

Fig. 3 Detection of solar-like oscillations in HD 221416 (TESS Object of Interest 197, TOI-197), the first *TESS* asteroseismic exoplanet host star. Left: power spectrum and echelle diagram of the *TESS* time series after removing the planetary transits. Right: Phase folded transit light curve and radial velocity follow-up observations using six different instruments. The combination of asteroseismology, transits and RV measurements constrained the density of the planet to $\sim 15\%$, making the planet one of the best characterized Saturn-sized planets to date. From Huber et al. [40]

2.3 Bright Benchmark Stars and Galactic Archeology

The *TESS* Mission is expected to yield detections of oscillations in many bright solar-type stars, which are benchmarks for testing stellar physics and stellar populations. The first example is ν Indi, one of the brightest metal-poor ([Fe/H] < -1.4 dex) stars in the solar neighborhood [17]. Oscillations had previously been detected using ground-based radial velocity observations, which enabled an age estimate based on the large frequency separation [5]. *TESS* significantly improved the SNR and window function of the power spectrum based on a single sector of observations, allowing an unambiguous mode identification (Fig. 4). Interestingly the identified frequencies from the *TESS* data are consistent with the mode identification by Bedding et al. [5], providing important confirmation of the reliability of asteroseismic results from early ground-based radial velocity observations.

The detection of oscillations in ν Indi enabled a powerful age constraint on the Gaia-Enceladus merger, a collision of the Milky Way with a dwarf galaxy which resulted in a population of accreted stars [37]. Through the combination of asteroseismic, spectroscopic, astrometric and kinematic observations, Chaplin et al. [17] show ν Indi is likely an in-situ member of our galaxy which has been dynamically heated by Gaia-Enceladus merger, and measure its age to be 11.0 ± 0.7 (stat) ± 0.8 (sys) Gyr. This age implies that the earliest the Gaia-Enceladus merger could have begun was 11.6 and 13.2 billion years ago (at 68% and 95% confidence). The recent first detections of solar-like oscillations in red giants by *TESS* [61] furthermore confirm the strong potential of *TESS* asteroseismology for galactic archaeology.

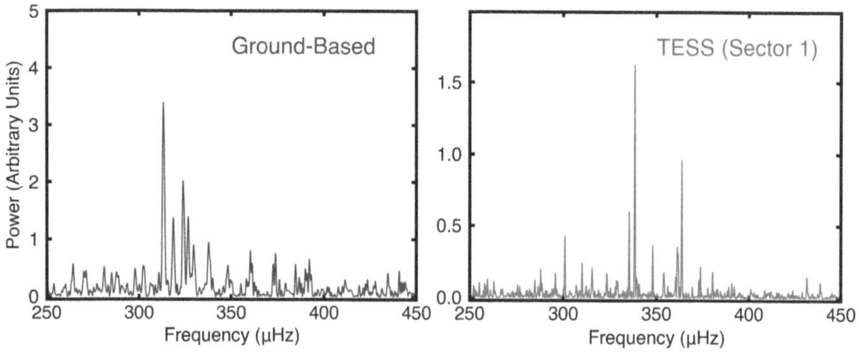

Fig. 4 Power spectrum showing the detection of solar-like oscillations in the bright metal-poor subgiant ν Indi ($V = 5.2$ mag) using ground-based radial velocity observations [left, 5] and a single sector of observations *TESS* [right, 17]. The continuous time series from *TESS* allowed an unambiguous mode identification, which was used to constrain the age of ν Indi to 10% and thus the earliest time of the Gaia-Enceladus merger

3 The Future: PLATO, WFIRST and Beyond

The asteroseismology revolution is set to continue in this decade (Fig. 5). Most importantly, the European-led PLATO mission will perform dedicated asteroseismology of main-sequence stars, with an expected yield of nearly 100,000 dwarfs and subgiants [55]. PLATO is expected to allow a systematic application of asteroseismology to characterize exoplanet host stars (rather than the largely serendipitous synergy which emerged from *Kepler*), providing unprecedented insights into the ages of small exoplanets. In addition to dwarfs, PLATO has enormous potential for galactic archaeology using red-giant asteroseismology, with yields expected to reach millions of stars [51, 52]. The detections will complement the deep, narrow red giant sampled probed by the K2 Galactic Archaeology program [66] and by *TESS*.

Another space-based telescope with strong potential to advance asteroseismology is the Wide-Field Infrared Telescope [WFIRST, 62]. WFIRST was the top-ranked space mission in the 2010 US Decadal Survey, and is slated for launch in the mid 2020s. Using a 2.3 m telescope over a 5 year nominal mission, WFIRST is expected to conduct a supernova and weak lensing survey as part of its dark energy program, and collect H-band time series photometry in a wide 2.8 square degree field toward the Galactic bulge to detect exoplanets using gravitational microlensing [53]. The planned 15-min cadence and 70 day long observing campaigns will yield sufficient precision to detect red-giant oscillations to the red clump, resulting in an expected 1 million detections [28]. Importantly, the H-band observations by WFIRST will extend galactic archaeology out to typical distances of 8 kpc and

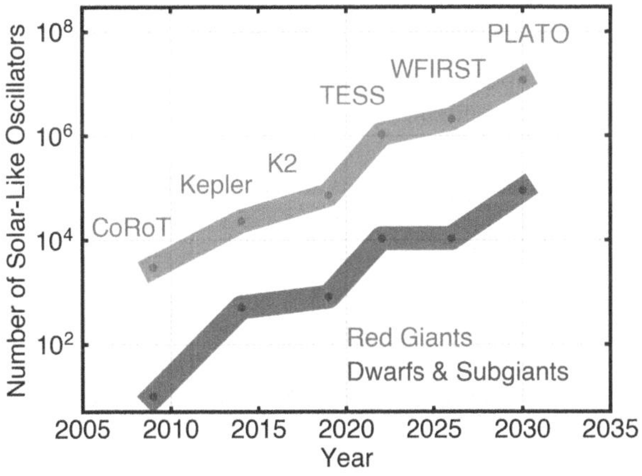

Fig. 5 Detection yield of solar-like oscillators by space-based telescopes, separated into dwarfs and subgiants (which require faster than 30 minute sampling) and red giants. Current and future space-based missions such as *TESS*, PLATO and WFIRST are expected to provide a several orders of magnitude yield increase in this decade, providing unprecedented opportunities for stellar physics, exoplanet science and galactic astronomy

provide the first probes of stellar populations using asteroseismology towards the galactic bulge.

In addition to space-based photometry, ground-based transient surveys such as ATLAS, ASAS-SN, ZTF and the Vera Rubin Observatory (LSST) will provide all-sky photometry at nightly cadence. While the precision and cadence will be insufficient to detect oscillations in main-sequence and subgiant stars, the detection of oscillations in M giants ascending to the tip of the red-giant branch are easily accessible using ground-based photometry [e.g. 67]. Indeed, comparisons of light curves obtained by ATLAS, ASAS-SN and *Kepler* show that a systematic detection of oscillations for M giants is feasible, opening the door for systematic and precise asteroseismic distance measurements beyond 5 kpc, providing invaluable information to study the outer galaxy and galactic halo [2].

An exciting and challenging new frontier for asteroseismology in the 2020s will be the systematic detection of solar-like oscillations in K-M dwarfs. Such stars are ubiquitous in our galaxy, yet our understanding of their structure and evolution remains a challenging problems. In particular, models frequently underestimate radii of K-M dwarfs at fixed temperature compared to empirical radii from interferometry and eclipsing binaries [45, 7]. While theories such as convective suppression by close binaries or magnetic fields can reduce the discrepancy [21], a universal explanation remains elusive. This is particularly important because these stars have become a primary targets for exoplanet science: both the NASA *TESS* Mission and JWST preferentially aim to characterize potentially habitable planets around cool dwarfs [e.g. 25, 58].

Asteroseismology can provide a unique solution to these problems. However, due to their low luminosities, oscillation amplitudes in dwarfs cooler than the Sun are small and difficult to detect, even with high-precision photometry from *Kepler*. To date, only a handful of stars cooler than the Sun have detected oscillations, and none cooler than 5000 K (Fig. 6). A solution is to use radial velocities (RV), which are less affected by granulation noise than photometry [33] and thus allow higher S/N detections in cool stars. Since there are no stellar noise sources on timescales shorter than oscillations, dedicated ground-based observations using small telescopes such as the Stellar Observations Network Group [SONG, 30] can build up sufficient S/N to detect oscillations in cool dwarfs. Additionally, high-precision Doppler spectrographs on large telescopes such as the Keck Planet Finder [KPF, 24] allow high cadence RVs for a selected number of high priority stars. Simulations demonstrate that a modest number of nights per target with KPF would allow the systematic application of asteroseismology in stars cooler than the Sun (Fig. 6), and similar results can be expected using larger time investments from a network of dedicated telescopes such as SONG.

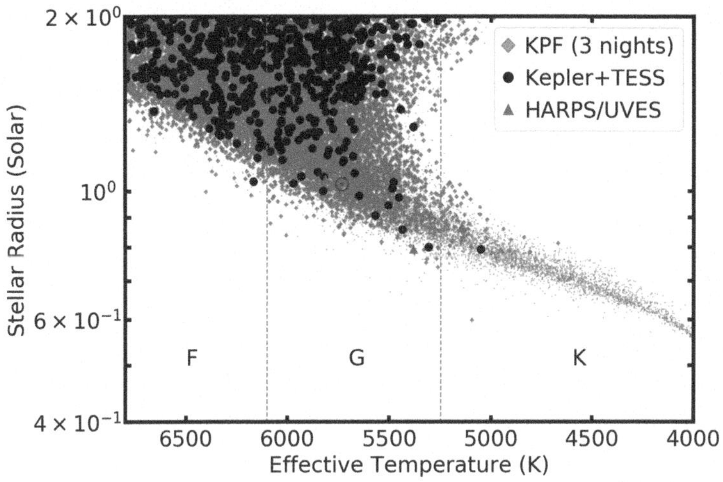

Fig. 6 H-R diagram of nearby field stars derived using Gaia parallaxes (grey) highlighting asteroseismic detections from space (*Kepler* & *TESS*, black) and stars with simulated S/N>5 detections from three nights of ground-based radial-velocity (RV) observations using the Keck Planet Finder (red). Next generation RV facilities such as KPF and dedicated networks such as SONG could allow the first systematic application of asteroseismology to stars cooler than the Sun

4 Conclusions

Asteroseismology of solar-like oscillators has undergone an exciting revolution over the past decades, building on the foundation of helioseismology established by pioneers such as the late Michael Thompson. The *TESS* Mission is currently continuing this revolution, and future space- and ground-based facilities such as PLATO, WFIRST and transient surveys are expected to expand the reach of asteroseismology to a galactic scale. An exciting challenge will be the extension of asteroseismology to cool dwarfs, which can be achieved through the strategic use of current and future ground-based radial velocity facilities. Overall, the future of asteroseismology and its synergy to fields such as exoplanet science and galactic archaeology will continue to bloom over the coming decades.

Acknowledgments I thank Jørgen Christensen Dalsgaard and the entire SOC and LOC for organizing this conference in memory of Michael Thompson, whose pioneering work helped launch and inspire careers in asteroseismology for numerous scientists of my generation. I thank Tim Bedding for providing the ground-based radial velocity data for ν Indi. I acknowledge support from the Alfred P. Sloan Foundation, the National Aeronautics and Space Administration (80NSSC18K1585, 80NSSC19K0379), and the National Science Foundation (AST-1717000).

References

1. Appourchaux, T., Michel, E., Auvergne, M., Baglin, A., Toutain, T., Baudin, F., et al. (2008). *The American Academy of Pediatrics, 488*, 705.
2. Auge, C., Huber, D., Heinze, A., Shappee, B. J., Tonry, J., Chakrabarti, S., et al. (2020). Beyond Gaia: Asteroseismic distances of M giants using ground-based transient surveys. *The Astronomical Journal, 160*(1), 11.
3. Barban, C., Matthews, J. M., De Ridder, J., Baudin, F., Kuschnig, R., Mazumdar, A., et al. (2007). *The American Academy of Pediatrics, 468*, 1033.
4. Bedding, T. R., Butler, R. P., Kjeldsen, H., Baldry, I. K., O'Toole, S. J., Tinney, C. G., et al. (2001). *The Astrophysical Journal Letters, 549*, L105.
5. Bedding, T. R., Butler, R. P., Carrier, F., Bouchy, F., Brewer, B. J., Eggenberger, P., et al. (2006). *The Astrophysical Journal, 647*, 558.
6. Bouchy, F., & Carrier, F. (2001). *The American Academy of Pediatrics, 374*, L5.
7. Boyajian, T. S., McAlister, H. A., van Belle, G., et al. (2012). *The Astrophysical Journal, 746*, 101.
8. Brown, T. M., Gilliland, R. L., Noyes, R. W., & Ramsey, L. W. (1991). *The Astrophysical Journal, 368*, 599.
9. Bruntt, H., Kjeldsen, H., Buzasi, D. L., & Bedding, T. R. (2005). *The Astrophysical Journal, 633*, 440.
10. Butler, R. P., Bedding, T. R., Kjeldsen, H., McCarthy, C., O'Toole, S. J., Tinney, C. G., et al. (2004). *The Astrophysical Journal Letters, 600*, L75.
11. Campante, T. L., Schofield, M., Kuszlewicz, J. S., Bouma, L., Chaplin, W. J., Huber, D., et al. (2016). *The Astrophysical Journal, 830*, 138.
12. Campante, T. L., Corsaro, E., Lund, M. N., Mosser, B., Serenelli, A., Veras, D., et al. (2019). *The Astrophysical Journal, 885*, 31.
13. Carrier, F., & Bourban, G. (2003). *The American Academy of Pediatrics, 406*, L23.
14. Carrier, F., Bouchy, F., Kienzle, F., Bedding, T. R., Kjeldsen, H., Butler, R. P., et al. (2001). *The American Academy of Pediatrics, 378*, 142.
15. Chaplin, W. J., Kjeldsen, H., Bedding, T. R., Christensen-Dalsgaard, J., Gilliland, R. L., Kawaler, S. D., et al. (2011). *The Astrophysical Journal, 732*, 54.
16. Chaplin, W. J., Basu, S., Huber, D., Serenelli, A., Casagrande, L., Silva Aguirre, V. (2014). *The Astrophysical Journal Supplement, 210*, 1.
17. Chaplin, W. J., Serenelli, A. M., Miglio, A., Morel, T., Ted Mackereth, J., Vincenzo, F., et al. (2020). *Nature Astronomy, 7*, 382–389
18. De Ridder, J., Barban, C., Carrier, F., Mazumdar, A., Eggenberger, P., Aerts, C., et al. (2006). *The American Academy of Pediatrics, 448*, 689.
19. De Ridder, J., Barban, C., Baudin, F., Carrier, F., Hatzes, A. P., Hekker, S., et al. (2009). *Nature, 459*, 398.
20. Edmonds, P. D., & Gilliland, R. L. (1996). *The Astrophysical Journal Letters, 464*, L157.
21. Feiden, G. A., & Chaboyer, B. (2013). *The Astrophysical Journal, 779*, 183.
22. Frandsen, S., Carrier, F., Aerts, C., Stello, D., Maas, T., Burnet, M., et al. (2002). *The American Academy of Pediatrics, 394*, L5.
23. Gandolfi, D., Barragán, O., Livingston, J. H., Fridlund, M., Justesen, A. B., Redfield, S., et al. (2018). *The American Academy of Pediatrics, 619*, L10.
24. Gibson, S. R., Howard, A. W., Marcy, G. W., Edelstein, J., Wishnow, E. H., & Poppett, C. L. (2016). *Society of Photo-Optical Instrumentation Engineers (SPIE) Conference Series* (Vol. 9908, p. 990870). KPF: Keck Planet Finder.
25. Gilbert, E. A., Barclay, T., Schlieder, J. E., Quintana, E. V., Hord, B. J., Kostov, V. B., et al. (2020). e-prints. arXiv:2001.00952.
26. Gilliland, R. L. (2008). *The Astronomical Journal, 136*, 566.
27. Gilliland, R. L., McCullough, P. R., Nelan, E. P., Brown, T. M., Charbonneau, D., Nutzman, P. et al. (2011). *The Astrophysical Journal, 726*, 2.

28. Gould, A., Huber, D., Penny, M., & Stello, D. (2015). *Journal of Korean Astronomical Society, 48*, 93.
29. Grunblatt, S. K., Huber, D., Gaidos, E., Hon, M., Zinn, J. C., & Stello, D. (2019). *The Astronomical Journal, 158*, 227.
30. Grundahl, F., Fredslund Andersen, M., Christensen-Dalsgaard, J., Antoci, V., Kjeldsen, H., Handberg, R., et al. (2017). *The Astrophysical Journal, 836*, 142.
31. Guenther, D. B., Kallinger, T., Gruberbauer, M., Huber, D., Weiss, W. W., Kuschnig, R., et al. (2008). *The Astrophysical Journal, 687*(1448).
32. Guzik, J. A., Houdek, G., Chaplin, W. J., Smalley, B., Kurtz, D. W., Gilliland, R. L., et al. (2016). *The Astrophysical Journal, 831*, 17.
33. Harvey, J. W. (1988). In J. Christensen-Dalsgaard & S. Frandsen (Eds.), *Advances in helio- and asteroseismology. IAU symposium* (Vol. 123, p. 497).
34. Hekker, S., & Christensen-Dalsgaard, J. (2017). *Astronomy and Astrophysics Reviews, 25*, 1.
35. Hekker, S., Kallinger, T., Baudin, F., De Ridder, J., Barban, C., Carrier, F., et al. (2009). *The American Academy of Pediatrics, 506*, 465.
36. Hekker, S., Gilliland, R. L., Elsworth, Y., Chaplin, W. J., De Ridder, J., Stello, D., et al. (2011). *Monthly Notices of the Royal Astronomical Society, 414*, 2594.
37. Helmi, A., Babusiaux, C., Koppelman, H. H., Massari, D., Veljanoski, J., Brown, A. G. A., et al. (2018). *Nature, 563*, 85.
38. Huang, C. X., Burt, J., Vanderburg, A., Günther, M. N., Shporer, A., Dittmann, J. A., et al. (2018). *The Astrophysical Journal Letters, 868*, L39.
39. Huber, D., Bedding, T. R., Arentoft, T., Gruberbauer, M., Guenther, D. B., Houdek, G., et al. (2011). *The Astrophysical Journal, 731*, 94.
40. Huber, D., Chaplin, W. J., Chontos, A., Kjeldsen, H., Christensen-Dalsgaard, J., Bedding, T. R., et al. (2019). *The Astronomical Journal, 157*, 245.
41. Kallinger, T., Guenther, D. B., Weiss, W. W., Hareter, M., Matthews, J. M., Kuschnig, R., et al. (2008). *Communications in Asteroseismology, 153*, 84.
42. Kjeldsen, H., & Bedding, T. R. (1995). *The American Academy of Pediatrics, 293*, 87.
43. Kjeldsen, H., Bedding, T. R., Viskum, M., & Frandsen, S. (1995). *The Astronomical Journal, 109*, 1313.
44. Kjeldsen, H., Bedding, T. R., Butler, R. P., Christensen-Dalsgaard, J., Kiss, L. L., McCarthy, C., et al. (2005). *The Astrophysical Journal, 635*, 1281.
45. Kraus, A. L., Tucker, R. A., Thompson, M. I., Craine, E. R., & Hillenbrand, L. A. (2011). *The Astrophysical Journal, 728*, 48.
46. Leighton, R. B., Noyes, R. W., & Simon, G. W. (1962). *The Astrophysical Journal, 135*, 474.
47. Matthews, J. M. (2007). *Communications in Asteroseismology, 150*, 333.
48. Matthews, J. M., Kusching, R., Guenther, D. B., Walker, G. A. H., Moffat, A. F. J., Rucinski, S. M., et al. (2004). *Nature, 430*, 51.
49. Metcalfe, T. S., Creevey, O. L., & Davies, G. R. (2015). *The Astrophysical Journal Letters, 811*, L37.
50. Michel, E., Baglin, A., Auvergne, M., Catala, C., Samadi, R., Baudin, F., et al. (2008). *Science, 322*, 558.
51. Miglio, A., Chiappini, C., Mosser, B., Davies, G. R., Freeman, K., Girardi, L., et al. (2017). *Astronomische Nachrichten, 338*, 644.
52. Mosser, B., Michel, E., Samadi, R., Miglio, A., et al. (2019). *The American Academy of Pediatrics, 622*, A76.
53. Penny, M. T., Gaudi, B. S., Kerins, E., Rattenbury, N. J., Mao, S., Robin, A. C., et al. (2019). *The Astrophysical Journal Supplement, 241*, 3.
54. Petigura, E. A., Sinukoff, E., Lopez, E. D., Crossfield, I. J. M., Howard, A. W., Brewer, J. M., et al. (2017). *The Astronomical Journal, 153*, 142.
55. Rauer, H., Catala, C., Aerts, C., Appourchaux, T., Benz, W., Brandeker, A., et al. (2014). *Experimental Astronomy, 38*(1–2), 249–330.
56. Retter, A., Bedding, T. R., Buzasi, D. L., Kjeldsen, H., & Kiss, L. L. (2003). *The Astrophysical Journal Letters, 591*, L151.

57. Ricker, G. R., Winn, J. N., Vanderspek, R., Latham, D. W., Bakos, G. Á., Bean, J. L., et al. (2014). In *Society of Photo-Optical Instrumentation Engineers (SPIE) Conference Series* (Vol. 9143, p. 20).
58. Rodriguez, J. E., Vanderburg, A., Zieba, S., Kreidberg, L., Morley, C. V., Kane, S. R., et al. (2020). e-prints. arXiv:2001.00954.
59. Schofield, M., Chaplin, W. J., Huber, D., Campante, T. L., Davies, G. R., Miglio, A., et al. (2019). *The Astrophysical Journal Supplement Series, 241*, 12.
60. Schou, J., & Buzasi, D. L. (2001). ESA special publication. In A. Wilson & P. L. Pallé (Eds.), *SOHO 10/GONG 2000 Workshop: Helio- and Asteroseismology at the Dawn of the Millennium* (Vol. 464, pp. 391–394)
61. Silva Aguirre, V., Stello, D., Stokholm, A., Mosumgaard, J. R., Ball, W., Basu, S., et al. (2019). e-prints. arXiv:1912.07604.
62. Spergel, D., Gehrels, N., Baltay, C., Bennett, D., Breckinridge, J., Donahue, M., et al. (2015). e-prints. arXiv:1503.03757.
63. Stello, D., & Gilliland, R. L. (2009). *The Astrophysical Journal, 700*, 949.
64. Stello, D., Bruntt, H., Preston, H., & Buzasi, D. (2008). *The Astrophysical Journal Letters, 674*, L53.
65. Stello, D., Huber, D., Bedding, T. R., Benomar, O., Bildsten, L., Elsworth, Y. P., et al. (2013). *The Astrophysical Journal Letters, 765*, L41.
66. Stello, D., Zinn, J., Elsworth, Y., Garcia, R. A., Kallinger, T., Mathur, S., et al. (2017). *The Astrophysical Journal, 835*, 83.
67. Tabur, V., Bedding, T. R., Kiss, L. L., Giles, T., Derekas, A., Moon, T. T. (2010). *Monthly Notices of the Royal Astronomical Society, 409*, 777.
68. Tarrant, N. J., Chaplin, W. J., Elsworth, Y., Spreckley, S. A., & Stevens, I. R. (2007). *Monthly Notices of the Royal Astronomical Society, 382*, L48.
69. Thompson, M. J., Toomre, J., Anderson, E. R., Antia, H. M., Berthomieu, G., Burtonclay, D., et al. (1996). *Science, 272*, 1300.
70. Thompson, M. J., Christensen-Dalsgaard, J., Miesch, M. S., & Toomre, J. (2003). *Annual Review of Astronomy and Astrophysics, 41*, 599.
71. Ulrich, R. K. (1970). *The Astrophysical Journal, 162*, 993.
72. Walker, G., Matthews, J., Kuschnig, R., Johnson, R., Rucinski, S., Pazder, J., et al. (2003). *Publications of the Astronomical Society of the Pacific, 115*, 1023.
73. Yu, J., Huber, D., Bedding, T. R., Stello, D., Murphy, S. J., Xiang, M., et al. (2016). *Monthly Notices of the Royal Astronomical Society, 463*(2), 1297–1306.

Oblique Pulsation: New, Challenging Observations with TESS Data

D. W. Kurtz and D. L. Holdsworth

Abstract A major result from the first cycle of the TESS mission shows that the well-studied roAp star HD 6532 has a completely different mode geometry when compared to the published extensive ground-based *B* observations taken in 1984, 1985 and 1994. This is a major challenge to the widely used and theoretically explored oblique pulsator model. There are three prime hypotheses: (1) The oblique pulsator model is wrong. If that is shown, 40 years of work on these stars will need reinterpretation. (2) The star has dramatically changed pulsation axes. No pulsating star has previously even been suspected of doing this. We do not know if a pulsating star *can* do this. (3) Theory and extensive spectroscopic observations show that the pulsation mode geometry in roAp stars is sensitive to atmospheric depth. Subsequently, in these stars, we can see pulsation amplitude, phase and geometry in 3D. It is possible that the ground-based *B* observations and the TESS red bandpass observations are seeing a very different pulsation structure at different depths. If this can be shown, it will be novel and unexpected with implications for stellar pulsation in general. The test of these hypotheses requires simultaneous ground-based observations in *UBV* at the time of new observations of HD 6532 by TESS. This problem is novel, has wider implications for asteroseismology, and is testable by our method.

1 Scientific Justification

Magnetically chemically peculiar stars of spectral type A (i.e. Ap stars) constitute approximately 7% of all A-type stars [24]. These stars have enhanced abundances of Cr, Eu, Si, or Sr and host a large-scale magnetic fields ranging in strength from a few hundred gauss to of order 30 kG (e.g., [3, 2, 7]). Amongst the Ap stars are

D. W. Kurtz (✉) · D. L. Holdsworth
Jeremiah Horrocks Institute, University of Central Lancashire, Preston, UK
e-mail: dwkurtz@uclan.ac.uk; dlholdsworth@uclan.ac.uk

M. J. P. F. G. Monteiro et al. (eds.), *Dynamics of the Sun and Stars*, Astrophysics and Space Science Proceedings 57,
https://doi.org/10.1007/978-3-030-55336-4_43

a rare subgroup called rapidly oscillating Ap (roAp) stars [15] which pulsate with periods between 5 and 24 min [20, 1, 8]. Currently, only 76 rapidly oscillating Ap (roAp) stars have been discussed in the literature [25, 8, 4, 11], which were mostly discovered via ground-based campaigns of individual stars [9, 14].

The quality of ground-based photometry is typically inferior compared to data from space telescopes such as *Kepler* and K2, such that only the highest amplitude roAp pulsators have likely been detected from the ground. Now with TESS, unlike with *Kepler*, we are able to conduct the first full sky survey for pulsations in Ap stars with a precision of a few micro-magnitudes. The first results of this work were published by Cunha et al. [8], in a follow up paper by Balona et al. [4], and an upcoming end of TESS cycle 1 paper (Holdsworth et al. in prep.).

One of the most significant and puzzling results so far are from the observations of HD 6532—a previously known and well studied roAp star. HD 6532 (V = 8.4; T = 8.3 mag) was first reported as an roAp star by Kurtz and Kriedl [16] and has been the subject of multi-site ground-based photometric observations totalling 188 h. Analysis of those observations led to the conclusion that the star was pulsating in a distorted dipole mode [17]. This conclusion was drawn directly from two observables: the pulsation amplitude and phase.

The pulsation axis in the roAp stars is inclined to the rotation axis which leads to *oblique pulsation*, resulting in a modulation of both the pulsation amplitude and phase over the rotation cycle of the star. These varying amplitudes and phases manifest themselves in a multiplet in a Fourier transform with a central peak being the actual pulsation mode frequency, and $2\ell + 1$ (where ℓ is the degree of the mode) frequencies split by *exactly* the stellar rotation frequency. These frequencies are the Fourier description of the mode amplitude and phase variation with rotation. So, for a pure dipole mode, where $\ell = 1$, we expect to observe a triplet, and a quintuplet for a quadrupole mode (where $\ell = 2$).

In the case of a dipole mode, the pulsational equator of the star is a node. When the equatorial node crosses the line of sight, the pulsation amplitude appears to be zero (as the two hemispheres act to cancel each other) and a π-rad change in the pulsation phase is observed. This was seen in HD 6532 in the ground-based B observations, as is shown in Fig. 1.

The TESS cycle 1 data for HD 6532 also shows these expected variations (as shown in red in Fig. 1). The pulsation phase variations in the TESS data mimic those seen in the B data. This implies we are still looking at the same dipole mode. The pulsation amplitude variations, however, are very different, and it is the amplitude variations that the oblique pulsator model [15, 23, 5, 6] uses to infer the geometry of the pulsation mode. As stated above, the pulsation amplitude variation results in a multiplet in a Fourier transform of the light curve. We show the schematic results in Fig. 2 of the multiplet observed in B and TESS data. In the case of a dipole mode, the ratio of the sum of the amplitudes of the $\pm\nu_{\rm rot}$ sidelobes to the amplitude of the central peak is the key value. We see from Fig. 2 that this ratio is very different from the B and TESS observations. It is the astrophysical origin of this difference that is the puzzle we address in this paper.

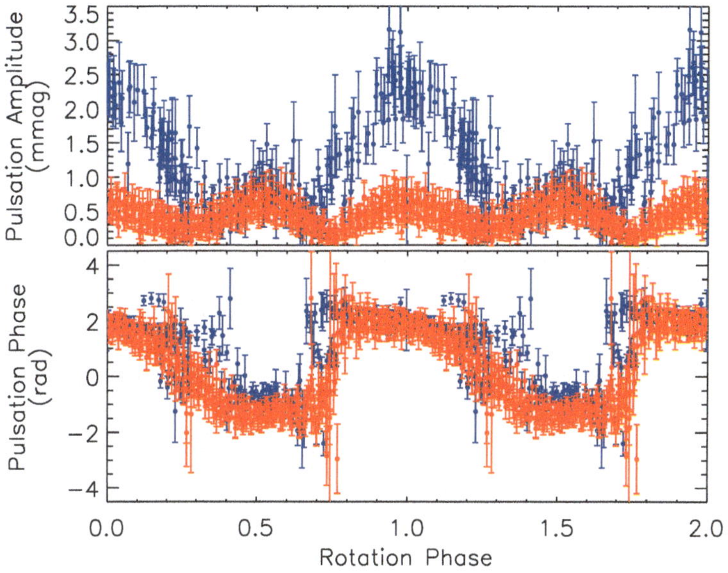

Fig. 1 Top: The pulsation amplitude as a function of rotation phase for HD 6532. Bottom: The pulsation phase as a function of rotation phase for HDS 6532. The blue points represent the ground-based B data, with the red points representing the TESS data. The expected π-rad phase change occurs in both data sets, but the amplitude variations have very different characteristics between the two. Two rotation periods are shown for clarity. The rotation zero-point has been chosen, independently for the two sets, to be at pulsation maximum amplitude

We have three hypotheses why the mode geometry is significantly different in the two sets of observations:

(1) **The star has dramatically changed pulsation axes.** No other pulsating star has been suspected of doing this; we do not know if a star *can* do this. The TESS and ground-based B data are separated in time by nearly 30 years, thus drawing conclusions from direct comparison to the two data sets is dubious. This significant time span may be sufficient for the atmospheric structure (or possibly even the magnetic field) of the star to change, thus shifting the pulsation axis of the mode. If this is the case, we have the opportunity to observe stellar evolution in action. The fact that the pulsational phase variation with rotation is so similar in the B and TESS data sets suggests that this hypothesis is unlikely, since most changes of pulsation axis would have also changed the pulsation phase versus rotation variation.

(2) **The oblique pulsator model is wrong.** The initial oblique pulsator model was proposed by Kurtz [15] and has been subsequently revised and improved many times (e.g., 16, 17, 18). Before the launch of the *Kepler* mission, the model was primarily developed and implemented with B observations. This results from the fact that the best signal-to-noise ratios for pulsations in roAp stars are attained through a Strömgren v filter, or the more commonly available Johnson-

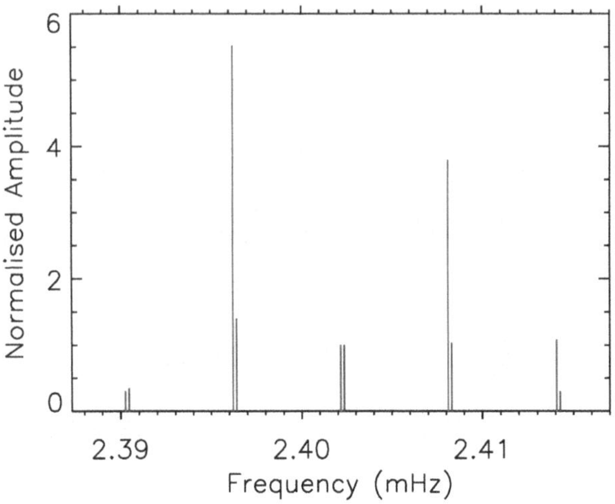

Fig. 2 Schematic of the pulsation quintuplet in HD 6532. The blue lines represent the ground-based *B* results, with the red lines representing the TESS results. The amplitudes of the quintuplet have been normalised to the amplitude of their pulsation mode, i.e. the central peak. The ground-based results have been shifted to higher frequency for clarity. Although the mode is a dipole, a quintuplet is seen as a result of distortion of the mode. The relative amplitudes of the multiplet are significantly different in the two data sets

Cousins *B* filter [21]. That is a consequence of those filters being close to the peak of the spectral energy distribution for A stars. *Kepler* observations provided the first instance for testing the model with white-light data, which provided some interesting results, such as the roAp star KIC 10195926 (discovered in the *Kepler* data) which was postulated to have two pulsation axes [19], and the most distorted roAp pulsation mode in HD 24355 as revealed by K2 [12]. TESS is just now—at the time of this writing, and after nearly 4 decades of using the oblique pulsator model—providing the first detailed light curves of the roAp stars in an almost entirely red bandpass, with *no* contribution from the traditional blue wavelengths, to confront the model. If the model is wrong, or only applicable to observations at certain wavelengths, this has significant implications to the field of roAp stars, with potentially 40 years of observations needing reinterpretation. Furthermore, a new model may need to be created to explain *all* TESS observations of the roAp stars.

(3) **The pulsation geometry in roAp stars changes as a function of depth in the atmosphere.** This has been shown spectroscopically and theoretically [13, 18, 22], but has yet to be explored photometrically. Due to the strong magnetic fields in Ap stars, convection is suppressed around the magnetic poles which allows for the stratification of chemical elements in the stellar

atmosphere through atomic diffusion. The roAp stars have strongly stratified atmospheres with some rare earth elements floating in "cirrus clouds" high in the atmospheres at continuum optical depths as low as 10^{-5} [10]. This makes it possible, when considering time-resolved spectroscopic observations, to measure the pulsation properties of a star at different atmospheric heights by studying spectral lines that originate at different heights in the atmosphere. It may now be possible, thanks to TESS, to perform this kind of depth analysis photometrically. Because of the higher opacity in B observations, they probe smaller geometric distances than the longer wavelength TESS observations, i.e., with B data we are looking higher in the stellar atmosphere, and with TESS we are probing closer to the continuum. This allows us to see how the geometry of the mode changes with atmospheric height, revealing information of the stellar atmosphere such at the temperature gradient. To study this from the ground alone is complicated by the low pulsation amplitude at long wavelengths. Thus TESS observations are required to obtain data with high signal-to-noise ratio to extract the low amplitudes found in the red for the pulsation frequencies.

This is our preferred hypothesis, although all three hypotheses will be tested by the multi band simultaneous observations planned. Within this hypothesis, the higher amplitude of the B observations at pulsation maximum seen in Fig. 1 arises because the B observations are sampling a layer higher in the atmosphere where the pulsation amplitude is greater [22]. This particular rotation phase presents the pulsation pole close to the line of sight, so we are looking to very different atmospheric depths in B and in TESS red. The secondary minimum is at a rotation phase where the pole is far from the line of sight and we are looking obliquely into the atmosphere, hence the difference in atmospheric depth of the B and TESS red observations is less, and the pulsation amplitudes are similar.

All three of these hypotheses are testable by obtaining simultaneous TESS and UBV observations of HD 6532. If the new UBV plus TESS red observations show the same pulsation geometry within the oblique pulsator model, then we will know that the mode axis did change over the years between the data sets discussed in this paper. On the other hand, if the UBV and TESS red observations are in agreement with the observations presented here, but still not in agreement with each other, then we consider that the third hypothesis is most viable, and will work on understanding and modelling the pulsation geometry as a function of atmospheric depth. If that should fail, then our understanding the oblique pulsator model will need to be revisited. Each of these possibilities leads to new astrophysical inference and understanding of these most peculiar and complex stars (Fig. 3).

Fig. 3 Michael Thompson was a friend, colleague and mentor, in whose company we always took great pleasure. There were so many times of shared personal and scientific time. Here is a picture that was typical of Michael, Kate and Robin—sharing their home with a dinner for friends. In this picture, from left-to-right around the table are: Juri Toomre, Robin Thompson, Rosanne Gough, Jesper Schou, June Kurtz, Douglas Gough, Michael at the head of the table, Linda Toomre, Stuart Jefferies, Don Kurtz, and Kate Thompson

References

1. Alentiev, D., Kochukhov, O., Ryabchikova, T., Cunha, M., Tsymbal, V., & Weiss, W. (2012). *Monthly Notices of the Royal Astronomical Society, 421*, L82.
2. Aurière, M., et al. (2004). *IAU Symposium* (Vol. 224, p. 633).
3. Babcock, H. W. (1960). *The Astrophysical Journal, 132*, 521.
4. Balona, L. A., Holdsworth, D. L., & Cunha, M. S. (2019). *Monthly Notices of the Royal Astronomical Society, 487*, 2117.
5. Bigot, L., & Dziembowski, W. A. (2002). *Astronomy and Astrophysics, 391*, 235.
6. Bigot, L., & Kurtz, D. W. (2011). *Astronomy and Astrophysics, 536*, A73.
7. Buysschaert, B., Neiner, C., Martin, A. J., Aerts, C., Bowman, D. M., Oksala, M. E., & Van Reeth, T. (2018). *Monthly Notices of the Royal Astronomical Society, 478*, 2777.
8. Cunha, M. S., Antoci, V., Holdsworth, D. L., Kurtz, D. W., Balona, L. A., Bognár, Zs., et al. (2019). *Monthly Notices of the Royal Astronomical Society, 487*, 3523.
9. Elkin, V. G., Kurtz, D. W., & Mathys, G. (2005). *Monthly Notices of the Royal Astronomical Society, 364*, 864.
10. Freyhammer, L. M., Kurtz, D. W., Elkin, V. G., Mathys, G., Savanov, I., Zima, W., et al. (2009). *Monthly Notices of the Royal Astronomical Society, 396*, 325.
11. Hey, D. R., Holdsworth, D. L., Bedding, T. R., Murphy, S. J., Cunha, M. S., Kurtz, D. W., et al. (2019). *Monthly Notices of the Royal Astronomical Society, 488*, 18.
12. Holdsworth, D. L., Kurtz, D. W., Smalley, B., Saio, H., Handler, G., Murphy, S. J., & Lehmann, H. (2016). *Monthly Notices of the Royal Astronomical Society, 462*, 876.
13. Kochukhov, O., & Ryabchikova, T. (2001). *Astronomy and Astrophysics, 374*, 615.

14. Kochukhov, O., Alentiev, D., Ryabchikova, T., Boyko, S., Cunha, M., Tsymbal, V., & Weiss, W. (2013). *Monthly Notices of the Royal Astronomical Society, 431*, 2808.
15. Kurtz, D. W. (1982). *Monthly Notices of the Royal Astronomical Society, 200*, 807.
16. Kurtz, D. W., & Kreidl, T. J. (1985). *Monthly Notices of the Royal Astronomical Society, 216*, 987.
17. Kurtz, D. W., Martinez, P., Koen, C., & Sullivan, D. J. (1996). *Monthly Notices of the Royal Astronomical Society, 281*, 883.
18. Kurtz, D. W., Elkin, V. G., Cunha, M. S., Mathys, G., Hubrig, S., Wolff, B., & Savanov, I. (2006). *Monthly Notices of the Royal Astronomical Society, 372*, 286.
19. Kurtz, D. W., Cunha, M. S., Saio, H., Bigot, L., Balona, L. A., Elkin, V. G., et al. (2011). *Monthly Notices of the Royal Astronomical Society, 414*, 2550.
20. Martinez, P., & Kurtz, D. W. (1994). *Monthly Notices of the Royal Astronomical Society, 271*, 118.
21. Medupe, R., & Kurtz, D. W. (1998). *Monthly Notices of the Royal Astronomical Society, 299*, 371.
22. Quitral-Manosalva, P., Cunha, M. S., & Kochukhov, O. (2018). *Monthly Notices of the Royal Astronomical Society, 480*, 1676.
23. Shibahashi H., & Takata M. (1993). *ASP Conference Series* (Vol. 563, p. 40).
24. Sikora, J., David-Uraz, A., Chowdhury, S., Bowman, D. M., Wade, G. A., Khalack, V., et al. (2019). *Monthly Notices of the Royal Astronomical Society, 487*, 4695.
25. Smalley, B., Niemczura, E., Murphy, S. J., Lehmann, H., Kurtz, D. W., Holdsworth, D. L., et al. (2015). *Monthly Notices of the Royal Astronomical Society, 452*, 3334.

Accounting for Asphericity

Kyle C. Augustson

Abstract Most global-scale simulations of stars neglect perturbations to its spherical geometry arising form external or internal forces, such as rotation or tides. To take these into account, one can map from the distorted geometry back onto the sphere, with the addition of a few new geometrical terms in the operators of a given set of equations. These are discussed here in the context of a magnetic field.

1 Introduction

The geometrical and indirect effects of internal and external forces such as the Lorentz force within stars, even complex ones, can play subtle dynamical roles in systems with the Eddington-Sweet circulation being an example. A relatively simple way to deal with these in a perturbative way, and yet still utilize existing numerical tools, is to map the distored geometry back into a sphere. The result is that the operators of the original set of equations are expanded to include new geometrical terms that are proportional to the map and its derivatives. This map and its implications are discussed here.

2 Equations of Motion

The perturbative classification and nondimensionalization of the various terms in the nondissipative MHD equations of motion can be made with the independent

K. C. Augustson (✉)
AIM, CEA, CNRS, Université Paris-Saclay, Université Paris Diderot, Gif-sur-Yvette Cedex, France
e-mail: kyle.augustson@cea.fr

M. J. P. F. G. Monteiro et al. (eds.), *Dynamics of the Sun and Stars*, Astrophysics and Space Science Proceedings 57,
https://doi.org/10.1007/978-3-030-55336-4_44

perturbative control parameter

$$\delta = \frac{B^2 R^4}{4\pi G M^2}. \tag{1}$$

where the characteristic value for the magnetic field is given by B, the radius of the star is R, and its mass is M, and where the gravitational constant is G. The eigenfunctions with zero eigenfrequency (the time-independent background state) may be linearized as $p = p_0 + \delta p_B$, $\rho = \rho_0 + \delta\rho_B$, $\Phi = \Phi_0 + \delta\Phi_B$, capturing the effect of the Lorentz force and its modification of the background stratification, where p is the pressure, ρ is the density, and Φ is the gravitational potential. This yields the following perturbative equations $\nabla p_0 + \rho_0 \nabla \Phi_0 = 0$ and $\nabla^2 \Phi_0 = 4\pi\rho_0$ for the zeroth-order spherically symmetric background and

$$\begin{aligned} &\nabla p_B + \rho_0 \nabla \Phi_B + \rho_B \nabla \Phi_0 - (\nabla \mathbf{x} \mathbf{B}) \mathbf{x} \mathbf{B} = 0, \\ &\nabla^2 \Phi_B = 4\pi \rho_B, \end{aligned} \tag{2}$$

for the magnetic perturbations labeled with an underscore B, where $\mathbf{B}$ is the magnetic field. This system of equations is completed with the boundary conditions that ρ, p, and Φ must be well-behaved as $r \to 0$, and that ρ, p, $\Phi \to 0$ as $r \to \infty$.

3 Asphericity and Coordinate Maps

To account for the asphericity of the volume and the photospheric boundary, the coordinate system can be adjusted to accommodate the magnetically induced spherical symmetry breaking. While the photosphere is mapped to the same level sets as in the hydrostatic case, this approximation neglects the higher order changes due to the modification of the energy flux balance. The coordinate map from the spherical coordinates of the case with a magnetic field (r, θ, ϕ) that describe an aspherical volume to the zeroth-order nonmagnetic coordinate system (r_0, θ_0, ϕ_0) that describe a spherical volume, yielding a new effective radius $r(r_0, \theta_0, \phi_0)$ which may be expressed to first-order in δ as

$$r = r_0 + \delta \sum_{\ell,m} h^m_{B,\ell}(r_0)\, Y^m_\ell(\theta_0, \phi_0), \tag{3}$$

where h_B encapsulates the local changes in the map due to the magnetic field and the angular coordinates remain the same as expressed in the spherical harmonics Y^m_ℓ. The solutions for h_B depend in turn upon the modified density and pressure [e.g., 3, 2, 1]. There is ambiguity in this choice of coordinates as it is linked to the spherical state and one may make one of several choices regarding which level set to choose, as long as they still satisfy the regularity condition at the origin and the outer boundary condition. The relevant level sets are isopycnal [2], isobaric [1], or equipotential. This is addressed in Appendix. Yet, since the horizontal coordinates remain unchanged, spherical harmonics still form an orthonormal basis on the chosen level set. This coordinate mapping then introduces several new terms due to the modification of the basis vectors and operators. To begin, note that the spherical coordinates of both the spherical and the aspherical volume are standard, being right-handed orthonormal coordinate systems with orthonormal moving frames and one-form frames in their own volumes. For compactness, the aspherical map and its derivatives will be written as

$$h = \sum_{\ell,m} h_{B,\ell}^{m} Y_\ell^m, \quad h_r = \sum_{\ell,m} \partial_{r_0} h_{B,\ell}^{m} Y_\ell^m, \quad h_{rr} = \sum_{\ell,m} \partial_{r_0}^2 h_{B,\ell}^{m} Y_\ell^m, \tag{4}$$

$$h_\theta = \sum_{\ell,m} \frac{h_{B,\ell}^{m}}{r_0} \partial_{\theta_0} Y_\ell^m, \quad h_\phi = \sum_{\ell,m} \frac{h_{B,\ell}^{m}}{r_0 \sin\theta_0} \partial_{\phi_0} Y_\ell^m, \tag{5}$$

and the basis vectors $\hat{\mathbf{e}}_\mathbf{i}$ will be refered to as a hatted coordinate, e.g. $\hat{\mathbf{r}} = \hat{\mathbf{e}}_\mathbf{r}$. Note the definition of the spherical Jacobian and its inverse, both of which are evaluated in the nonmagnetic coordinates to avoid implicit functions, which are

$$J = \frac{\partial \mathbf{r}}{\partial \mathbf{r}_0} = \begin{bmatrix} 1+\delta h_r & \delta h_\theta & \delta h_\phi \\ 0 & 1 & 0 \\ 0 & 0 & 1 \end{bmatrix}, \quad J^{-1} = \frac{\partial \mathbf{r}_0}{\partial \mathbf{r}} = \begin{bmatrix} \frac{1}{1+\delta h_r} & -\frac{\delta h_\theta}{1+\delta h_r} & -\frac{\delta h_\phi}{1+\delta h_r} \\ 0 & 1 & 0 \\ 0 & 0 & 1 \end{bmatrix}. \tag{6}$$

In addition to the coordinates, the basis vectors, or the moving frame, and operators in the aspherical volume may be expanded in the spherical volume. The angular coordinates are identical, and so are their basis vectors, therefore one need only deal with the change in the radial basis vector noting that under a coordinate transformation and a normalization

$$\hat{\mathbf{r}} = \frac{\sum_{i=1}^{3} J_{ri}^{-1} \hat{\mathbf{x}}_\mathbf{i}}{\left(\sum_{i=1}^{3} \left(J_{ri}^{-1}\right)^2\right)^{1/2}} = \frac{\hat{\mathbf{r}}_0 - \delta h_\theta \hat{\boldsymbol{\theta}}_0 - \delta h_\phi \hat{\boldsymbol{\phi}}_0}{\sqrt{1+\delta^2 h_\theta^2 + \delta^2 h_\phi^2}}, \tag{7}$$

which to first order in δ yields

$$\hat{\mathbf{r}} = \hat{\mathbf{r}}_0 - \delta h_\theta \hat{\boldsymbol{\theta}}_0 - \delta h_\phi \hat{\boldsymbol{\phi}}_0 + O\left(\delta^2\right). \tag{8}$$

With these quantities in hand, one may then compute the modifications to the spherical operators that account for the asphericity of the stellar volume. The gradient of a scalar function f may be defined as

$$\begin{aligned}
\boldsymbol{\nabla} f &= \partial_r f\hat{\mathbf{r}} + \frac{1}{r}\partial_\theta f\hat{\boldsymbol{\theta}} + \frac{1}{r\sin\theta}\partial_\phi f\hat{\boldsymbol{\phi}},\\
&= \frac{\partial_{r_0} f}{1+\delta h_r}\frac{\hat{\mathbf{r}}_0 - \delta h_\theta\hat{\boldsymbol{\theta}}_0 - \delta h_\phi\hat{\boldsymbol{\phi}}_0}{\sqrt{1+\delta^2 h_\theta^2+\delta^2 h_\phi^2}} + \frac{\partial_{\theta_0} f\hat{\boldsymbol{\theta}}_0}{r_0\left(1+\delta h/r_0\right)} + \frac{\partial_{\phi_0} f\hat{\boldsymbol{\phi}}_0}{r_0\sin\theta_0\left(1+\delta h/r_0\right)},\\
&= \boldsymbol{\nabla}_0 f - \delta\partial_{r_0} f\boldsymbol{\nabla}_0 h - \delta\frac{h}{r_0}\boldsymbol{\nabla}_{\perp,0} f + O\left(\delta^2\right),
\end{aligned} \tag{9}$$

where $\boldsymbol{\nabla}_\perp$ is just the angular part of the gradient. In a similar vein, the divergence of a vector field $\boldsymbol{\xi}$ may be expressed as

$$\begin{aligned}
\boldsymbol{\nabla}\cdot\boldsymbol{\xi} &= \partial_r\left(\hat{\mathbf{r}}\cdot\boldsymbol{\xi}\right) + \frac{2}{r}\left(\hat{\mathbf{r}}\cdot\boldsymbol{\xi}\right) + \frac{\partial_\theta\left(\sin\theta\xi_\theta\right)}{r\sin\theta} + \frac{\partial_\phi\xi_\phi}{r\sin\theta},\\
&= \frac{1}{1+\delta h_r}\partial_{r_0}\left(\frac{\xi_{r_0} - \delta h_\theta\xi_{\theta_0} - \delta h_\phi\xi_{\phi_0}}{\sqrt{1+\delta^2 h_\theta^2+\delta^2 h_\phi^2}}\right) + \frac{2}{r_0\left(1+\delta h/r_0\right)}\left[\frac{\xi_{r_0} - \delta h_\theta\xi_{\theta_0} - \delta h_\phi\xi_{\phi_0}}{\sqrt{1+\delta^2 h_\theta^2+\delta^2 h_\phi^2}}\right]\\
&\quad + \frac{\partial_{\theta_0}\left(\sin\theta_0\xi_{\theta_0}\right)}{r_0\sin\theta_0\left(1+\delta h/r_0\right)} + \frac{\partial_{\phi_0}\xi_{\phi_0}}{r_0\sin\theta_0\left(1+\delta h/r_0\right)},\\
&= \boldsymbol{\nabla}_0\cdot\boldsymbol{\xi} - \delta\frac{h}{r_0}\boldsymbol{\nabla}_{\perp,0}\cdot\boldsymbol{\xi} - \delta\left(\partial_{r_0}\boldsymbol{\xi}\right)\cdot\boldsymbol{\nabla}_0 h - \delta\frac{\boldsymbol{\xi}}{r_0^2}\cdot\left[\partial_{r_0}\left(r_0^2\boldsymbol{\nabla}_{\perp,0} h\right) + 2h\hat{\mathbf{r}}_0\right] + O\left(\delta^2\right).
\end{aligned} \tag{10}$$

Likewise, the directional derivative of a scalar function f with respect to a vector field $\boldsymbol{\xi}$ can be written as

$$\begin{aligned}
\boldsymbol{\xi}\cdot\boldsymbol{\nabla} f &= \xi_r\partial_r f + \frac{\xi_\theta}{r}\partial_\theta f + \frac{\xi_\phi}{r\sin\theta}\partial_\phi f,\\
&= \frac{\xi_{r_0} - \delta h_\theta\xi_{\theta_0} - \delta h_\phi\xi_{\phi_0}}{\left(1+\delta h_r\right)\left(1+\delta^2 h_\theta^2+\delta^2 h_\phi^2\right)^{1/2}}\partial_{r_0} f + \frac{\xi_{\theta_0}\partial_{\theta_0} f}{r_0\left(1+\delta h/r_0\right)} + \frac{\xi_{\phi_0}\partial_{\phi_0} f}{r_0\sin\theta_0\left(1+\delta h/r_0\right)},\\
&= \boldsymbol{\xi}\cdot\boldsymbol{\nabla}_0 f - \delta\frac{h}{r_0}\boldsymbol{\xi}\cdot\boldsymbol{\nabla}_{\perp,0} f - \delta\left(\partial_{r_0} f\right)\boldsymbol{\xi}\cdot\boldsymbol{\nabla}_0 h + O\left(\delta^2\right),
\end{aligned} \tag{11}$$

which agrees with the above definition of the gradient and the basis vector to first order. And finally, the Laplacian of a scalar may be written as

$$\begin{aligned}
\nabla^2 f &= \frac{\partial_{r_0}^2 f}{\left(1+\delta h_r\right)^2} + \left[\frac{2}{r_0\left(1+\delta h_r\right)\left(1+\delta h/r_0\right)} - \frac{\delta h_{rr}}{\left(1+\delta h_r\right)^3}\right]\partial_{r_0} f + \frac{1}{\left(1+\delta h/r_0\right)^2}\nabla_{\perp,0}^2 f,\\
&= \left(1-2\delta\frac{h}{r_0}\right)\nabla_0^2 f - \delta\left[2r_0\partial_{r_0}\left(\frac{h}{r_0}\right)\partial_{r_0}^2 f + h_{rr}\partial_{r_0} f\right] + O\left(\delta^2\right).
\end{aligned} \tag{12}$$

The curl follows in a similar manner.

4 Nonaxisymmetric Thermodynamic Perturbations

It is now useful to define the nonaxisymmetric components of the perturbative thermodynamic quantities. For this one may return to Eq. (2). In order to preserve the outer boundary condition on the eigenfunctions, as in [1], the isobars are chosen to be fixed which requires $p_B = 0$, which then specifies the aspherical coordinate mapping functions. First, expand the operators into the new coordinate system and the thermodynamic perturbations on spherical harmonics, which yields

$$\rho_0 \mathbf{\nabla}_0 \Phi_B - \rho_0 \mathbf{\nabla}_B \Phi_0 - \mathbf{\nabla}_B p_0 + \rho_B \mathbf{\nabla}_0 \Phi_0 - (\mathbf{\nabla}_0 \times \mathbf{B}) \times \mathbf{B} = 0, \tag{13}$$

$$\nabla_0^2 \Phi_B - \nabla_B^2 \Phi_0 = 4\pi \rho_B. \tag{14}$$

The latter equation eliminates ρ_B from the suite of equations for h and Φ_B, after integrating over angle against $Y_{\ell_\rho}^{m_\rho,*}$, $Y_{\ell_\Phi}^{m_\Phi,*}$, and $Y_{\ell_h}^{m_h}$ respectively, leaving first the density perturbation in terms of the coefficients of the gravitational potential perturbations $\Phi_{B,\ell}^m$ and the height function $h_{B,\ell}^m$ as

$$\begin{aligned} 4\pi \rho_{B,\ell_\rho}^{m_\rho} = {} & \frac{1}{r_0^2} \partial_{r_0} \left(r_0^2 \partial_{r_0} \Phi_{B,\ell_\rho}^{m_\rho} \right) - \frac{\ell_\rho (\ell_\rho + 1)}{r_0^2} \Phi_{B,\ell_\rho}^{m_\rho} - 8\pi \rho_0 \frac{h_{B,\ell_\rho}^{m_\rho}}{r_0} \\ & - 2 r_0 \partial_{r_0} \left(\frac{h_{B,\ell_\rho}^{m_\rho}}{r_0} \right) \partial_{r_0}^2 \Phi_0 - \partial_{r_0}^2 h_{B,\ell_\rho}^{m_\rho} \partial_{r_0} \Phi_0. \end{aligned} \tag{15}$$

Noting that the two $\mathbf{\nabla}_B$ terms in Eq. (13) vanish given that in the spherical reference state $\partial_{r_0} p_0 + \rho_0 \partial_{r_0} \Phi_0 = 0$, that the ℓ_h and ℓ_Φ components equivalent to the above equation can be substituted into it, and after taking the inner product with $\mathbf{Y}_{\ell_h,\ell_{h'}}^{m_h,*}$, one can see that

$$\begin{aligned} & \rho_0 G_{0,\ell_h,\ell_{h'}} \Phi_{B,\ell_h}^{m_h} + \sum_{\ell_{h''}=\ell_h-1}^{\ell_h+1} R_{\ell_h,\ell_{h'}} \rho_{B,\ell_h}^{m_h} \partial_{r_0} \Phi_0 = \\ & \sum_{\substack{j_J,\ell_J,m_J \\ j_B,\ell_B,m_B}} \sum_{\ell_{J'}=\ell_J-1}^{\ell_J+1} \left(C_{\ell_J,\ell_{J'}}^{1,j_J} B_{j_J,\ell_J'}^{m_J} \right) \mathcal{J}_{\ell_J,m_J,j_J}^{\ell_{h'},m_h,\ell_h}{}_{\ell_B,m_B,j_B} B_{j_B,\ell_B}^{m_B} = F_{L,\ell_h,\ell_{h'}}^{m_h}, \end{aligned} \tag{16}$$

where F_L is a SVH series coefficient for the Lorentz force and

$$\mathbf{Y}_{j,\ell}^m \cdot \hat{\mathbf{r}}_0 = \left[-\sqrt{\frac{j+1}{2j+1}} \delta_{\ell,j+1} + \sqrt{\frac{j}{2j+1}} \delta_{\ell,j-1} \right] Y_j^m = R_{j,\ell} Y_j^m. \tag{17}$$

Therefore, adding and subtracting these equations for $\ell_{h'} = \ell_h - 1$ and $\ell_{h'} = \ell_h + 1$ (and translating indices from ℓ_h to ℓ_Φ for the first equation), one has that

$$\left[\left(\sqrt{\ell_\Phi + 1} + \sqrt{\ell_\Phi}\right)\partial_{r_0} + \sqrt{\ell_\Phi\left(\ell_\Phi + 1\right)}\left(\sqrt{\ell_\Phi + 1} - \sqrt{\ell_\Phi}\right)\frac{1}{r_0}\right]\Phi^{m_\Phi}_{B,\ell_\Phi}$$

$$= \sqrt{2\ell_\Phi + 1}\left[F^{m_\Phi}_{L,\ell_\Phi,\ell_\Phi - 1} - F^{m_\Phi}_{L,\ell_\Phi,\ell_\Phi + 1}\right], \tag{18}$$

$$\left[\frac{8\pi\rho_0}{r_0} + 2\partial^2_{r_0}\Phi_0\left(\partial_{r_0} - \frac{1}{r_0}\right) + \partial_{r_0}\Phi_0\partial^2_{r_0}\right]h^{m_h}_{B,\ell_h}$$

$$= \frac{1}{r_0^2}\partial_{r_0}\left(r_0^2\Phi^{m_h}_{B,\ell_h}\right) - \frac{\ell_h\left(\ell_h + 1\right)}{r_0^2}\Phi^{m_h}_{B,\ell_h}$$

$$- \frac{1}{2\sqrt{\ell_h + 1} - 2\sqrt{\ell_h}}\left\{\left[\left(\sqrt{\ell_h} - \sqrt{\ell_h + 1}\right)\partial_{r_0}\right.\right.$$

$$\left. + \sqrt{\ell_h\left(\ell_h + 1\right)}\left(\sqrt{\ell_h} + \sqrt{\ell_h + 1}\right)\frac{1}{r_0}\right]\Phi^{m_h}_{B,\ell_h}$$

$$\left. - \sqrt{2\ell_h + 1}\left(F^{m_h}_{L,\ell_h,\ell_h + 1} + F^{m_h}_{L,\ell_h,\ell_h - 1}\right)\right\}. \tag{19}$$

5 Conclusions

We have made explict the perturbative contributions to the geometrical distortion of the star and a means of mapping it back into spherical coordinates such that these effects can be incorporated into global-scale numerical simulations. Note that this treatment is in fact general such that any external force impacting the first-order thermodynamics may be treated with the above equations. For rotation, this is particularly simple, with the need to include only a few terms of low angular degree. Likewise, for a distant tidal partner, the effects are of low angular degree and easily incorporated.

Acknowledgments K. C. Augustson acknowledges support from the ERC SPIRE 647383 grant and PLATO CNES grant at CEA/DAp-AIM.

References

1. Gough, D. O., & Thompson, M. J. (1990). *Monthly Notices of the Royal Astronomical Society, 242*, 25. https://doi.org/10.1093/mnras/242.1.25
2. Lebovitz, N. R. (1970). *Astrophysical Journal, 160*, 701 (1970). https://doi.org/10.1086/150463
3. Simon, R. (1969). *Astronomy and Astrophysics, 2*, 390.

A Comparison of Global Helioseismic-Instrument Performances: Solar-SONG, GOLF and VIRGO

S. N. Breton, R. A. García, P. L. Pallé, S. Mathur, F. Hill, K. Jain, A. Jiménez, S. C. Tripathy, F. Grundahl, M. Fredslund-Andersen, and A. R. G. Santos

Abstract The SONG spectrograph has recently demonstrated its ability to perform solar radial velocity measurement during the first test run of the Solar-SONG initiative. A preliminary assessment of its performance is carried out here by comparing the results of Solar-SONG during the summer 2018 test run, with GOLF and VIRGO/SPM taken as reference instruments.

1 Solar-SONG Performances Over the Summer 2018 Test Run

We consider 30 continuous days of observation of the first Solar-SONG observing run [2] and two sets of contemporaneous VIRGO [3] and GOLF [4] subseries, all sampled at 60s. This implies that the original Solar-SONG series is interpolated to this cadence (and filtered below 800 μHz). In order to homogenize the series we multiply each one by the common observational window function and we

---

S. N. Breton (✉) · R. A. García
IRFU, CEA, Université Paris-Saclay, Gif-sur-Yvette, France

AIM, CEA, CNRS, Université Paris-Saclay, Université Paris Diderot, Gif-sur-Yvette, France
e-mail: sylvain.breton@cea.fr

P. L. Pallé · S. Mathur · A. Jiménez
Instituto de Astrofísica de Canarias, La Laguna, Spain

Universidad de La Laguna, Dpto. de Astrofísica, La Laguna, Spain

F. Hill · K. Jain · S. C. Tripathy
National Solar Observatory, Boulder, CO, USA

F. Grundahl · M. Fredslund-Andersen
Stellar Astrophysics Centre, Aarhus University, Aarhus C, Denmark

A. R. G. Santos
Space Science Institute, Boulder, CO, USA

M. J. P. F. G. Monteiro et al. (eds.), *Dynamics of the Sun and Stars*, Astrophysics and Space Science Proceedings 57,
https://doi.org/10.1007/978-3-030-55336-4_45

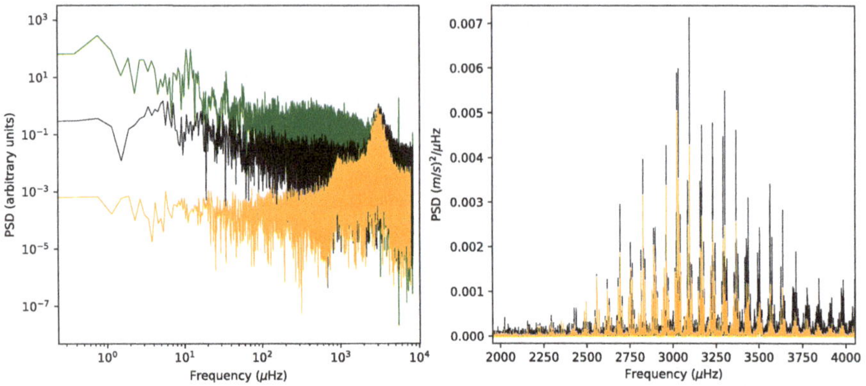

Fig. 1 Solar-SONG (orange), GOLF (black) and VIRGO (green) 30-days PSD during the 2018 summer. Solar-SONG signal has been filtered below 800 μHz using an ad-hoc non phase distortion FIR filter

Table 1 $\Delta\nu$ and ν_{max} as computed by the A2Z pipeline analysis and RMS p-mode and noise amplitudes [1] for GOLF, VIRGO and Solar-SONG 30-day series

Instrument	Duty cycle (%)	$\Delta\nu$ (μHz)	ν_{max} (μHz)	RMS value (m/s)	Noise amplitude (m/s)
GOLF	99.9	136.4 ± 2.6	3229.2 ± 150.7	–	–
	49.9	135.6 ± 2.9	3226.8 ± 153.8	0.69	0.28
VIRGO	99.9	134.4 ± 3.2	3158.7 ± 111.2	–	–
	49.9	134.4 ± 2.7	3117.1 ± 120.8	–	–
Solar-SONG	49.9	136.0 ± 2.8	3133.2 ± 136.6	0.60	0.15
	49.9	136.0 ± 2.8	3133.1 ± 136.2	0.60	0.15

compute the Power Spectrum Density (PSD) which is shown in Fig. 1. Global seismic parameters of the solar acoustic modes ($\Delta\nu$ and ν_{max}) are determined by applying the A2Z pipeline [5]. Table 1 summarizes the results. For the velocity instruments (GOLF and Solar-SONG) we include a determination of the RMS and noise amplitude in m/s.

References

1. Appourchaux, T., Boumier, P., Leibacher, J. W., & Corbard, T. (2018). *Astronomy & Astrophysics, 617*, A108.
2. Fredslund Andersen, M., Pallé, P. L., Jessen-Hansen, J., Wang, K., Grundahl, F., Bedding, T. R., et al. (2019). *Astronomy & Astrophysics, 623*, L9.
3. Fröhlich, C., Romero, J., Roth, H., Wehrli, C. Andersen, B. N., Appourchaux, T., et al. (1995). *Solar Physics, 162*(1–2), 101.
4. García, R. A., Turck-Chièze, S., Boumier, P., Robillot, J. M., Bertello, L., Charra, J., et al. (2005). *Astronomy & Astrophysics, 442*(1), 385.
5. Mathur, S., García, R. A., Régulo, C., Creevey, O. L., Ballot, J., Salabert, D. et al. (2010). *Astronomy & Astrophysics, 511*, A46.

Open Discussion

R. A. García, S. Mathur, M. J. P. F. G. Monteiro, J. Christensen-Dalsgaard, and S. W. McIntosh

Abstract During the last morning of the conference, a one-hour open discussion allowed the participants to debate some of the "hot" topics presented all along the meeting as well as on some of the key issues in the field mostly related with the work Prof. Michael J. Thompson studied during his carrier. The discussion covered theory and methods, current and future modeling efforts, observations, and future instrumentation. At the end, Dr. Robin Thompson discussed about the use of inversion methods in his current research, of particular interest these days, about the control of infectious disease outbreaks.

R. A. García (✉)
AIM, CEA, CNRS, Université Paris-Saclay, Université Paris Diderot, Gif-sur-Yvette, France
e-mail: rafael.garcia@cea.fr

S. Mathur
Instituto de Astrofísica de Canarias, La Laguna, Spain

M. J. P. F. G. Monteiro
Instituto de Astrofísica e Ciências do Espaço, Universidade do Porto, CAUP, Porto, Portugal

Departamento de Física e Astronomia, Faculdade de Ciências, Universidade do Porto, Porto, Portugal

J. Christensen-Dalsgaard
Stellar Astrophysics Centre (SAC), Department of Physics and Astronomy, Aarhus University, Aarhus C, Denmark

S. W. McIntosh
High Altitude Observatory, National Center for Atmospheric Research, Boulder, CO, USA

M. J. P. F. G. Monteiro et al. (eds.), *Dynamics of the Sun and Stars*, Astrophysics and Space Science Proceedings 57,
https://doi.org/10.1007/978-3-030-55336-4_46

1 Theory and Methods

1.1 Inversions

The discussion started with one of Michael's seminal works: Solar and stellar inversions. One of the current problems while inverting real data is the proper treatment of jumps, whether it is in space or in time. This is one of the current challenges that scientists must face to extract the best of the internal structure and dynamics. We need to find a way to deal with the inversions as a whole. A solution could be to make custom target kernels, which could allow us to deal with jumps for instance, or a clever way to place the nodes. Sharp changes like the tachocline or near the core are one of the most complicated features to deal with and the community expressed the need to improve their treatment.

Another important "worry" concerns the systematic errors in the input data and how they can be solved. Indeed their treatment is very important and they have a larger influence on some of the methods, in particular, for the optimally localized averaging (OLA) inversion technique. Moreover, the way in which over-resolution can be used properly and a better way to improve the real data prior to make the inversions are still open issues. A-priory information should not be used to force any given inversion.

One objective of the community is to move towards 3D inversions where the new dimension could be not only space but rather time. The question is how to do it.

What we are looking for is a quantitative response from the inversions and not a subjective one. Indeed, in general the inversion techniques usually assume a smooth profile, in particular when no information at different depths is available. A trade-off needs to be reached between data-driven inversions and our own prejudice.

Finally, and mostly for the solar case, it was pointed out that better high-frequency modes are required. The community is facing this issue and several groups are currently working to extract better high-frequency modes but we need to wait until this is done. These high-frequency modes imply that non-adiabatic calculations should also be improved on the theoretical side. This path has already been taken with averaged 3D Radiative Hydrodynamical simulations of the stellar surface (e.g. [7]) that can replace standard near-surface layers in 1D stellar structure models.

1.2 Surface Effects: From Machine Learning Tools to Open-Source Codes

This brings us to the (in)famous surface effects. The key point is how to extract them from simulations. The good news is that several groups are working on it but there are still important problems to solve and we need to know them. For example, in the tachocline, is there a dynamo operating there? Can the highly turbulent convective zones build activity? Maybe not in the way we want. . . We also do not know how boundary layers work and how they affect the surface effects and the dynamics below them. We also need to understand how magnetism impacts surface effects and move from studying snapshots towards a continuous time evolution of the different physical effects in play.

Can Machine Learning (ML) techniques contribute to better understand surface effects? A hot and animated debate followed this last question and got far beyond the framework of surface effects. Indeed, ML techniques are more and more often implemented in all disciplines and helio- and astero-seismology also follow this trend. ML techniques can be considered as highly non-linear mapping. So, they seem to be well adapted to the surface effect problem. However, it is important to not use these ML techniques for problems that can be solved by classical methods. ML are tools that should be better understood and we need to learn to interpret them in the best possible way. There is a balance to be found.

This discussion opened a second hot topic about open-source codes. The astronomy community in general is adopting these codes very fast but the seismic community is lagging behind. It is true that there are some codes such as the Modules for Experiments in Stellar Astrophysics (MESA, [5]) or several global seismic pipelines available but they are just exceptions and not the norm. The debate concerned the use of these programs as a "black box" with some examples of publications that contain important errors due to the bad use of these packages. Of course it has been pointed out that good open-source codes need good support behind and there is often a lack of funding by the institutions to give that to the community. Fortunately, there are a few people deeply involved on this path and more support is required from the community as well as feedback to the developers in order to improve the codes.

2 Current and Future Modeling

The general feeling of the community is that we need to go beyond the standard 1-D stellar evolution codes and generalize the use of 2-D codes, such as the Evolution STellaire en Rotation (ESTER, [2]), or even 3-D codes.

A question was raised whether it is possible to do structure evolution and model convection in a 2-D code with the same time scales as evolution. Nowadays, 2-D convection models can be done for short time scales, too short compared to evolution. To go beyond that, a large effort should be undertaken towards a fully parallelization of the codes, which requires also a deep investment in manpower.

A final question was raised concerning the differences found between 2-D and 3-D convection models. This implies that the community needs to understand those differences first, before moving forward.

3 Observations

While modes located near the frequency of maximum power ($\nu_{\max}$) are "easy" to characterize, the current challenges for helio- and asteroseismic observations are to go lower in frequency. Because these modes have long lifetimes they are more precisely determined and because they are less affected by the surface effects and activity, they can be used in theory with less assumptions and are better suited to be compared with theoretical frequencies. As a consequence, they allow us to obtain better inversion results close or near the core.

In the solar case, a large effort is given to extract a larger number of precise high-degree modes to better probe the sub-surface layers of the Sun.

For astero-seismology, high-frequency modes are extremely important to properly define the surface effect correction. Moreover in many cases, it was pointed out that the *ultimate* precision cannot be reached for other stars because of a lack of high-precision spectroscopic observations. The community needs to make some extra effort in acquiring complementary parameters to properly feed the evolution codes. Moreover, a better understanding of the atmospheric spectrum is required to improve the precision and accuracy of the spectroscopic parameters measured. Combining seismology, spectroscopy, and astrometry with stellar evolution codes will provide this *ultimate* precision. The PLAnetary Transits and Oscillations of stars (PLATO, [6]) community is trying to face some of these challenges to get the maximum from this mission.

In main-sequence stars, including the Sun, the core is still inaccessible using p modes only. G modes are needed to progress in the understanding of the innermost layers of these stars. Only when this is achieved will we be able to start to answer some of the most burning questions regarding the angular momentum transport in the deep layers of a solar-like star as well as the existence and characterization of a magnetic field in the core.

Finally, the question about what is going on in the solar poles will be soon addressed thanks to the Solar Orbiter mission [4]. The Japanese Hinode mission [3] already provided some glimpse on it. On the stellar side, with hundreds of thousands of stars observed by space missions, we have in hand a broad range of inclination angles (of the rotation axis compared to the line-of-sight), meaning that we should have a non-negligible sample of stars observed from the poles. The community should take advantage of those observations to study stellar poles. However, the advantage of the solar observations is that the solar poles will be spatially resolved, giving more information than what can be obtained from other stars.

4 Future Instrumentation

A discussion was engaged on what would be the future missions that would lead us to make a quantum leap in our knowledge of solar and stellar physics, compared to what we have today or what is planned in the near future (e.g. Solar Orbiter,

PLATO). A consensus was reached that this big step forward will come with a Stellar Imager in space (e.g. [1]). Even the observation of a handful of stars, pushing the degree of the modes would improve the resolution of the inversions not only close to the surface but also inside (e.g. the tachocline of stars) and better resolve internal kernels. Higher degree modes would also imply a much better latitudinal inversions.

For the Sun, it is necessary to think about new instrumentation capable of reducing the leakage of modes and move towards multi-band observations that could reduce the convective noise and increase the odds of detecting individual g modes with enough precision to be used in the inversions.

Finally all the audience agreed that an imperative requirement of future instrumentation for helio- and asteroseismic observations is to be able to provide decades monitoring for the former and as long as possible for the latter.

5 Concluding Remarks

To end the discussion on a different perspective a question was addressed to Dr. Robin Thompson, the son of Michael and his wife Kate, about the use of inversions in his research on epidemiology. Although this would be a very interesting technique to implement in such field, there are other issues to be addressed before. For example, there is a need in the current data sets to get to the origin of the pandemics. These data sets are in general noisy and come from multiple time series which complicates the modeling. One idea prior to use inversion techniques would be to do forward-backward modeling as it is also commonly done in solar and stellar physics.

At the time of writing this summary the world is facing the COVID-19 pandemic. Maybe in a few years from now, and using all the data sets that will be available, the inversions on this field will also be commonly used.

Acknowledgments The authors of this work would like to thank all the participants of the "Dynamics of the Sun & stars: Honoring the life & work of Michael J. Thompson" for their enthusiastic participation and friendly debates that we are sure Michael would have enjoyed.

References

1. Christensen-Dalsgaard, J., Carpenter, K. G., Schrijver, C. J., Karovska, M., Si Team (2011). *Journal of Physics Conference Series, 271*(1), 012085. https://doi.org/10.1088/1742-6596/271/1/012085
2. Espinosa Lara, F., & Rieutord, M. (2007). *Astronomy & Astrophysics 470*(3), 1013. https://doi.org/10.1051/0004-6361:20077263
3. Kosugi, T., Matsuzaki, K., Sakao, T., Shimizu, T., Sone, Y., Tachikawa, S., et al. *Solar Physics, 243*(1), 3 (2007). https://doi.org/10.1007/s11207-007-9014-6
4. Müller, D., Marsden, R. G., St. Cyr, O. C., & Gilbert, H. R. (2013). *Solar Physics, 285*(1–2), 25. https://doi.org/10.1007/s11207-012-0085-7

5. Paxton, B., Cantiello, M., Arras, P., Bildsten, L., Brown, E. F., Dotter, A., et al. (2013). *The Astrophysical Journal Supplement, 208*, 4. https://doi.org/10.1088/0067-0049/208/1/4
6. Rauer, H., Catala, C., Aerts, C., Appourchaux, T., Benz, W., Brandeker, A., et al. (2014). *Experimental Astronomy, 38*, 249. https://doi.org/10.1007/s10686-014-9383-4
7. Trampedach, R., Aarslev, M. J., Houdek, G., Collet, R., Christensen-Dalsgaard, J., Stein, R. F., et al. (2017). *Monthly Notices of the Royal Astronomical Society, 466*(1), L43. https://doi.org/10.1093/mnrasl/slw230

Contemplating the Future

Jørgen Christensen-Dalsgaard

Abstract The prospects for the further development of stellar astrophysics remain golden. Much can still be done based on the *Kepler* data, the TESS mission is providing data over most of the sky, at a 10 min cadence in the extended phase, PLATO will provide data comparable with those of *Kepler* for relatively near-by stars, ground-based observations yield data of even higher quality for selected stars, and Gaia is revolutionizing the determination of global stellar properties. On the modelling side ever more realistic simulations are becoming possible of dynamic phenomena in stellar interiors, and new techniques, known as artificial intelligence or deep learning, are helping the analysis of the huge amounts of data that are becoming available. Much of this effort is aimed at applications of stellar physics in other areas of astrophysics, such as exoplanets or Galactic archaeology. In the excitement of these developments, in the spirit of Michael we should not forget the basic goal of improving our physical understanding of stars, combining human intelligence and intuition with mathematical analysis and modelling, observations, and data interpretation. This should be kept in mind in our work and, even more importantly, in the education of coming generations of stellar astrophysicists.

1 Introduction

Michael Thompson was a close friend and for many years my closest collaborator. He had his first postdoc position with me in Aarhus, as my first postdoc after I took up my position there. Our collaboration continued with mutual visits and, in particular, through summers at the High Altitude Observatory in Boulder, as summer visitors and later, with Michael in a position at HAO. This combined

---

J. Christensen-Dalsgaard (✉)
Stellar Astrophysics Centre, Department of Physics and Astronomy, Aarhus University, Aarhus C, Denmark
e-mail: jcd@phys.au.dk

M. J. P. F. G. Monteiro et al. (eds.), *Dynamics of the Sun and Stars*, Astrophysics and Space Science Proceedings 57,
https://doi.org/10.1007/978-3-030-55336-4_47

intense work with a great deal of, perhaps sometimes a little noisy, fun, making colleagues wonder what all this fun was about. As mentioned also elsewhere in these proceedings, in this collaboration Michael played a seminal role in the development and application of helioseismology, and in the broader organization of the field. In later years, as Michael to a larger extent moved into administrative activities our contact weakened, and my yearly visits to HAO stopped. It is tragic that he died at a time where he was returning to scientific work. He was much needed in helioseismology and is sorely missed.

This is a time where there would have been a huge need for Michael's scientific qualities, in a combination of physical insight and overview, mathematical talent and computational expertise. Helioseismic data of exquisite quality continue to be accumulated, but the interpretation in terms of improved understanding of the solar interior is not keeping up, and asteroseismology has been completely revolutionized in the past decade by photometric observations from space. This has also spawned new techniques for the analysis of the data, with some emphasis on computational procedures for seeking patterns in the data. Developments in computational resources have also further strengthened the possibilities for carrying out detailed simulations of conditions and processes in stellar interiors. Much of this development was discussed at the conference and is reflected in these proceedings.

Here I consider some aspects of the outcome of the conference and provide personal thoughts about the coming development. This is undoubtedly the view of what some might call a dinosaur, with a substantially longer past experience than future prospect, but I hope that it may stimulate further thought and discussion on these issues, and perhaps inspire new developments, in the spirit of Michael and his work.

2 Goals of Solar and Stellar Physics

The study of the Sun and stars plays a key role in astrophysics, and, in the solar case, beyond. Characterization of stellar properties is central to investigations of extra-solar planetary systems: determination of the properties of the planets depends on knowledge of the mass and radius of the host star, and the only way to infer the age of a planetary system beyond the solar system is to constrain the age of the star. Studies of the structure and evolution of the Milky Way Galaxy, which in itself is a benchmark for the general understanding of galactic evolution, are to a large extent dependent on determining the location and motion of stars in the Galaxy, as well as their age and chemical composition. Owing to its proximity the Sun is obviously a benchmark for the study of stars. Furthermore, the Sun through its magnetic activity has a more direct influence on the Earth in what is known as space weather, with solar eruptions having potentially serious effects on our technological society, and longer-term solar variations may affect the climate on Earth, and hence have potential relevance also for the study of infectious-disease epidemiology (see the paper by Robin Thompson et al., these proceedings). In addition, solar observations have played a key role in developing the understanding of neutrinos and have been used to place constraints on the properties of dark-matter particles.

At a more basic level, however, the goal of solar and stellar astrophysics is, with a reference to Eddington [5], to understand so simple a thing as a star. This was the theme of the present conference, which clearly demonstrated both progress towards this goal and the fact that we are still far from reaching it. There remain key uncertainties in areas central to the conference, and to Michael's work, involving hydrodynamical and magneto-hydrodynamical processes in stellar interiors. Convection and dynamics beyond convective regions remain major uncertainties in stellar modelling and the stellar pulsations. Rotation, and the associated mixing processes, remain inadequately dealt with, and there are serious limitations in our understanding of the evolution of stellar rotation, starting with the helioseismically inferred relatively slow and uniform rotation of the solar interior. Magnetic fields undoubtedly play a major role in the interiors of many stars, yet are generally ignored, while the dynamo processes generating the field in the Sun and Sun-like stars are still not fully understood.

The challenge that we face is to build a coherent picture on a strong physical basis, which is consistent with the observational results in a solid statistical manner, of the structure and evolution of stars.

3 New Possibilities

There is excellent potential, in terms of observational data, for advances in these areas. The Solar Dynamics Observatory and the GONG and BiSON networks continue to obtain high-quality data on solar oscillations, with large amounts of data still awaiting full analysis. With the recent launch of Solar Orbiter there are prospects for helioseismic data addressing the dynamics of the solar poles within the coming decade. Much of the asteroseismic data from the *Kepler* mission still await detailed analysis, while data of lower quality, but for relatively nearby stars in fields covering most of the sky, are being obtained from the TESS mission. Data of a quality matching the *Kepler* data are expected from the PLATO mission after its launch in 2026. In addition to the seismic data, precise astrometry for a huge number of stars is being obtained by the Gaia mission, while large-scale spectroscopic surveys are providing extensive information about stellar composition.

In addition to these observational possibilities, the continuing development of computing resources allows new possibilities for data analysis, as well as for the modelling of stellar interiors.

The utilization of these possibilities requires extensive efforts, including the development of new data-analysis techniques. An interesting development in helioseismology is the use of new combinations of inferred properties in structure inversions (Buldgen et al. [3]; Basu, these proceedings). Following a suggestion and early development by Michael a concerted effort is underway to determine the solar internal rotation, using the huge amount of data that have been accumulated (Howe, these proceedings). Local helioseismology is being based on increasingly realistic representations of the sensitivity of the observations to subsurface solar features, with emphasis on the meridional flow and flows around sunspots (see papers by Birch and by Rajaguru, these proceedings). A very interesting development, based on earlier work by Schad et al. [7] and Woodard et al. [8] is the use of mode-coupling analysis to infer flows deep in the convection zone. As reported by Shravan Hanasoge at the conference the results show much weaker flows in the deeper parts of the convection zone than predicted by simple mixing-length models or detailed three-dimensional hydrodynamical simulations. This convective conundrum was discussed by Rast (these proceedings).

Asteroseismology provides a much broader range of challenges and possibilities. We are only beginning to address the deeper aspects of detailed probing of the structure of stellar interiors, and the underlying physics, where for the time being the *Kepler* data will be the main resource. Here, also, further developments of the analysis techniques are required. Application of inversion techniques, although in a far more limited sense than for the Sun, is being developed and is providing interesting discrepancies with otherwise best-fitting stellar models (Bellinger et al., these proceedings). Also, analysis of particular properties of the frequency spectra is providing interesting diagnostics of specific features in stellar internal structure, in particular the so-called glitches, where the structure varies on a scale small compared with the mode wavelength (Cunha, these proceedings). A rich picture is emerging of stellar rotation in different phases of stellar evolution, although the underlying physics, in terms of angular momentum loss and transport, is far from fully understood [1]. As noted in several papers in these proceedings, it is becoming increasingly clear that magnetic fields play an central role in some cases. As discussed by Santos (these proceedings), an important issue is to relate the inferred properties of solar-like stars, including their structure and magnetic activity, to the properties of the Sun and in this way, one may hope, get an improved understanding of solar evolution on both shorter and longer timescales.

An important development has been the huge increase in the general use of machine learning, including applications in the physical sciences [for a brief review with some examples, see 4]. This involves 'teaching' of algorithms to recognize specific patterns or relations between variables, extending to the so-called 'deep learning' where the information is passed through multiple layers in a neural network. These techniques have major strengths in image recognition and classification but can also be used in more direct analysis of observational data. An example is provided by the asteroseismic analysis implemented by Bellinger et al. [2] (see Fig. 1).

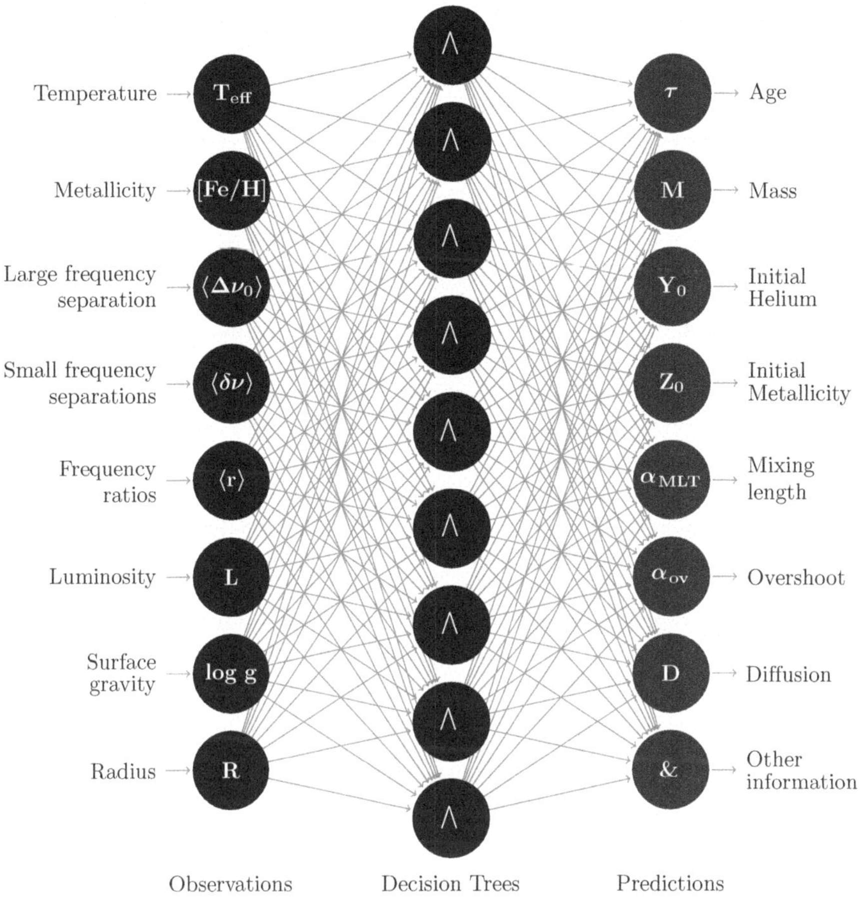

Fig. 1 Application of machine learning for the analysis of stellar data, including asteroseismic observations, to infer stellar properties. The so-called decision trees in the intermediate layer combine the input data with selection rules that are defined using a training set of models. The outcome of the analysis is a suitable average of the results from the individual decision trees. (From Bellinger et al. [2])

4 How do We Proceed?

In the following I mainly have the asteroseismic case in mind (having, as Jesper Schou pointed out many years ago, moved to 'the dark side'). However, several of the points raised are equally applicable to solar studies.

In the asteroseismic analyses two largely distinct procedures can be identified. One is what has been called 'boutique analysis', where single stars undergo detailed investigation, using different techniques and often involving independent analysis and modelling procedures. This clearly involves intensive human interaction, to optimize the analyses and obtain the best and most reliable inferences from the data. Such detailed analysis may be justified, for example, to infer accurate properties, including a reliable age, for an exoplanet host star, and it is required for detailed inferences of stellar internal properties and tests of the physical assumptions in stellar modelling.

The second class of analysis is ensemble analyses of large groups of stars, typically to determine global stellar parameters. In this case the observed quantities are often global seismic parameters, such as the frequency at maximum power and the large frequency separation. This type of analysis is becoming increasingly relevant with the accumulation of asteroseismic data from the TESS mission, where tens of thousands of red giants will be analysed, primarily for use in galactic archaeology. This is an area where automated analysis is inevitable, and where artificial intelligence, in some form of machine learning, may be highly relevant.

With the PLATO mission an intermediate case is becoming important. In the mission asteroseismology is an integral part in the analysis of exoplanetary systems, and hence automated seismic analysis of large numbers of stars will be required, at a level of detail going beyond the ensemble analysis and perhaps approaching the boutique analysis. The development and verification of the required techniques is an ongoing challenge. This should undoubtedly include tests for problematic cases, flagging failures of the automated analysis for further investigation.

In contemplating these issues related to the analysis and interpretation it is perhaps useful to keep in mind the well-known statement

The object of computing is insight, not numbers.

Given the crucial importance of computing in current data analysis the first part of the following addition is perhaps not needed, but let me make the following amendment of the saying:

The object of computing ***and observations*** *is insight* ***and understanding****, not numbers.*

Here the added 'understanding' is undoubtedly a somewhat vague concept, which nevertheless can perhaps be seen as the ultimate goal of our efforts, in the sense of the Eddington quote alluded to earlier.

To these considerations, let me add a statement which I attribute to Douglas Gough:

Never carry out a calculation before you know the answer

'knowing', in this sense, obviously meaning having an idea about the magnitude or behaviour of the result, based on simplified considerations, and hence providing a sanity check of the results. If they deviate from expectations they may either be very interesting, or possibly wrong!

This description of the modelling and analysis is the background for Fig. 2, which can be applied both to purely theoretical modelling and to the interpretation of observational data, the data being involved both in the analytical insight and the numerical computations. The emphasis is here on the interaction between all three aspects, certainly a feature which was emphasized in Michael's work. Helio- and asteroseismic investigations are obvious examples of this interplay, with the

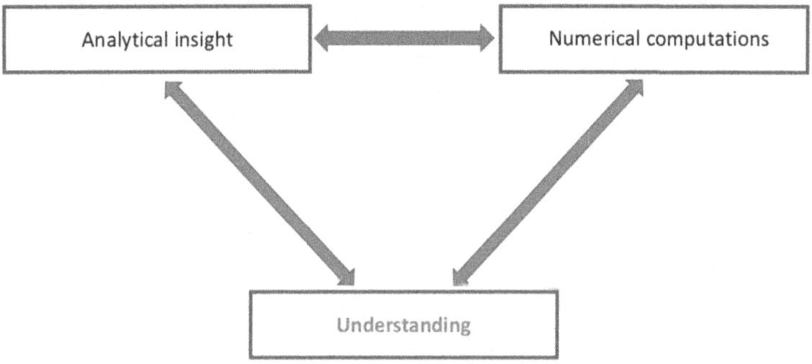

Fig. 2 A schematic illustration of the interaction between different modes of working, perhaps with specific reference to theoretical investigations, although analysis of observational data could also be involved. Analytical insight primarily refers to working with pen and paper, on simplified descriptions that can provide inspiration to the numerical analysis and contribute to understanding, while the latter obviously also inspires development of both analytical and numerical work

asymptotic description of the oscillations playing a major role as analytical insight in the interpretation of the observations and in the numerical computations, but where computations can also inspire deeper insight, for example in the more detailed description of the effect of glitches.

From my dinosaurian viewpoint I see a tendency that is not entirely consistent with the above way of working, in what might be regarded as an increasing application of black boxes, as schematically (and perhaps not completely fairly) illustrated in Fig. 3. The availability of comprehensive tools is undoubtedly essential

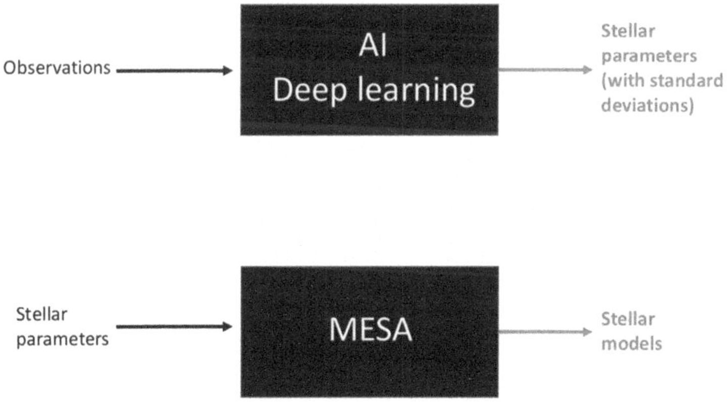

Fig. 3 Schematic illustration of ways of working that are undoubtedly convenient and efficient, but carry some risk if not properly applied. In the case of the analysis of observations machine learning is singled out here, but uninformed use of any statistical package could have been shown instead. Similarly, MESA [e.g., 6] is shown as an example of a stellar modelling code, certainly with major advantages in terms of flexibility and relative simplicity of use, including options for fitting observed asteroseismic data

to the progress of science and allows interesting results to be obtained without always starting from a very basic level. This is particularly important in the analysis of the large amounts of high-quality data coming from space observations. Also, such tools allow convenient exploration of broad parameter sets, which in itself certainly can contribute to the understanding. The risk is that they can be applied, and produce results, in the worst case without any sense of underlying assumptions or limitations, and hence making the results meaningless or, even worse, having apparent but entirely erroneous meaning. This is certainly a case where having a good idea about the result in advance can be very helpful. Also, the mounting pressure for publications undoubtedly motivates liberal use of existing tools, with insufficiently critical evaluation by the users of the results.

A crucial requirement for the further development of these fields is the involvement of new generations of scientists with new ideas. A very encouraging aspect of this and many other recent conferences is the participation of young scientists at the PhD and postdoc levels with very exciting contributions, enthusiastically presented. This bodes well for the future. Indeed, the continuing training of these new generations is an important tasks. This obviously involves teaching of the required techniques in modelling and observations, including a good understanding of the statistical properties of data and of the assessment of the significance or otherwise of results. It also involves training in creative analytical work, an area that Michael excelled in; a concern, at least in the Danish school system before university, is an increasing move from using paper and pencil towards symbolic manipulation on laptops, surely not conducive to creativity at this level. However, beyond these technical skills we should aim for a broader understanding of the underlying background and how the work on the specific projects fits into our broader picture of the Universe and the physics that determines its properties. The ultimate goal, which of course is rarely achieved, is true deep learning.

An important part of this is to inspire young people considering a scientific career about the interest of astrophysics, and the excitement of stellar studies. Thus public outreach is an significant contribution to the development of the field.

An unavoidable concern for the future development is the required funding. Development of the research infrastructure in terms of observational facilities on the ground and in space is proceeding, as is the continuing development of high-performance computing facilities. A more problematic issue is funding for the archiving and, in particular, long-term preservation of research data, including the meta-data required to make the data useful over several decades. Much of the data produced by missions like *Kepler* or TESS will not be superseded in the foreseeable future and hence should be kept in a form where it can be used, for a very long time. However, in many ways the real bottleneck in these projects are the people required to carry out and continue the work. Support for PhD students and postdocs is essential and is available from a number of sources. However, there is a shortage of permanent positions at academic or research institutions, with strong and obvious competition from other fields, limiting the possibilities for long-term continuity in the efforts and of course restricting the career possibilities for young scientists.

5 Concluding Remarks

We are witnessing a period of great promise for a better understanding of the physics of stars and their evolution. The observational basis for this work is outstanding and promises to improve further, and the tools and techniques for data analysis and modelling are developing rapidly, in terms of computational speed and programmatic sophistication. The conference provided a strong impression of these prospects and set out directions for further development. This is a situation that Michael would have relished, and where his contributions in terms of inspiration, insight and development would have been fundamental.

While missing him greatly we must carry on the work in Michael's spirit, instilling that spirit also in future generations, with emphasis on physical insight, the enjoyment of science and all the other aspects that life offers, in terms of music, books, nature and human interaction.

Acknowledgments I thank the High Altitude Observatory for hosting the conference, with general thanks to NCAR and the funding support from NSF, the Local Organizing Committee for the flawless organization in all regards, and in particular Sheryl Shapiro for her huge effort in bringing it all together. I am very grateful to the Scientific Organizing Committee for their efforts in establishing what I think was a very interesting programme, and to the speakers for making it so successful. Funding for the Stellar Astrophysics Centre is provided by The Danish National Research Foundation (Grant DNRF106).

References

1. Aerts, C., Mathis, S., & Rogers, T. M. (2019). Angular momentum transport in stellar interiors. *The Annual Review of Astronomy and Astrophysics, 57*, 35–78.
2. Bellinger, E. P., Angelou, G. C., Hekker, S., Basu, S., Ball, W. H., & Guggenberger, E. (2016). Fundamental parameters of main-sequence stars in an instant with machine learning. *The Astrophysical Journal, 830*(1–20), 31.
3. Buldgen, G., Salmon, S. J. A. J., Noels, A., Scuflaire, R., Montalban, J., Baturin, V. A., et al. (2019). Combining multiple structural inversions to constrain the solar modelling problem. *Astronomy & Astrophysics, 621*(1–16), A33.
4. Carleo, G., Cirac, I., Cranmer, K., Daudet, L., Schuld, M., Tishby, N., et al. (2019). Machine learning and the physical sciences. *Reviews of Modern Physics, 91*(1–39), 045002.
5. Eddington, A. S. (1926). *The internal constitution of the stars*. Cambridge: Cambridge University Press.
6. Paxton, B., Bildsten, L., Dotter, A., Herwig, F., Lesaffre, P. & Timmes, F. (2011). Modules for Experiments in Stellar Astrophysics (MESA). *The Astrophysical Journal Supplement, 192*(1–35), 3.
7. Schad, A., Timmer, J., & Roth, M. (2011). A unified approach to the helioseismic inversion problem of the solar meridional flow from global oscillations. *The Astrophysical Journal, 734*(1–8), 97.
8. Woodard, M., Schou, J., Birch, A. C., & Larson, T. P. (2013). Global-oscillation eigenfunction measurements of solar meridional flow. *Solar Physics, 287*, 129–147.

The manufacturer's authorised representative in the EU is Springer Nature Customer Service Centre GmbH, Europaplatz 3, 69115 Heidelberg, Germany. If you have any concerns regarding our products, please contact ProductSafety@springernature.com

Printed and bound by CPI Group (UK) Ltd, Croydon, CR0 4YY
30/09/2026
02225695-0011